# Geothermal Energy Engineering: Exploration, Extraction and Usage

# Geothermal Energy Engineering: Exploration, Extraction and Usage

Edited by Kale Stewart

New York

Published by Syrawood Publishing House,
750 Third Avenue, 9ᵗʰ Floor,
New York, NY 10017, USA
www.syrawoodpublishinghouse.com

**Geothermal Energy Engineering: Exploration, Extraction and Usage**
Edited by Kale Stewart

International Standard Book Number: 978-1-68286-462-3 (Hardback)

**Cataloging-in-publication Data**

Geothermal energy engineering : exploration, extraction and usage / edited by Kale Stewart.
    p. cm.
Includes bibliographical references and index.
ISBN 978-1-68286-462-3
1. Geothermal engineering. 2. Geothermal resources. I. Stewart, Kale.
TJ280.7 .G46 2017
621.44--dc23

Printed in the United States of America.

# TABLE OF CONTENTS

**Permissions**

**List of Contributors**

**Index**

# PREFACE

This book provides comprehensive insights into the field of geothermal energy engineering. It explores all the important aspects of this field in the present day scenario. Geothermal energy is one of the most important forms of alternative energy present today. It is generated within earth's crust because of radioactive decay of materials and is stored in the same. It is the most reliable, cost-effective and environmental friendly power. This text will give detailed analysis about the use and extraction of geothermal energy. It will also discuss the engineering methods required for its exploration. Different approaches, evaluations, methodologies and advanced studies in this area have been included in the text. Students, researchers, experts and all associated with the field of geothermal energy engineering will benefit alike from this text.

This book has been an outcome of determined endeavour from a group of educationists in the field. The primary objective was to involve a broad spectrum of professionals from diverse cultural background involved in the field for developing new researches. The book not only targets students but also scholars pursuing higher research for further enhancement of the theoretical and practical applications of the subject.

It was an honour to edit such a profound book and also a challenging task to compile and examine all the relevant data for accuracy and originality. I wish to acknowledge the efforts of the contributors for submitting such brilliant and diverse chapters in the field and for endlessly working for the completion of the book. Last, but not the least; I thank my family for being a constant source of support in all my research endeavours.

**Editor**

# Effectiveness of acidizing geothermal wells in the South German Molasse Basin

S. Schumacher and R. Schulz

Leibniz Institute for Applied Geophysics, Stilleweg 2, 30655 Hanover, Germany

*Correspondence to:* S. Schumacher (sandra.schumacher@liag-hannover.de)

**Abstract.** In Germany, many hydro-geothermal plants have been constructed in recent years, primarily in the region of Munich. As the host formation here mainly consists of carbonates, nearly all recently drilled wells have been acidized in order to improve the well yield. In this study, the effectiveness of these acid treatments is analyzed with respect to the amount of acid used and the number of acid treatments carried out per well. The results show that the first acid treatment has the largest effect, while subsequent acidizing improves the well only marginally. Data also indicate that continued acidizing can lead to degradation of the well. These findings may not only be important for geothermal installations in Germany but also for projects, for example, in Austria, France or China where geothermal energy is produced from carbonate formations as well.

## 1 Introduction

### 1.1 Geothermal energy in Germany

The growing need for energy and the rising prices of conventional energy sources such as oil, gas and coal have led to increased interest and investment in environmentally friendly, renewable energy in Germany. One such renewable resource is geothermal energy, which, compared to other more widespread resources such as wind or solar energy, has the tremendous advantage of being able to deliver heat and electrical energy independent of weather conditions and time of day.

In Germany, large geothermal plants require the use of deep geothermal energy, which so far has only been used in the form of hydro-geothermal energy (Schellschmidt et al., 2010). For this kind of energy, certain requirements have to be fulfilled, such as the existence of an aquifer of hot water. Hence, the generation of hydro-geothermal energy in Germany is mainly confined to three areas: the North German Basin, the Upper Rhine Graben and the South German Molasse Basin. The Molasse Basin (specifically the region of Munich) has become the center of geothermal energy production in Germany within the last few years due to favorable geological conditions.

### 1.2 Geological setting

In the Molasse Basin, hot water can be found in the stratigraphic unit of the Malm aquifer (Upper Jurassic). The sedimentary layers of the karstic Malm aquifer primarily consist of carbonate rocks, namely small-pored white limestones as well as fine-to-coarse crystalline dolomites (e.g., Wolfgramm et al., 2007). For the well Pullach Th2, pure limestones as well as limestone layers that contain clay in varying degrees have been described (Böhm et al., 2010), while Reinhold (1998) identifies oolitic platform sands with associated mounts of microbe–siliceous sponges in the Upper Jurassic of the Swabian Alb (Swabian facies). These variations in material can be explained by the different facies that can be encountered in the Malm aquifer. Facies found in the Malm aquifer encompass the Swabian, the Franconian, and the Helvetic facies.

The Helvetic facies developed under distal conditions and as a consequence contains not only carbonates but also a significant amount of marls (Meyer and Schmidt-Kaler, 1996). Therefore, it exhibits only very small transmissivities (Villinger, 1988; Bayerisches Staatsministerium für Wirtschaft, Infrastruktur, Verkehr und Technologie, 2010). Although the Helvetic facies is restricted to the southwestern part of the basin, small transmissivities are not restricted to the Helvetic facies itself but are also found in areas adjacent

**Figure 1.** A map of southern Germany and geothermal installations within the Malm aquifer. Main uses are district heating (red dots), power generation (blue dots), and spas (gray dots).

to it (Stober and Villinger, 1997), which can be explained by a lack of karstification (Villinger, 1997).

The Franconian facies, found in the region of Munich, developed in a shelf environment. It contains basin facies and reef facies. While the basin facies consists of banked limestones, which were deposited in shallow waters between reefs, the carbonates of the reef facies are built up by reef detritus. Thus, the matrix porosity of the reef facies tends to be higher than that of the other facies. As a result, recrystallization of limestone into dolomite primarily takes place in the reef facies (e.g., Bausch, 1963). Dolomites are characterized by a predominantly good porosity caused by the reduction in volume due to the dolomitization of limestone (Koch, 1997), which can result in an increase in porosity by up to 13 % (Böhm et al., 2011). Therefore, the reef facies in the Molasse Basin is of special interest for the planning of geothermal projects.

In addition to matrix porosity, pathways for fluid flow that have been created by karstification play a dominant role within the Malm aquifer. As a result, transmissivity is higher in the north of the basin than in the south, since rocks in the north underwent a higher rate of karstification than those in the south, where the transmissivity depends mainly on the matrix porosity of the carbonates (Koch and Sobott, 2005) and on fissures (Wolfgramm et al., 2009). Analysis of pump tests shows that transmissivities within the Malm aquifer vary by more than seven orders of magnitude (Birner et al., 2012). However, in the region of Munich, which is of special interest for this study due to the high number of geothermal wells in this area, transmissivities change by only two orders of magnitude (Birner et al., 2012).

It is also important to note that the water within the Malm aquifer shows only low mineralization (e.g., Prestel, 1990;

Wolfgramm and Seibt, 2008) despite the fact that this water has been buried to significant depths for long periods. Prestel (1990), for example, gives an age of 6700 to 10 400 yr for water encountered in the well Saulgau GB3 (for approximate location see Fig. 1, "Saulgau"), dating it to the Holocene. Bertleff and Watzel (2002) state that water in the basin's center is of Pleistocene age. In general, water tends to be younger toward the borders of the basin, but it has been found that the mineral load is low even at the center. Because of this, water within the Malm aquifer is a perfect repository for hydrogeothermal energy use.

## 1.3  Technical details

The geologic setting is important for the exploitation of geothermal energy in the Molasse Basin as not only is a high transmissivity needed in order to operate geothermal power plants economically, but high temperatures are also essential. As a general trend, it can be observed that with increasing depth of the Malm aquifer to the south, temperatures found within this layer also increase. This, however, is contrary to the behavior of the transmissivity, which tends to decrease to the south (Birner et al., 2012).

In the region of Munich, both parameters were assumed to exhibit values suitable for the exploitation of geothermal energy for district heating and/or electricity generation. Therefore, in recent years many geothermal plants have been constructed in this area. In order to increase the yield of the wells and therefore their economic efficiency, acidizing has been performed on nearly all of these wells. The primary acid used was hydrochloric acid in varying dilutions, but in some cases small amounts of citric and acetic acid were also used as admixing. In this work the effectiveness of these acid treatments has been analyzed based on data from 17 wells.

## 1.4 Data origin

In Germany, operators of geothermal plants are legally obligated to pass on information about stimulation procedures and pump tests to the responsible state geological survey. However, they are not required to go into detail or to give access to the original measurements. In general, the results of pump test analyses done by service companies are reported. Depending on the service company, these reports include details about the duration of the stimulation measure or the part of the well acidized, but often only the bare minimum (such as flow rate and specific capacity, i.e., flow rate divided by drawdown) is included.

As part of the research project "Geothermal Information System for Germany" (GeotIS), access to the information of the state geological surveys was granted, which enabled this metaanalysis of acidizing data. However, this meant that none of the original pump test measurements and only limited information about their analysis was available. Therefore, this paper deals only with the most basic data, which are routinely acquired by operators of geothermal plants. This has the advantage that future geothermal projects do not need to deviate from standard procedures in data acquisition and analysis in order to use the insights gained from this work.

## 2 Theory of laminar and turbulent flow

Fluid flow into a well is defined by Darcy's law.

$$Q = k \cdot A \cdot i \tag{1}$$

with $Q$ being the flow rate [$m^3 s^{-1}$], $k$ the coefficient of permeability or hydraulic conductivity [$m s^{-1}$], $A$ the surface area through which the fluid passes [$m^2$], and $i$ is the hydraulic gradient [1]. Under the assumption that the water enters the well perpendicular to the well's axis through parts of the uncemented regions of the borehole, Eq. (1) can be rewritten as

$$Q = 2\pi r H k \frac{dh}{dl}, \tag{2}$$

where $r$ is the radius of the borehole [m], $H$ the thickness of the aquifer [m], and $dh/dl$ is the gradient of the water table or piezometric surface [1] (Hamill and Bell, 1986; Hölting and Coldewey, 2009). It is therefore apparent that the flow rate depends linearly on the hydraulic conductivity of the surrounding rock. In the case of a karst aquifer, such as the Malm aquifer, the hydraulic conductivity of the immediate surroundings may be small due to a lack of fissures and fractures, and therefore impeding a high flow rate. The overall hydraulic conductivity of the aquifer, however, may be comparatively high and well-suited for geothermal exploitation. Thus, the idea of acidizing is to improve the connection of

the well to fissures and fractures in its immediate surroundings, and to widen those in order to enhance the well's flow rate.

The theory of fluid dynamics has shown that the transition from laminar to turbulent flow in a pipe occurs at a Reynolds number $Re$ between about 2200 (Turcotte and Schubert, 2002) and 2300 (Schlichting and Gersten, 2001), with $Re$ [1] being defined as

$$Re = \frac{\rho \cdot v \cdot D}{\mu}, \tag{3}$$

where $\rho$ is the fluid's density [$kg\, m^{-3}$], $D$ the diameter of the borehole [m], and $\mu$ the fluid's dynamic viscosity [Pa s]. The critical velocity for the onset of turbulence for boreholes considered in this study can be calculated to be in the range of about 0.002 to 0.010 m s$^{-1}$. This is far below the velocity of about 1.4 m s$^{-1}$ that is obtained for a flow rate of 50 L s$^{-1}$, which is a typical flow rate for the wells considered in this analysis. Fully developed turbulent flow can be assumed to take place for $Re > 5000$ (Spurk, 2006). As the Reynolds numbers for all wells are larger than 400 000, it can be concluded that all wells operate within the fully turbulent regime.

Pump tests are a combination of an aquifer test and a well test. Flow within the aquifer is assumed to be laminar, while it is assumed to be turbulent within the well and its immediate surroundings. In previous modeling of the Malm aquifer in the region of Munich, it has been shown that the pressure conditions within the aquifer can be simulated using a porous matrix with laminar flow instead of a karst model (Bartels et al., 2012; Bartels and Wenderoth, 2012), which can also exhibit turbulent flow. The porous matrix model has been calibrated against well test data as well as data from operating geothermal plants, and a good correlation between measured and modeled data has been achieved. Thus, the assumption of laminar flow within the Malm aquifer for large-scale considerations seems justified. The assumption of turbulent flow within the well and its surroundings has been shown above to be correct for the wells considered in this analysis. Therefore, the equation proposed by, for example, Jacob (1947) and Hamill and Bell (1986) for the drawdown in the well can describe this behavior mathematically:

$$s = BQ + CQ^2, \tag{4}$$

where $s$ is the drawdown [m], $Q$ the flow rate [$m^3 s^{-1}$], and $B$ and $C$ coefficients with units of [$s\, m^{-2}$] and [$s^2 m^{-5}$], respectively. The drawdown can be described either by the height difference of the water column in the well or by the pressure difference that is caused by changes in water column height and that is measured by a pressure gauge within the well. In the latter case, the unit of drawdown changes to [MPa], with the units of $B$ and $C$ changing accordingly. The term $BQ$ describes the aquifer loss, while $CQ^2$ is the well loss. However, this equation has been derived from experimental data, and Rorabaugh (1953) argues that the exponent of $CQ^2$ can vary as it depends on the well efficiency.

In this case, the well loss can be more generally expressed as $CQ^p$, where $p$ is the exponent as determined by well efficiency, and varies between 1.5 and 3.5. The specific capacity $S_c$ in [m$^2$ s$^{-1}$] or [m$^3$ s$^{-1}$ MPa$^{-1}$] can then be determined by the following equation:

$$S_c = \frac{Q}{s} = \frac{1}{B + CQ^{(p-1)}}. \qquad (5)$$

It becomes apparent that the specific capacity of a well decreases with increasing flow rate independent of the mathematical approach used for the well loss. The reason for this is that the water loses kinetic energy due to friction that already occurs in the laminar regime. Turbulence introduces yet another element into the analysis of the fluid flow. As long as the well operates within the laminar regime, the viscosity of the water mainly depends on its temperature. In the turbulent regime, however, eddies within the water lead to the so-called "eddy viscosity", which describes the apparently higher viscosity because of a transfer of kinetic energy to internal energy of the fluid (Schlichting and Gersten, 2001). Because of this and other smaller effects, turbulence introduces more complexity into the calculations and has to be considered in separate terms.

## 3   Data set and normalization

For many wells acidizing took place in several steps with interposed pump tests. Moreover, a large number of pump tests were carried out as step drawdown tests. The result of this combination can be seen in Fig. 2, where all available data for step drawdown tests are displayed.

Figure 2 shows that the specific capacity is reduced if the flow rate is increased. In this and all following figures, the same symbols and the same well numbers signify the same borehole, while the same colors are indicators of the same acidizing step. The reduction of the specific capacity occurs because a higher flow rate also means a higher velocity of the water within the well and its surroundings, which leads to turbulence for all wells considered in this analysis.

The decrease in specific capacity makes it harder to compare the results of pump tests. Most pump tests were done at different flow rates (see Fig. 2), so due to the effect described above, their results cannot be directly compared, but need to be normalized. In order to achieve this, a linear equation was established for each borehole and acidizing step, describing the decrease in specific capacity with increasing flow rate. The validity of assuming a linear equation is shown in Fig. 2, where especially the long-time pump tests indicate a linear decrease of the specific capacity for the range of flow rates analyzed. For the area of Munich, 23 data sets were available, which contained more than one pump test per acidizing step and for which a linear equation could be found. The resulting slopes of these equations were then averaged so as to

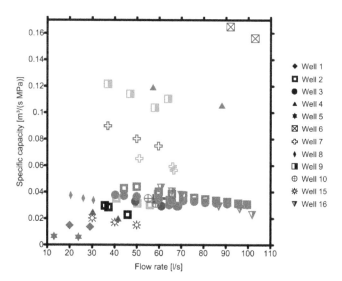

**Figure 2.** Data used to calculate linear equation for the reduction of the specific capacity with increasing flow rate. Colors indicate the time when the pump tests were performed: before first acid treatment (black), after first acid treatment (green), after second acid treatment (blue), after third acid treatment (red), after fourth acid treatment (orange), after fifth acid treatment (pink), after sixth acid treatment (brown), and long-time pump tests (gray).

use a common value for all further calculations. The resulting linear equation for normalization is

$$S_{c_{norm}} = -0.438\,\text{MPa}^{-1} \cdot Q + y, \qquad (6)$$

where $S_{c_{norm}}$ is the normalized specific capacity and $y$ the intercept with the ordinate, which has been determined for each acidizing step and borehole beforehand. In principle, it would have been better if each borehole and acidizing step were normalized based on its own linear equation. This, however, was not feasible due to lack of data since for some boreholes and acidizing steps the pump test was only carried out for one flow rate. As a result, no linear equation could be obtained for four boreholes (boreholes 11, 12, 13 and 14). Thus, an averaged slope was used for normalization in order to use a standardized method for all boreholes and to be able to increase the number of boreholes analyzed. The variance $\sigma^2$ of the slope of Eq. (6) is 0.066. Even though using an averaged slope leads to some minor distortion regarding the absolute values of specific capacity, it does not influence the values for the relative improvement of the specific capacity from one acidizing step to the next, which are of far more interest for this study.

Since for most boreholes and acidizing steps more than one specific capacity value was obtained from pump tests, the existing specific capacity and flow rate values of each acidizing step and each borehole were averaged to generate mean values. The resulting mean values for each acidizing step then served as starting points for the normalization. The result of this normalization can be seen in Fig. 3.

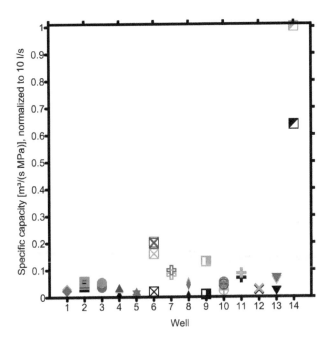

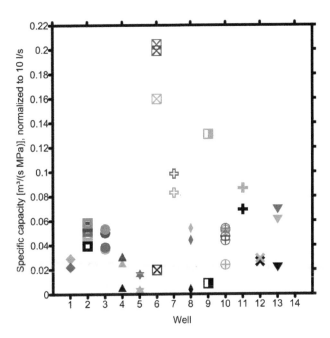

**Figure 3.** Specific capacity for different wells before and after acidizing (for details see Fig. 4). Colors indicate the time when the pump tests were performed: before first acid treatment (black), after first acid treatment (green), after second acid treatment (blue), after third acid treatment (red), after fourth acid treatment (orange), and after fifth acid treatment (pink).

**Figure 4.** Specific capacity for different wells before and after acidizing (extract of Fig. 3). Colors indicate the time when the pump tests were performed: before first acid treatment (black), after first acid treatment (green), after second acid treatment (blue), after third acid treatment (red), after fourth acid treatment (orange), and after fifth acid treatment (pink).

The results of all pump tests were normalized to a flow rate of $10\,\mathrm{L\,s^{-1}}$ by using Eq. (6). A rate of $10\,\mathrm{L\,s^{-1}}$ is very low, much lower than the usual flow rate for most of the wells. It was chosen for two reasons. First of all, extrapolation to higher flow rates of, for example, $100\,\mathrm{L\,s^{-1}}$ can be problematic as it is unclear up to which flow rate the inferred linear equation holds true. As discussed above (Sect. 2), high flow rates are connected to the turbulent regime. Turbulence, however, implies that the exponential term of Eq. (5) comes to bear and, subsequently, the linear equation can no longer be used. The second reason for using a rather low value was the fact that some wells exhibit very low specific capacities. Extrapolating the linear equation to high flow rates would lead to negative values for these wells. As this is physically infeasible, it had to be avoided. Thus, a flow rate of $10\,\mathrm{L\,s^{-1}}$ met all requirements with regard to validity of extrapolation.

## 4 Results

In Fig. 3 the results of 14 boreholes are displayed, for which data for more than one acidizing step could be obtained. It should be noted that two boreholes of Fig. 2, wells 15 and 16, are not part of the group of these 14 boreholes for which all subsequent calculations and considerations are done.

In Fig. 3 some wells (namely wells 6, 9 and 14) exhibit an enormous increase in the specific capacity due to acidizing, while for the majority of wells acidizing only led to a

marginal improvement. This latter observation becomes especially apparent in Fig. 4, which displays the same data as Fig. 3 with the exception of data for well 14. Thus, it is possible to show the small increases in specific capacity for all other wells in more detail. The fact that acidizing has only a small impact on some wells is especially pronounced for wells 2, 3 and 10. For these wells there are plenty of data for different acidizing steps. All these data show that the effect of the acidizing is diminished with increasing number of acidizing steps.

However, the improvement that is generated by acidizing depends not only on the number of acidizing steps, but also on the amount of acid used for each step. For some wells, the amount used was kept constant for different acidizing steps but some wells exhibit great variations (Table 1). In order to test how far the improvement of the specific capacity correlates with the amount of acid used, the specific capacity improvement was normalized to the amount of acid (15 % HCl) used. The results of this are shown in Fig. 5. In this figure, the changes in the specific capacity after an acidizing step, as compared to before the acidizing step, are displayed. Thus, only the relative improvement and not the absolute improvement is shown. This approach is used for all further comparisons. As could be expected, the improvement of the specific capacity is large for well 14 even after normalization to the amount of acid used. Moreover, it is again made evident that the effectiveness of the acidizing is reduced with every

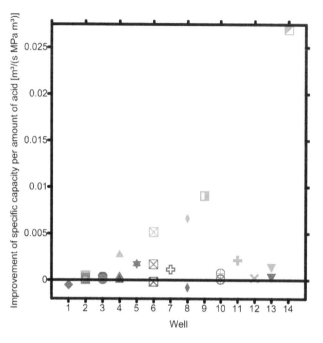

**Figure 5.** Improvement of the specific capacity normalized to the amount of acid used (for details see Fig. 7). Colors indicate the acidizing step that causes the improvement: first (green), second (blue), third (red), fourth (orange), and fifth (pink).

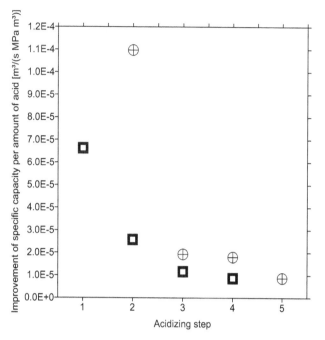

**Figure 6.** Improvement of the specific capacity normalized to the amount of acid used for two wells: well 2 (black squares) and well 10 (black crossed circles).

further acidizing undertaken. Thus, the trend that could already be observed in Fig. 4 holds true even if the amount of acid used is taken into account.

It becomes even clearer in Fig. 6, where the increase in specific capacity per acidizing step has been plotted for two wells. Wells 2 and 10 are those that exhibit the highest number of acidizing steps undertaken and monitored, and they therefore provide the best database. However, even for those two wells the data are not perfect. For well 10 the specific capacity of the unstimulated well has not been measured, and thus the improvement due to the first acid treatment is unknown. Still, the trend of initially high gains due to acidizing with subsequent stagnation at low levels is apparent.

Except for well 14, there are some other wells that exhibit a significant improvement of the specific capacity. These are primarily wells that already displayed a large improvement in Fig. 4; however, well 8 is an exception to this rule. For this well, the first acid treatment was extremely effective, while the second acid treatment shows a surprising result: the specific capacity decreased. Thus, the second acid treatment worsened the well compared to the situation after the first acid treatment. If this were the only well for which such an observation could be made, an error in the recorded pump test data or a mistake in its analysis would have been a likely explanation for this unexpected result. However, three other wells, namely wells 1, 4 and 6, exhibit a similar behavior. This can be observed in Fig. 7, which is an extract of Fig. 5 and shows the results in more detail.

In this figure it is very obvious that for some wells continued acidizing results in a reduction of the specific capacity. The first and maybe second acid treatment lead to an improvement of the specific capacity, but subsequent acidizing worsens the results. In all cases in which the initial specific capacity is known, acidizing was successful (i.e., the specific capacity after the last acidizing step was higher than initially), even for wells that show a reduction after the first or second acidizing step. However, in some cases the result could have been better if less acid treatments had been applied.

In Fig. 8 the improvement of the specific capacity in percent is shown. Again it is normalized to the amount of acid used, and in this case also to a flow rate of $10 \, \text{L s}^{-1}$. The latter normalization is necessary as percentages are based on absolute values and therefore need to have the same reference value if they are to be compared.

As in all other figures, the improvement is not related to the initial specific capacity but to that of the acidizing step before. In the case of percentages, this means that some wells exhibit extremely high improvements of more than 25 % per $m^3$ acid. However, in general, these wells were suffering from very low initial specific capacities so that even moderate improvements translated into high percentage gains.

Nevertheless, it is noteworthy that wells that show a reduction in specific capacity with consecutive acidizing (wells 4, 6 and 8) are among the group of wells that show high initial percentage gains. Unfortunately, there are no data for wells 1, 5 or 9, which could indicate if this is a pattern or a mere

**Table 1.** Listing of the different sorts, concentrations and amounts of acids used for acidizing.

| Well number | Acidizing step | Acid | Amount |
|---|---|---|---|
| 1 | 1 | 15 % HCl | 90 m$^3$ |
|  | 2 | 15 % HCl | 90 m$^3$ |
| 2 | 1 | 20 % HCl | 100 m$^3$ |
|  | 2 | 15 % HCl, 3 % $C_2H_4O_2$ | 130 m$^3$ |
|  | 3 | 15 % HCl | 320 m$^3$ |
|  | 4 | 15 % HCl, 6 % $C_2H_4O_2$, 1 % $C_6H_8O_7$ | 320 m$^3$ |
| 3 | 1 | 20 % HCl | 100 m$^3$ |
|  | 2 | 20 % HCl | 150 m$^3$ |
|  | 3 | 20 % HCl | 200 m$^3$ |
|  | 4 | 20 % HCl | 200 m$^3$ |
| 4 | 1 | 7.5 % HCl | 100 m$^3$ |
|  | 2 | 7.5 % HCl | 200 m$^3$ |
|  | 3 | 7.5 % HCl | 200 m$^3$ |
| 5 | 1 | 15 % HCl | 50 m$^3$ |
|  | 2 | 7.5 % HCl | 100 m$^3$ |
| 6 | 1 | 15 % HCl | 180 m$^3$ |
|  | 2 | 15 % HCl | 180 m$^3$ |
|  | 3 | 15 % HCl | 180 m$^3$ |
| 7 | 1 | 7.5 % HCl | 100 m$^3$ |
|  | 2 | 7.5 % HCl | 190 m$^3$ |
| 8 | 1 | 15 % HCl | 50 m$^3$ |
|  | 2 | 15 % HCl | 80 m$^3$ |
| 9 | 1 | 7.5 % HCl | 180 m$^3$ |
| 10 | 1 | 15 % HCl | 180 m$^3$ |
|  | 2 | 15 % HCl | 180 m$^3$ |
|  | 3 | 15 % HCl | 180 m$^3$ |
|  | 4 | 15 % HCl | 270 m$^3$ |
|  | 5 | 15 % HCl | 180 m$^3$ |
| 11 | 1 | 15 % HCl, 6 % $C_2H_4O_2$ | 50 m$^3$ |
| 12 | 1 | 15 % HCl | 75 m$^3$ |
| 13 | 1 | 15 % HCl | 200 m$^3$ |
|  | 2 | 15 % HCl | 200 m$^3$ |
| 14 | 1 | 7.5 % HCl | 180 m$^3$ |

coincidence. Either no pump test was performed before the first acid treatment (wells 1 and 5), or only one acidizing step was carried out (well 9).

The temperature dependence of the reaction of hydrochloric acid with limestone has also been analyzed. No clear correlation between temperature and effectiveness of acidizing could be observed. Therefore, this effect has not been taken into account for the comparison of wells in Figs. 3 to 8 (see Sect. 5.3).

## 5  Discussion

The analysis of data from 14 wells in the region of Munich shows that acidizing with hydrochloric acid can significantly improve a well drilled into carbonate rock. The improvement can be well over 10 % per m$^3$ of 15 % HCl used. However, the data also indicate that the increases in specific capacity result primarily from the first acid treatment. All subsequent acidizing – especially everything above two treatments – does not have a significant impact any more. This holds true even for

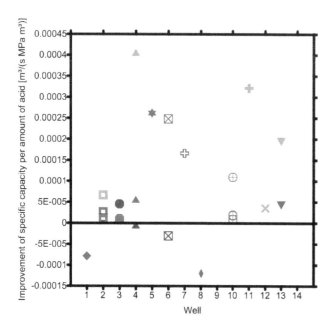

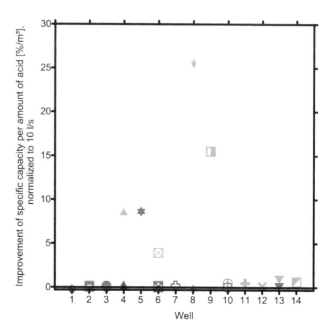

**Figure 7.** Improvement of the specific capacity normalized to the amount of acid used (extract of Fig. 5). Colors indicate the acidizing step that causes the improvement: first (green), second (blue), third (red), fourth (orange), and fifth (pink).

**Figure 8.** Improvement of the specific capacity normalized to the amount of acid used and a flow rate of $10 \, \text{L s}^{-1}$. Colors indicate the acidizing step that causes the improvement: first (green), second (blue), third (red), fourth (orange), and fifth (pink).

wells for which an initial pump test was performed in which the well should have been pumped clean of drilling residue. The results of the first acid treatment indicate that these initial cleaning measures did not have the desired effect and that only acidizing can effectively remove fine material from the well's walls. Thus, the primary effect of acidizing is to clean the well and not to improve the reservoir by widening already existing fissures or generating new pathways for water to circulate. As a consequence, only the first and sometimes second acid treatment have a significant impact on the specific capacity.

## 5.1 Amount of acid

The amounts of acid used are well below the recommended values for the acidizing of wells in carbonate rock that have been published by Gdanski (2005). This author gives about $1 \, \text{m}^3$ of 15 % HCl per meter of acidized borehole as a rule of thumb for matrix acidizing. In such a case, the injection pressure of the acid is below the fracture pressure of the rock. If acid fracturing is assumed (i.e., the rock is first fractured with high pressure and then treated with acid), even higher amounts of acid are recommended. Thus, the amounts actually used (see Table 1) were always on the lower side of acid volumes suggested for acidizing. However, it has to be clearly stated that acidizing of the wells analyzed was performed as matrix acidizing and not as acid fracturing even though matrix acidizing of the karstified Malm aquifer might have required larger amounts of acid than matrix acidizing of, for example, tight carbonates.

## 5.2 Injection rate

On the other hand, it can be argued that geothermal wells do not require the same amount of acidizing as oil wells, since water is less viscous than oil and can therefore far better percolate through small fractures than oil. As a consequence, geothermal wells might rely more heavily on the penetration depth of the acid than on the widening effect of existing fractures. The penetration depth of the acid depends primarily on the injection rate as a higher injection rate means higher pressure within the well. So for a given porosity, a higher injection rate means that the acid will penetrate deeper into the matrix before it is spent. As a rule of thumb, for matrix acidizing conditions Gdanski (2001, 2005) gives about $0.1 \, \text{barrel min}^{-1}$ for each foot of penetration depth, which translates into about $0.7 \, \text{L s}^{-1}$ per meter of penetration depth. Given the known pump rates for some of the analyzed wells, penetration depths according to this rule should have been about 10 to 40 m. It is therefore obvious that no wide-ranging effects can be expected, but that merely the immediate surroundings of the well are treated. This adds to the observation that the effect of acidizing is strongest for the first treatment, which primarily removes the damage resulting from the drilling process in the immediate surrounding of the well.

## 5.3 Temperature dependence

Another factor that can significantly influence the acidizing effect is the temperature dependence of the acid spending rate. According to, for example, Lund et al. (1974) and Allen and Roberts (1989), the acid spending rate increases significantly over the temperature interval encountered. Gdanski (2005), however, has shown that for the temperature range that can be encountered in the wells in the area of Munich ($\approx 75$–$145\,°C$), the reactivity of limestone does not vary much and is moreover similar to that of dolomite. The temperature dependence within the wells of this analysis has been examined, but no clear correlation between temperature and effectiveness of acidizing could be observed. This finding is therefore in good agreement with recent research. As the temperature dependence plays a minor role, it has not been considered further.

## 5.4 Reasons for well deterioration

It is important to note that the analysis indicates that continued acidizing can worsen the well. Overall, acidizing improved all wells considered in this analysis, but for some wells the final result could have been better if the acidizing had been stopped at an earlier point in time. Moreover, this effect cannot simply be explained by faulty measurements or erroneous pump test analysis as nearly one-third of all wells show this unexpected behavior. Although two wells exhibit reductions in the specific capacity of about 3 % compared to the results directly before the acid treatment in question, two wells suffer from a deterioration of about 18 % and 24 %, respectively. Therefore, this effect is not negligible and can seriously affect the yield and thus the profitability of a well.

The reason for this unexpected behavior is unclear. It is impossible to determine a single property all the affected wells have in common, except the fact that they seem to react extremely well to the first acid treatment, if the improvement in percentage is considered. This might provide a tool for companies conducting the acidizing to estimate the risk of worsening the well, but unfortunately it does not indicate why the well reacts like this.

It can be argued that hydrochloric acid reacts with silicate minerals and causes them to increase their volume by up to five times the original size (Hamill and Bell, 1986). These particles then may cause the fine fissures within the rock to become blocked, thereby also reducing the yield of the well (Hamill and Bell, 1986). Another possibility is that continued acidizing releases insoluble particles, which then clog parts of the well.

In the case of well 4, clogging due to the precipitation of silicate minerals could be an explanation. For this well an injection test with several thousand cubic meters of freshwater during wintertime occurred between acidizing steps two and three, and directly before the decrease in specific capacity was observed. It is therefore possible that due to the injection of huge amounts of cold water the temperature inside the well was lowered significantly. At temperatures of $50\,°C$ or even less, the solubility of silica is already markedly lower than, for example, $100\,°C$ (e.g., Morey et al., 1964; Fournier and Rowe, 1977; Chigira and Watanabe, 1994), which could

lead to the precipitation of silicate minerals and subsequently a blocking of fluid pathways if the well is cooled down. Another factor underlines this possibility. The specific capacity for well 4 was low even after acidizing, which indicates a rather tight rock matrix. Thus, it is conceivable that precipitation of minerals led to a significant closure of not that abundant and/or small fissures. If this was the case, the decrease in specific capacity was not the result of continued acidizing but of injections tests between acid treatments. However, this explanation can only be employed for well 4. The decrease in specific capacity for well 6 requires an alternative explanation as no injection of huge amounts of freshwater in addition to the acid treatment happened here. For wells 1 and 8, there are no data regarding activities between acidizing steps, so based on the available information, worsening of these wells due to injection tests can also be ruled out.

As an additional explanation, the precipitation of ferric iron, which can lead to plugged fractures, is highly unlikely. The mineralization of waters from the Malm aquifer is low, and iron does not play a role (Wolfgramm and Seibt, 2008). Moreover, for one well some acidizing steps included acetic and citric acid, which react very slowly and therefore result in a low pH level over a prolonged period of time. Those acids chelate iron, which might exist in the aquifer, and prohibit its precipitation (Allen and Roberts, 1989). Nevertheless, the use of these acids did not improve the acidizing result. It can thus be concluded that at least iron precipitation is not a problem that has to be considered for wells in the area of Munich.

As the effect of the worsening of the specific capacity was only observed for wells that initially reacted very well to acidizing, a purely mechanical explanation is also possible. The first acid treatment might have etched sufficient flow channels into the formation to increase the specific capacity significantly. Subsequent acidizing then destroyed the pillars and posts necessary to keep these flow channels open so that the pressure of the overburden or tectonic stresses led to a partial closure of the flow channels (Allen and Roberts, 1989). However, due to lack of information it cannot be deduced which of these processes led to the observed results or if perhaps even another mechanism was responsible.

## 5.5 Relevance for other geothermal areas

Although only wells in the area of Munich have been considered for this study, the results can be transferred to other geothermal projects in different regions as the setting of geothermal wells in a karstic limestone is not unique to the South German Molasse Basin.

In the western part of Austria, the Malm aquifer also forms a huge repository for geothermal energy, which has already been tapped by several projects. It is likely that new projects will be developed in the next years for which acidizing of the wells will be considered.

For the Paris Basin especially the limestones of the Dogger have been identified as potential geothermal reservoirs

(Ungemach et al., 2005) and are currently used by more than 30 geothermal plants with new projects in planning.

In China, waters from the Tianjin geothermal field, which consists of sandstones and fractured and karstic limestones, as well as from the Xiaotangshan geothermal field, which consists of limestones and dolomites, have been used for hundreds of years. In both areas development of new projects and the improvement of existing plants for, for example, space heating is under way (Duan et al., 2011).

Therefore, the findings of this study are not only relevant to geothermal projects planned in Germany but also to projects worldwide.

## 6 Conclusions

Acidizing of geothermal wells in carbonate rock in the Malm aquifer in the area of Munich/Germany can lead to significant increases in production even if less acid is used than recommended in, for example, the treatment of hydrocarbon wells. The results, however, show that the first acid treatment of a well is the most successful one, while the increases in specific capacity are lower with every further acid treatment. This indicates that the acidizing primarily removes the damage caused by the drilling process and does not generate a substantial number of new flow channels.

Moreover, about a quarter of all analyzed wells suffer from a decrease in specific capacity after continued acidizing. Although the first treatment leads to an improvement in specific capacity for all wells, subsequent treatments can result in a deterioration of the well. This seems especially true if the first acid treatment of a given well was extremely successful. The reason for this, however, is unclear and several explanations are possible.

The data show that the acidizing of geothermal wells in carbonate rock generally leads to an increase in specific capacity. However, it becomes also clear that more is not always better as continued acidizing results in only marginal improvements, and in some cases even a deterioration of the well's productivity.

**Acknowledgements.** This work was funded by the German Federal Ministry for the Environment, Nature Conservation, and Nuclear Safety (BMU) under project number 0327542A. The data were collected within the Geothermal Information System for Germany (GeotIS) project, which can be accessed for free via http://www.geotis.de. However, site-specific data such as those on acidizing are confidential and are therefore not published within GeotIS. We thank all colleagues who provided the data that made this work possible.

The authors also thank Gioia Falcone and Wolfgang Wirth for their helpful reviews.

## References

Allen, T. O. and Roberts, A. P.: Production Operations, Vol. 2, Oil & Gas Consultants International, Inc., Tulsa, Oklahoma, 3rd Edn., 1989.

Bartels, J. and Wenderoth, F.: Numerische thermisch-hydraulische 3D-Modellierung für den Großraum München, Neubrandenburg and Berlin, 1st Edn., 2012.

Bartels, J., Wenderoth, F., Fritzer, T., Huber, B., Dussel, M., Lüschen, E., Thomas, R., and Schulz, R.: A new simulation model to evaluate interaction between neighbouring hydrogeothermal installations developing the deep Malm aquifer in the Munich region, Geophys. Res. Abstr., Vol. 14, EGU2012–9157, European Geophysical Union, 2012.

Bausch, W. M.: Der Obere Malm an der unteren Altmühl – Nebst einer Studie über das Riff-Problem, Erlanger Geologische Abhandlungen, 49, 38 pp., 1963.

Bayerisches Staatsministerium für Wirtschaft, Infrastruktur, Verkehr und Technologie, Bayerischer Geothermieatlas, 2010.

Bertleff, B. and Watzel, R.: Tiefe Aquifersysteme im südwestdeutschen Molassebecken – Eine umfassende hydrogeologische Analyse als Grundlage eines zukünftigen Quantitäts- und Qualitätsmanagements, Abhandlungen des Landesamts für Geologie, Rohstoffe und Bergbau Baden-Württemberg, 15, 75–90, 2002.

Birner, J., Fritzer, T., Jodocy, M., Savvatis, A., Schneider, M., and Stober, I.: Hydraulische Eigenschaften des Malmaquifers im Süddeutschen Molassebecken und ihre Bedeutung für die geothermische Erschließung, Zeitschrift für geologische Wissenschaften, 40, 133–156, 2012.

Böhm, F., Koch, R., Höferle, R., and Baasch, R.: Der Malm in der Geothermiebohrung Pullach Th2 – Faziesanalyse aus Spülproben (München, S-Deutschland), Geologische Blätter für Nordost-Bayern, 60, 17–49, 2010.

Böhm, F., Birner, J., Steiner, U., Koch, R., Sobott, R., Schneider, M., and Wang, A.: Tafelbankiger Dolomit in der Kernbohrung Moosburg SC4: Ein Schlüssel zum Verständnis der Zuflussraten in Geothermiebohrungen des Malmaquifers (Östliches Molasse-Becken, Malm $\delta - \zeta$; Süddeutschland), Z. Geol. Wissenschaft., 39, 117–157, 2011.

Chigira, M. and Watanabe, M.: Silica precipitation behavior in a flow field with negative temperature gradients, J. Geophys. Res., 99, 15539–15548, 1994.

Duan, Z., Pang, Z., and Wang, X.: Sustainability evaluation of limestone geothermal reservoirs with extended production histories in Beijing and Tianjin, China, Geothermics, 40, 125–135, doi:10.1016/j.geothermics.2011.02.001, 2011.

Fournier, R. O. and Rowe, J. J.: The solubility of amorphous silica in water at high temperatures and high pressures, Am. Mineral., 62, 1052–1056, 1977.

Gdanski, R.: The symmetry of wormholing, E&P, 2001.

Gdanski, R.: Advances in carbonate stimulation, Encounter and Exhibition International of the Oil Industry, College of Mexican Petroleum Engineers, 2005.

Hamill, L. and Bell, F. G.: Groundwater Resource Development, Butterworths, London, 1st Edn., 1986.

Hölting, B. and Coldewey, W. G.: Hydrogeologie, Spektrum Akademischer Verlag, Heidelberg, 2nd Edn., 2009.

Jacob, C. E.: Drawdown test to determine effective radius of artesian well, Transactions, American Society of Civil Engineers, 112, 1047–1070, 1947.

Koch, R.: Daten zur Fazies und Diagenese von Massenkalken und ihre Extrapolation nach Süden bis unter die Nördlichen Kalkalpen, Geologische Blätter für Nordost-Bayern, 47, 117–150, 1997.

Koch, R. and Sobott, R.: Porosität in Karbonatgesteinen – Genese, Morphologie und Einfluss auf Verwitterung und Konservierungsmaßnahmen, Zeitschrift der Deutschen Gesellschaft für Geowissenschaften, 156, 33–50, doi:10.1127/1860-1804/2005/0156-0033, 2005.

Lund, K., Fogler, H. S., McCune, C. C., and Ault, J. W.: Acidization – II. The dissolution of calcite in hydrochloric acid, Chem. Eng. Sci., 30, 825–835, 1974.

Meyer, R. K. F. and Schmidt-Kaler, H.: Jura, in: Erläuterungen zur Geologischen Karte von Bayern 1:500.000, Bayerisches Geologisches Landesamt, München, 90–111, 1996.

Morey, G. W., Fournier, R. O., and Rowe, J. J.: The Solubility of Amorphous Silica at 25 °C, J. Geophys. Res., 69, 1995–2002, 1964.

Prestel, R.: Untersuchungen zur Diagenese von Malm-Karbonatgesteinen und Entwicklung des Malm-Grundwassers im süddeutschen Molassebecken, Ph.D. thesis, Institut für Geologie und Paläontologie der Universität Stuttgart, 1990.

Reinhold, C.: Multiple episodes of dolomitization and dolomite recrystallization during shallow burial in Upper Jurassic shelf carbonates: eastern Swabian Alb, southern Germany, Sediment. Geol., 121, 71–95, 1998.

Rorabaugh, M. I.: Graphical and theoretical analysis of step-drawdown test of artesian wells, Proceedings of the American Society of Civil Engineers, 79, 1–23, 1953.

Schellschmidt, R., Sanner, B., Pester, S., and Schulz, R.: Geothermal Energy Use in Germany, in: Proceedings World Geothermal Congress 2010, International Geothermal Association, 2010.

Schlichting, H. and Gersten, K.: Boundary Layer Theory, Springer, 8th Edn., 2001.

Spurk, J. H.: Fluid Mechanics, Springer, 2006.

Stober, I. and Villinger, E.: Hydraulisches Potential und Durchlässigkeit des höheren Oberjuras und des Oberen Muschelkalks unter dem baden-württembergischen Molassebecken, Jahreshefte des Geologischen Landesamts Baden-Württemberg, 37, 77–96, 1997.

Turcotte, D. L. and Schubert, G.: Geodynamics, Cambridge University Press, Cambridge, 2nd Edn., 2002.

Ungemach, P., Antics, M., and Papachristou, M.: Sustainable Geothermal Reservoir Management, in: Proceedings World Geothermal Congress 2005, International Geothermal Association, 2005.

Villinger, E.: Bemerkungen zur Verkarstung des Malms unter dem westlichen süddeutschen Molassebecken, Bulletin der Vereinigung schweizerischer Petroleum-Geologen und -Ingenieure, 54, 41–59, 1988.

Villinger, E.: Der Oberjura-Aquifer der Schwäbischen Alb und des baden-württembergischen Molassebeckens (SW-Deutschland), Tübinger Geowissenschaftliche Arbeiten, C34, 79–109, 1997.

Wolfgramm, M. and Seibt, A.: Zusammensetzung von Tiefenwässern in Deutschland und ihre Relevanz für geothermische Anlagen, in: Der Geothermiekongress 2008, GtV-Bundesverband Geothermie, 2008.

Wolfgramm, M., Bartels, J., Hoffmann, F., Kittl, G., Lenz, G., Seibt, P., Schulz, R., Thomas, R., and Unger, H. J.: Unterhaching geothermal well doublet: structural and hydrodynamic reservoir characteristics; Bavaria (Germany), in: Proceedings European Geothermal Congress 2007, German Geothermal Association, Berlin, 2007.

Wolfgramm, M., Obst, K., Beichel, K., Brandes, J., Koch, R., Rauppach, K., and Thorwart, K.: Produktivitätsprognosen geothermischer Aquifere in Deutschland, in: Der Geothermiekongress 2009, GtV-Bundesverband Geothermie, 2009.

# 2

# Convective, intrusive geothermal plays: what about tectonics?

**A. Santilano, A. Manzella, G. Gianelli, A. Donato, G. Gola, I. Nardini, E. Trumpy, and S. Botteghi**

National Research Council, Institute for Geosciences and Earth Resources, Pisa, Italy

*Correspondence to:* A. Santilano (alessandro.santilano@igg.cnr.it)

**Abstract.** We revised the concept of convective, intrusive geothermal plays, considering that the tectonic setting is not, in our opinion, a discriminant parameter suitable for a classification. We analysed and compared four case studies: (i) Larderello (Italy), (ii) Mt Amiata (Italy), (iii) The Geysers (USA) and (iv) Kizildere (Turkey). The tectonic settings of these geothermal systems are different and a matter of debate, so it is hard to use this parameter, and the results of classification are ambiguous. We suggest a classification based on the age and nature of the heat source and the related hydrothermal circulation. Finally we propose to distinguish the convective geothermal plays as volcanic, young intrusive and amagmatic.

## 1 Introduction

Geothermal energy is a renewable resource suitable for baseload power production. Various countries are working toward an increase of geothermal exploitation and research development. Although geothermal energy has been exploited for many decades in many countries, a clear and unique classification of geothermal systems has not been accepted worldwide, probably due to the strong variability of geological, geophysical and thermodynamic conditions. In the past, many authors proposed a classification of geothermal systems and resources, based mainly on temperature (e.g. Muffler, 1979; Sanyal et al., 2005). Recently Moeck et al. (2014) proposed an alternative scheme to classify geothermal systems, in the frame of "geothermal plays", based on geological characteristics. The "play" concept hails from oil and gas exploration and corresponds to a "…model in the mind of the geologist of how a number of geological factors might combine to produce petroleum accumulation in a specific stratigraphic level of a basin" (Allen P. A. and Allen J. R., 2005). It is hard to import this concept to geothermal exploration due to the possible development of geothermal systems in many geodynamic settings with extremely various geological characteristics worldwide. On the other hand, we agree with Moeck (2014) on the need for a clear and widely accepted new catalogue of geothermal plays to support geothermal exploration activities at least in their very first activities. By merging different opinions and scientific discussions during a recent workshop held by the International Geothermal Association (IGA) in Essen, Germany (IGA, 2013), classification of the geothermal plays has been attempted as follows:

1. convective, volcanic field, divergent margins;

2. convective, volcanic field, convergent margins;

3. convective, intrusive, extensional;

4. convective, intrusive, convergent;

5. convective, extensional domains fault controlled;

6. conductive, intracratonic basin;

7. conductive, foreland basin/orogenic belt;

8. conductive, basement (igneous and metamorphic).

In this paper we analyse and discuss the structural setting, the heat source and the reservoir characteristics of four important geothermal fields in exploitation over decades: (i) Larderello (Italy), (ii) Mt Amiata (Italy), (iii) The Geysers (USA) and (iv) Kizildere (Turkey). We classify them as convective and intrusive play types, and we stress the similar geological features that could depict this type of play.

The understanding of common features of convective and intrusive plays is important since they host some of the most productive geothermal fields in the world. The Larderello and Mt Amiata fields, both located in Tuscany (Italy), are two large convective geothermal systems with similarities but also many differences. Larderello is one of the few vapour-dominated systems worldwide, where the first geothermal power plant was installed in 1913. The Mt Amiata geothermal area is located close to the homonymous extinct volcano (0.3–0.2 Myr) and is characterized by a liquid-dominated system. The Geysers field is located in California (USA) close to the Clear Lake volcanic field and is the most productive vapour-dominated geothermal system in the world; it has been exploited since the 1960s. The Kizildere field, located in western Turkey, is a liquid-dominated system exploited for power production since 1984, the first field in Turkey. In our analysis we include also Kizildere, which has been differently classified (IGA and IFC, 2014, and references therein), since the aim of our paper is to open a discussion and comparison between the different proposed models.

We argue that the prospective resources are hardly classified on the basis of both tectonic setting and stratigraphic features, and we propose a new classification.

## 2   Structural setting

### 2.1   Larderello and Mt Amiata geothermal fields

The geothermal fields of Larderello and Mt Amiata (southern Tuscany, Italy) are located in the inner part of the Northern Apennines, a sector of the Apennine orogenic belt developed as a consequence of the Cenozoic collision between the European (Corso–Sardinian block) and the Adria plates (Boccaletti et al., 2011). Southern Tuscany is characterized by a shallow Moho discontinuity (20–25 km depth), a reduced lithosphere thickness due to uprising asthenosphere and the delamination of crustal lithosphere (Gianelli, 2008). Many authors proposed a tectonic evolution of the Northern Apennines due to two main deformational processes: (i) a first one related to eastward-migrating compressional tectonics and (ii) a subsequent extensional tectonics migrating eastward which has been affecting the inner part of the orogenic belt since at least the early Miocene (Carmignani et al., 1994; Jolivet et al., 1998; Brogi, 2006, and reference therein). Alternative models have been proposed to describe the tectonic evolution of the inner Northern Apennines (Boccaletti et al., 1997; Bonini and Sani, 2002). These studies revealed a complex tectonic evolution during the Miocene–Pleistocene with alternating compressive and extensional tectonics events but suggest a prevalent contribution of compressive tectonics till the Pleistocene epoch, in contrast with an uninterrupted regional extensional tectonics active since at least the early Miocene as suggested by other authors. After the Pleistocene, southern Tuscany is characterized by active extensional tectonics as inferred from borehole breakout analysis (Mon-

tone et al., 2012). Considering Quaternary tectonics, recent studies have suggested an important role of strike-slip faults and step-over zones controlling the magma emplacement in the inner Northern Apennines (Acocella et al., 2006) and in the Mt Amiata area (Brogi and Fabbrini, 2009). Batini et al. (1985) presented a seismological study of the Larderello area, showing an intense seismic activity of low magnitude, partially induced, that could be correlated with seismically active structures.

### 2.2   The Geysers geothermal field

The Geysers–Clear Lake geothermal field is located in northern California, between the San Andreas fault system and the Coast Range thrust (Stanley and Rodriguez al., 1995). This region belongs to the California Coastal Ranges, and its geological features are a consequence of the eastward subduction of the Farallon oceanic plate underneath the North America Plate since late Mesozoic times. The tectonic evolution of the region is quite complex. The late Mesozoic subduction system along western North America was replaced, in the Eocene period, by the Mendocino Triple Junction, which evolved in the San Andreas transform system (Stanley and Rodriguez, 1995).

The Geysers geothermal field is located between NW-trending right-lateral strike-slip faults that belong to the San Andreas fault system and exhibits normal and strike-slip faulting (Boyle et al., 2013). The analysis of seismicity (Oppenheimer, 1986; Boyle et al., 2013) indicates that most of the fault plane solutions show an extensional and strike-slip component. However, above 1 km depth, a reverse component is present.

### 2.3   Kizildere geothermal field

The Kizildere geothermal field is located in the Denizli and Aydin provinces of western Turkey in the easternmost part of the Büyük Menderes Graben. The western Anatolian horst-and-graben system forms the eastern boundary of the Aegean extensional system, which is one of the most active extensional regions in the world and is undergoing a N–S extension (Gürer et al., 2009, and reference therein). The west Anatolian–Aegean area underwent continental collisions which started in the Mesozoic with a collisional zone migrating southward, down to the present position of the Cyprus–Hellenic subduction zone. The differential velocity fields of plates involved in the western Anatolia–Aegean region may explain the opening of the Aegean extensional system (Doglioni et al., 2002). The Büyük Menderes Graben is about 140 km long and up to 14 km wide, and approximately trends E–W in the Kizildere area. The main normal fault, bounding the northern margin of the graben, terminates eastward close to the Kizildere geothermal field where a horse-tailing termination could facilitate the hydrothermal circulation (Faulds et al., 2009). The Büyük Menderes Graben is characterized by intense seismicity, mainly concentrated in the Çameli–Denizli district close to the geothermal field and dominated by low-magnitude seismic swarms (Süer et al., 2010).

## 3   Heat source and thermal regime

### 3.1   Larderello and Mt Amiata geothermal fields

The geodynamic setting and the magmatic activity produce a huge geothermal anomaly in southern Tuscany, with maximum peaks centred in the Larderello and Mt Amiata areas with values of heat flow up to $1000 \, \text{mW} \, \text{m}^{-2}$ (Baldi et al., 1994). The heat source of Larderello and Mt Amiata geothermal fields is related to shallow igneous intrusions belonging to the Tuscan Magmatic Province (TMP) according to many authors (see Gianelli, 2008, and references therein). Geophysical data (gravimetry, seismic reflection, seismology and MT) and thermal numerical modelling support the hypothesis of deep buried still molten igneous intrusions below the geothermal systems of southern Tuscany (Foley, 1992; Baldi et al., 1994; Batini et al., 1995; Bernabini et al., 1995; Gianelli et al., 1997a; Manzella et al., 1998; Mongelli et al., 1998; Gianelli, 2008). Various models relate the genesis of this magmatic activity in the inner part of the Northern Apennines to the west-dipping subduction, delamination and eastward rollback of the Adriatic lithosphere. Both the magmatism and the extensional tectonics migrated from west to east following the eastward migration of the collisional front. The Larderello intrusive bodies, cored in several deep wells, can be classified as two-mica granites ranging in composition from monzogranites to syeno-monzogranites, with ages ranging from 3.8 to 1.3 Myr (Dini et al., 2005). Gianelli and Puxeddu (1994) summarize the geophysical evidence of the batholith beneath Larderello area: (i) a Bouguer gravity low (20–25 mGal minimum peaks), (ii) a thermal anomaly (heat flow values $> 120 \, \text{mW} \, \text{m}^{-2}$) over an area of $600 \, \text{km}^2$, (iii) P wave delays (up to 1 s), (iv) lack of hypocentres below 7–8 km and (v) mineralogical evidence in well SP2 (post-tectonic occurrence of corundum, sanidine and biotite-tourmaline level).

Mt Amiata is a young (0.3–0.2 Myr) extinct volcano belonging to the TMP made up of trachytes, trachylatites and olivine-latites (Gianelli, 2008). The volcanic edifice hosts an important reservoir of cold and drinkable water and overlies impermeable, clayey units. As for Larderello, the high-temperature hydrothermal circulation occurs in two deep-seated non-volcanic reservoir. Major bodies of intrusive rocks were never crossed by deep wells in Mt Amiata, but the heat source may be related to shallow intrusions inferred from geophysical data (Bernabini et al., 1995; Manzella et al., 1998; Finetti, 2006). This allows us to consider this geothermal play as intrusive.

### 3.2   The Geysers geothermal field

The heat source of The Geysers geothermal field corresponds to a Quaternary pluton complex ($> 100 \, \text{km}^3$) of batholithic dimension known as "felsite" that occurs only in the subsurface and is clearly affiliated geochemically and mineralogically with the Cobb Mountain volcanic centre of the Clear Lake volcanic field (Hulen and Nielson, 1996; Dalrymple et al., 1999). Movement of the Mendocino Triple Junction is widely believed to be the cause of northward-migrating late Tertiary and Quaternary volcanism in the California Coast Ranges (Stanley and Rodriguez, 1995). A slab window is assumed to favour asthenosphere upwelling and basic magmas emplacement that in turn have fractionated, melted or assimilated continental crust, producing felsic magma (Hulen and Nielson, 1996). According to Dalrymple et al. (1999), The Geysers plutonic complex (GPC) crystallized at 1.18 Ma and suggests a further heat source, in addition to the intrusive mass of the GPC, to explain the observed thermal evolution of the complex. Based on deep-well data, Hulen and Nielson (1996) distinguished three type of rocks constituting the igneous body: (i) granite, (ii) microgranite porphyry and (iii) late granodiorite. The presence of batholith is supported by (i) a Bouguer gravity low ($-24 \, \text{mGal}$ minimum peaks), (ii) a thermal anomaly with heat flow values grater than $168 \, \text{mW} \, \text{m}^{-2}$ over an area of $750 \, \text{km}^2$ and values in the range $335$–$500 \, \text{mW} \, \text{m}^{-2}$ over an area of $75 \, \text{km}^2$ centred on the field, (iii) P wave delays (up to 1 s), (iv) lack of hypocentres below 5–7 km depth and (v) the occurrence of a thick aureole of biotite-tourmaline-rich hornfels around the felsite (Walters and Combs, 1989; Gianelli and Puxeddu, 1994; Nielson and Moore, 2000, and reference therein).

Both the huge vapour-dominated reservoir and the upper portion of the felsite are oriented NW–SE, sub-parallel to the right-lateral San Andreas Fault and related wrench faults. In fact, Hulen and Norton (2000) considered the emplacement of the felsite to be possibly related to pull-apart extension. The presence of a batholith or multiple silicic magma chambers at depth are supported by geophysical evidence, but a shallow intrusion cyclically replenished by new magma (at least 500 000 years each) is required to keep the present-day heat flow and thermal anomaly (Erkan et al., 2005).

### 3.3   Kizildere geothermal field

Surface heat flow in western Turkey depicts wide thermal anomalies with values up to $150 \, \text{mW} \, \text{m}^{-2}$ in the Menderes Massif area (Tezcan and Turgay, 1991). The exploration activities carried out in the Kizildere area were not able to clearly identify a shallow intrusion or a magmatic chamber as a possible heat source of the system.

Faulds et al. (2009) compared the western Turkey region to the western Great Basin undergoing significant extension and relatively sparse volcanism and excluded a magmatic heat source at upper crustal levels for the geothermal systems in the area. The primary control of structural features accommodating deep hydrothermal circulation is therefore suggested.

On the other hand, geochemical and isotopic analyses of C, S and B (Özgür, 2002; Simsek, 2003; Özgür and Karamenderesi, 2015) support the hypothesis of relatively shallow and recent magmatic intrusion. On the basis of helium isotopic data, Güleç and Hilton (2006) suggest the occurrence of plutonic activity underneath the Büyük Menderes Graben. This is consistent with the fact that during the late Miocene to Quaternary an oceanic-island basalt (OIB)-type volcanism occurs during the recent extensional phase in the Anatolian–Aegean region (Agostini et al., 2007). The most recent Quaternary volcanic products are found in the Kula region located about 65 km NW of Kizildere field.

The $C^{13}$ analysis indicates a substantial contribution of magmatic $CO_2$, although $CO_2$ mainly derives from decarbonatization processes (Simsek, 2003). By comparison, similar processes occurred at Larderello and Mt Amiata, where $CO_2$ is in part produced by high-temperature decarbonatization reaction of sediments under thermo-metamorphic conditions (Gianelli and Calore, 1996; Gianelli et al., 1997b; Orlando et al., 2010).

Considering that young intrusion is inferred in Kizildere and remembering that intrusive rocks below Larderello and The Geysers were disputed and not clearly proved until these rocks were reached by deep drilling, we include Kizildere in the discussion of intrusive plays.

## 4 Reservoir characteristics

### 4.1 Larderello and Mt Amiata geothermal fields

There are differences and similarities between the Larderello and Mt Amiata geothermal reservoirs. Both areas host two reservoirs, the shallow being hosted in sedimentary units and the deep in crystalline rocks. At Larderello superheated steam is present at depths over 3.5 km and with temperatures exceeding 350 °C, whereas the deep reservoir of the Mt Amiata geothermal fields is in a two-phase (liquid + vapour mixture) state with temperatures of 300–350 °C (Barelli et al., 2010). In the upper levels (shallow reservoir), the Larderello reservoir consists of several rock types: sandstone; marls; radiolarites; and, more commonly, Mesozoic micritic limestone and anhydrite dolostone. The deep reservoir consists of phyllite, micaschist, skarn, hornfelses and granite. Similar rocks form the reservoir of the Mt Amiata geothermal field: Mesozoic limestone and anhydrite dolostone (shallow reservoir), and phyllite, quartzite and dolomitic marbles (deep reservoir) (Pandeli et al., 1988). Strong reflectors in the metamorphic complexes have been explained with rock fracturing and the presence of fluids (Batini et al., 1983; Cameli et al., 1995). Gianelli and Bertini (1993) report the occurrence of a hydrothermal breccia at 1090 m depth and suggest that natural hydraulic fracturing could have occurred within the sys-

tem. Hydraulic fracturing may also be a present-day mechanism of rock fracturing at Larderello. Also at Mt Amiata, in the deep reservoir, the occurrence of hydrothermal breccias (Ruggieri et al., 2004) leads us to assume a similar process of permeability enhancement. Coupled with this process, it is clear that faults and densely fractured zones play a fundamental role in the permeability of the reservoir, considering that primary permeability is extremely low.

Barelli et al. (2010) highlight that the shallow and deep reservoirs of the Mt Amiata system are in piezometric equilibrium as pointed out by the hydrostatic pressure distribution.

Thermal springs and diffuse gas discharge are abundant in both Larderello and Amiata geothermal fields and surrounding areas, with fierce manifestations in Larderello (Duchi et al., 1986; Minissale et al., 1991, 1997; Frondini et al., 2009).

### 4.2 The Geysers geothermal field

At The Geysers the geothermal fluids are hosted principally by highly deformed late-Mesozoic-age subduction-trench-related metasedimentary and meta-igneous rocks of the Franciscan complex. The system is disrupted by high-angle, generally northwest-trending faults related to the still-active San Andreas Fault and low- to moderate-angle thrust faults. The Franciscan rocks at The Geysers are intruded by a northwest-trending Plio–Pleistocene multi-phase felsic pluton, which actually hosts a portion of the steam reservoir and underwent further mineral recrystallization due to the intrusion and related fluids. The configurations of the felsite and reservoir coincide, strongly suggesting that the intrusion critically influenced steam-field evolution (Hulen and Nielson, 1993). The two reservoirs (shallow and deeper "high-temperature zone") produce steam at temperatures in the range 235–342 °C at depth of approximately 500–2500 m b.g.l., and the permeability is mainly related to rock fractures. Recent experimental redrilling and deepening of an abandoned well were able to significantly increase the flow rate of a low-permeability level at 3350 m depth and 400 °C temperature, and, practically, create an enhanced geothermal system (EGS) demonstration project into the high-temperature zone (Garcia et al., 2012)

Geothermal surface manifestation are widely diffused counting several thermal springs in the surrounding area (Donnelly-Nolan et al., 1993).

### 4.3 Kizildere geothermal field

The Kizildere field is characterized by three different liquid-dominated reservoirs hosted in fractured sedimentary and metamorphic rocks. The first well (KD1) was completed in 1968, reaching 198 °C at 540 m b.g.l. in the first reservoir, constituted mainly of Neogene limestone and marls (Kindap et al., 2010). The second geothermal reservoir is hosted in Paleozoic metamorphic rocks (marble and quartzite schists) with a maximum temperature of 212 °C. A third reservoir was discovered in 1998, after the completion of the deep well R1, initially designed for re-injection and converted into a production well. The well R1 reached a temperature of 242 °C at 2261 m b.g.l., as presumed by previous geochemical studies (Serpen and Ugur, 1998, and reference therein). Serpen et al. (2000) do not consider the lithological differences enough to distinguish first and second reservoir but suggested only a single fractured reservoir, independent of stratigraphic features, located at 300–1000 m depth with an average temperature of 205 °C. The hydrothermal circulation in the reservoir is related to structural permeability, and primary porosity of rocks is low. Geothermal surface manifestation are widely diffused in the Kizildere field, counting several thermal springs in the surrounding area, with temperatures ranging from 37 to 88 °C (Özgür, 2002).

## 5 Discussion

In our opinion a worldwide-accepted temperature-based classification of geothermal resources is needed, because it provides a quantitative evaluation of power production. On the other hand, in agreement with the definition of geothermal play of Moeck (2014), in the first stage of exploration it is useful to take into account a catalogue of plays based on geological features. But we argue with the following question: are geological features clear enough for characterizing favourable conditions for geothermal resources?

The comparison of the main geological conditions among the Larderello and Mt Amiata (Italy), The Geysers (USA) and Kizildere (Turkey) geothermal fields led us to identify common features that may characterize the convective and intrusive play. The most important common feature of this type of play is the effectiveness of the heat source represented by shallow plutonic intrusions, although nowadays for Mt Amiata and Kizildere fields the magmatic contribution is only inferred, as it was inferred in Larderello and The Geysers at the beginning. This is a crucial point, and the term intrusive for a play is not so immediate to apply. For example, different models have been also proposed for Kizildere with the magmatic activity the main matter of debate, as in Larderello and The Geysers before drilling and coring granites and felsite. Another example is that recent acidic intrusions are considered to be the heat source of Larderello and Mt Amiata geothermal systems (Bertini et al., 2006; Gianelli, 2008), whereas in the conceptual model of the Larderello geothermal area of Brogi et al. (2003, and references therein) the geothermal area is located in a "basin and range"-like structure, the magmatic contribution as heat source being minimized and depicting an overall scenario of a fault-controlled system.

The efficacy of the heat source is a leading issue. In fact, with regard to the Italian and American fields the available information nowadays endorses the effectiveness of buried intrusion older than 1 Myr. Mathematical models exclude the possibility that intrusions of any reasonable, even large, size can supply enough heat and are able to feed large geothermal systems for more than 1 Myr (Norton and Knight, 1977; Calore et al. 1981; Cathless and Erendi, 1997). A continuous magma, and therefore heat, feeding is therefore necessary to maintain a geothermal system of the size of Larderello or The Geyser. In our opinion, the term "intrusive" should be accompanied by the term "Young". We can define young as an intrusion at least coeval, or younger than the last tectonic phase affecting the geothermal area, and if isotopic dating is available it should be younger than approximately 1 Myr.

The age of the magmatism is used also for the catalogue of geothermal play proposed by IGA and IFC (2014), which, however, lacks a clear distinction of volcanic and intrusive plays. In our opinion a system fed by young intrusions with the geothermal reservoir hosted in the associated volcanites has different features with respect to a "convective, intrusive" play, with the reservoir hosted in sedimentary and crystalline units. The geothermal plays characterized by intrusions approximately older than 1 Myr and without evidence of melt or partial melting in the upper crust could be included into a "amagmatic play", to be eventually sub-classified. Geochemical data on surface manifestations should be considered to support the cataloguing activities because they provide useful information about the hypothesis of magmatic contribution, helping in discriminating the intrusive and amagmatic systems.

Of course, large or composite batholiths are better heat sources than small dikes or laccoliths, which cannot induce thermal anomaly for a long period of time. The volume of the intrusion, however, is not a good discriminating parameter, because during the exploration it is difficult to define its size. Thus, further distinguishing the intrusive plays on the basis of the size of the intrusion is in our opinion not of practical use.

Another important issue is the convective heat transfer that implies the circulation of a thermovector fluid. This condition distinguishes the conventional system exploitable by current technologies from the conductive unconventional geothermal systems that require engineering stimulation. The four geothermal systems are classified as convective since they show a wide and effective hydrothermal circulation, even complex, considering the presence of more than one reservoir for each field. In our analysis we could count on geophysical data and well logs for fields in operation. Considering an initial stage of exploration for a play, without geophysical data, it is difficult to assess the regime of heat transfer at depth. A preliminary indication could be provided by the number and type of geothermal manifestations in the surrounding areas. Surface manifestations (e.g. hot springs and gas discharge) are common in the four fields of interest, disregarding the fluid phase in the reservoir (steam-dominated in Larderello and The Geysers, and liquid-dominated in Amiata and Kizildere). In the considered cases, the low-permeability layer acting as a cap rock is a low-permeability sedimentary or crystalline unit, and the abundance and distribution of natural manifestations, as well as the hydraulic head of steam or brine, are strictly related to the depth of the reservoir and the faults and fractures regime.

Attempts at evaluating the steam fraction in a geothermal reservoir have been proposed by D'Amore and Truesdell (1979), but its application before drilling and during the geochemical survey of natural manifestation is problematic. In any case the presence of steam phase is not an indication for intrusive or fault-controlled geothermal systems. For example Mt Amiata is an intrusive geothermal system with liquid-dominated reservoirs.

The comparison of these fields drove us to exclude lithological and stratigraphic conditions as key parameters to classify the plays. A geothermal reservoir can be hosted in various typologies of sedimentary and crystalline rocks. What really matters is the rock permeability.

What makes things even more complex is the geodynamic and structural setting, which may spatially vary in stress regime (from compressive to extensional or strike slip) and in time (polyphasic tectonic). In areas rich in data such as those we analysed the tectonic evolution is still under debate in the scientific community. We have shown that Larderello and Mt Amiata are located in the inner sector of an active orogenic belt that has undergone extensional tectonics since the Miocene or Pleistocene (non-univocal consensus about timing), and some authors have suggested the importance of recent strike-slip faults during the emplacement of plutons that represent the heat source.

The main elements in common in the four fields which we used for the classification are the hydrothermal circulation and the known or inferred plutonic heat source, respectively identifying the convective and intrusive terms. We observe that there are two other common parameters in the four areas: (i) relevant seismic activity and (ii) high heat flow. It is known that geothermal fields are common in tectonically active areas and earthquake swarms could be associated with areas of recent volcanic or geothermal activity (Sibson, 1996). The heat flow values depict huge thermal anomalies in the surrounding areas of the Larderello, Mt Amiata, The Geysers and Kizildere fields, with maximum values centred on the field in exploitation.

## 6   Conclusions

We compared the main geological features of the Larderello, Mt Amiata, The Geysers and Kizildere geothermal fields in order to describe the common elements that could be useful for the classification of geothermal plays based on the terminology proposed for the IGA workshop held in Essen, Germany (IGA, 2013). We classified these fields as convective and intrusive plays. The first term would indicate the presence of a reservoir suitable for economic exploitation with current technologies without engineering stimulation. The term intrusive is correlated with the plutonic heat source

that feeds wide and highly productive geothermal systems. We do not adhere to the proposal to split this play into different kinds depending on the tectonic setting. Considering that a play should be defined in an unambiguous way, and should help in classifying resources and planning exploration decisions, we conclude that recognized resources, such as those we analysed, and even more so the prospective resources, can hardly be classified on the basis of tectonic setting. We explained that geodynamic and structural setting are still debated in such well-known fields, and a tectonics-based classification of geothermal plays could not simplify the exploration planning. The structural survey remains a milestone in a geothermal exploration project to understand tectonics evolution and to assess the faults and fractures systems that control hydrothermal circulation.

With regard to the classification of geothermal plays we suggest simplifying the classification of the convective plays, distinguishing volcanic, young intrusive and amagmatic. Our classification reduces the emphasis on the tectonic settings, which can be subjective and therefore lead to ambiguous conceptual models. Besides highlighting the importance of geochemical data for inferring magmatic heat source, we identify two more features that are common in the four fields: (i) they are seismically active, and (ii) they show high heat flow values and wide thermal anomalies. These features, more than structural and tectonic features, might be used for a sub-classification.

**Acknowledgements.** The authors would like to thank the anonymous reviewers for their valuable comments and suggestions, which greatly contributed to improve the final version of the paper. A special acknowledgement also goes to Samuele Agostini for the useful discussion on the subject of this study.

## References

Acocella, V. and Funiciello, R.: Transverse systems along the extensional Tyrrhenian margin of central Italy and their influence on volcanism, Tectonics, 25, 1–24, doi:10.1029/2005TC001845, 2006.

Agostini, S., Doglioni, C., Innocenti, F., Manetti, P., Tonarini, S., and Savaşçin, M. Y.: The transition from subduction-related to intraplate Neogene magmatism in the Western Anatolia and Aegean area, in: Cenozoic Volcanism in the Mediterranean Area, edited by: Beccaluva, L., Bianchini, G., and Wilson, M., Geol. S. Am. S., 418, 1–15, doi:10.1130/2007.2418(01), 2007.

Allen, P. A and Allen, J. R.: The Petroleum Play, in: Basin Analysis-Principles and Applications, Blackwell Science Ltd., Oxford, UK, 405–494, 2005.

Baldi, P., Bellani, S., Ceccarelli, A., Fiordelisi, A., Squarci, P., and Taffi, L.: Correlazioni tra le anomalie termiche ed altri elementi geofisici e strutturali della Toscana meridionale, Studi Geologici Camerti, Volume Speciale 1994-1, 139–149, 1994.

Barelli, A., Ceccarelli, A., Dini, I., Fiordelisi, A., Giorgi, N., Lovari, F., and Romagnoli, P.: A Review of the Mt. Amiata Geothermal System (Italy), Proceedings of the World Geothermal Congress, Bali, Indonesia, 25–29 April 2010, 1–6, 2010.

Batini, F., Bertini, G., Gianelli,G., Pandeli, E., and Puxeddu, M.: Deep structure of the Larderello field: contribution from recent geophysical and geological data, Mem. Soc. Geol. It., 25, 219–235, 1983.

Batini, F., Console, R., and Luongo, G.: Seismological study of Larderello-Travale geothermal area, Geothermics, 14, 255–272, 1985.

Batini, F., Fiordelisi, A., Graziano, F., and Toksöz, M.N.: Earthquake Tomography in the Larderello Geothermal Area, Proceedings of the World Geothermal Congress, Florence, Italy, 18–31 May 1995, 1995-2, 817–820, 1995.

Bernabini, M., Bertini, G., Cameli, G. M., Dini, I., Orlando, L.: Gravity Interpretation of Mt. Amiata Geothermal Area (Central Italy), Proceedings of the World Geothermal Congress, Florence, Italy, 18–31 May 1995, 1995-2, 817–820, 1995.

Bertini, G., Casini, M.. Gianelli, G., and Pandeli E.: Geological structure of a long-living geothermal system, Larderello, Italy, Terra Nova, 18, 163–169. 2006.

Boccaletti, M., Gianelli, G., and Sani, F.: Tectonic regime, granite emplacement and crustal structure in the inner zone of the Northern Apennines (Tuscany, Italy): A new hypothesis, Tectonophysics, 270, 127–143, doi:10.1016/S0040-1951(96)00177-1, 1997.

Boccaletti, M., Corti, G., and Martelli, L.: Recent and active tectonics of the external zone of the Northern Apennines (Italy), Int. J. Earth Sci., 100, 1331–1348, doi:10.1007/s00531-010-0545-y, 2011.

Bonini, M. and Sani, F.: Extension and compression in the Northern Apennines (Italy) hinterland: Evidence from the late Miocene-Pliocene Siena-Radicofani Basin and relations with basement structures, Tectonics, 21, 1–33, doi:10.1029/2001TC900024, 2002.

Boyle, K. and Zoback, M.: Stress and fracture orientation in the northwest Geysers geothermal field, Proceedings of Thirty-Eighth Workshop on Geothermal Reservoir Engineering, Stanford University, Stanford, CA, USA, 11–13 February 2013, 7 pp., 2013.

Brogi, A., Lazzarotto, A., Liotta, D., and Ranalli, G.: Extensional shear zones as imaged by reflection seismic lines: the Larderello geothermal field (Central Italy), Tectonophysics, 363, 127–139. 2003.

Brogi, A.: Neogene extension in the Northern Apennines (Italy): insights from the southern part of the Mt. Amiata geothermal area, Geodin. Acta, 19/1, 1–9, doi:10.3166/ga.19.33-50, 2006.

Brogi, A. and Fabbrini, L.: Extensional and strike-slip tectonics across the Monte Amiata–Monte Cetona transect (Northern Apennines, Italy) and seismotectonic implications, Tectonophysics, 476, 195–209, doi:10.1016/j.tecto.2009.02.020, 2009.

Calore, C., Celati, R., Gianelli, G., Norton, D., and Squarci, P.: Studi sull'origine del sistema geotermico di Larderello. Atti II Seminario Informativo del Sottoprogetto Energia Geotermica, Progetto Finalizzato Energetica, Roma, Italy, 16–19 June 1981, PEG Editrice, 218–225, 1981.

Cameli, G. M., Batini, F., Dini, I., Lee, J. M., Gibson R. L., and Toksoz M. N.: Seismic delineation of a geothermal reservoir in the Monteverdi area from VSP data, Proceedings of World Geothermal Congress, Florence, Italy, 18–31 May 1995, 821–826, 1995.

Carmignani, L., Decandia, F. A., Fantozzi, P. L., Lazzarotto, A., Liotta, D., and Meccheri, M.: Tertiary extensional tectonics in Tuscany (Northern Apennines, Italy), Tectonophysics, 238, 295–315, 1994.

Cathless, L. M. and Erendi, A. H. J.: How long can a hydrothermal system be sustained by a single event?, Econ. Geol., 77, 1071–1084, 1997.

D'Amore, F. and Truesdell, A. H.: Models for steam chemistry at Larderello and The Geyser, Proc. 5th Workshop on Geothermal Reservoir Engineering, 12–14 December 1979, 283–297, Stanford, CA, USA, 1979.

Dalrymple, G. B., Grove, M., Lovera, O. M., Harrison, T. M., Hulen, J. B., and Lanphere, M. A.: Age and thermal history of the Geysers plutonic complex (felsite unit), Geysers geothermal field, California: a $^{40}$Ar/$^{39}$Ar and U–Pb study, Earth Planet. Sc. Lett., 173, 285–298, 1999.

Dini, A., Gianelli, G., Puxeddu, M., and Ruggieri, G.: Origin and evolution of Pliocene–Pleistocene granites from the Larderello geothermal field (Tuscan Magmatic Province, Italy), Lithos, 81, 1–31, doi:10.1016/j.lithos.2004.09.002, 2005.

Doglioni, C., Agostini, S., Crespi, M, Innocenti, F., Manetti, P., Riguzzi, F., and Savaşçin, Y.: On the extension in western Anatolia and the Aegean sea, in: Reconstruction of the evolution of the Alpine-Himalayan Orogen, edited by: Rosenbaum, G. and Lister, G. S., Journal of the Virtual Explorer, 8, 161–176, 2002.

Donnelly-Nolan, J. M., Burns, M. G., Goff, F. E., Peters, E. K., and Thompson, J. M.: The Geysers-Clear Lake area, California: thermal waters, mineralization, volcanism, and geothermal potential, Econ. Geol., 88, 301–316, doi:10.2113/gsecongeo.88.2.301, 1993.

Duchi, V., Minissale, A., and Rossi, R.: Chemistry of thermal springs in the Larderello-Travale geothermal region, southern Tuscany, Italy, Appl. Geochem., 1, 659–667, 1986.

Erkan, K., Blackwell D. D., and Leidig M. M.: Crustal thermal regime at The Geysesrs/Clear Lake Area, California, Proceedings World Geothermal Congress, Antalaya, Turkey, 24–29 April 2005, 1–9, 2005.

Faulds, J. E., Bouchot, V., Moeck, I., and Oguz, K.: Structural Controls on Geothermal Systems in Western Turkey: A Preliminary Report, Geoth. Res. T., 33, 375–381, 2009.

Finetti, I. R.: Basic regional crustal setting and superimposed local pluton-intrusion-related tectonics in the Larderello–Mt. Amiata geothermal province, from integrated CROP seismic data, Boll. Soc. Geol. Ital., 125, 117–146, 2006.

Foley, J. E., Toksoz, M. N., and Batini, F.: Inversion of teleseismic traveltime residuals for velocity structure in the Larderello geothermal system, Italy, Geophys. Res. Lett., 19, 5–8, 1992.

Frondini, F., Caliro, S., Cardellini, C., Chiodini, G., and Morgantini, N.: Carbon dioxide degassing and thermal energy release in the Monte Amiata volcanic-geothermal area (Italy), Appl. Geochem., 24, 860–875, 2009.

Garcia, J., Walters, M., Beall, J., Hartline, C., Pingol, A., Pistone, S., and Wright, M.: Overview of The Northwest Geysers EGS Demonstration Project, Proceedings of Thirty-Seventh Workshop on Geothermal Reservoir Engineering, Stanford, CA, USA, 30 January–1 February 2012, 11 pp., 2012.

Garg, S. K., Haizlip, J., Bloomfield, K. K., Kindap, A., Haklidir, F. S. T., and Guney, A.: A Numerical Model of the Kizildere Geothermal Field, Turkey, Proceedings of World Geothermal Congress, 19–25 April 2015, Melbourne, Australia, 1–15, 2015.

Gianelli, G.: A comparative analysis of the geothermal fields of Larderello and Mt. Amiata, Italy, in: Geothermal energy research trends, edited by: Ueckermann, H. I., Nova Science Publishers, New York, USA, 59–85, 2008.

Gianelli, G. and Bertini, G.: Natural hydraulic fracturing in the Larderello geothermal field: evidence from well MV5A, Boll. Soc. Geol. Ital., 112, 507–512, 1993.

Gianelli, G. and Puxeddu, M.: Geological comparison between Larderello and The Geysers geothermal fields, Mem. Soc. Geol. It., 48, 715–717, 1994.

Gianelli, G. and Calore, C.: Models for the origin of carbon dioxide in the Larderello geothermal field, Boll. Soc. Geol. Ital., 115, 75–84. 1996

Gianelli, G., Manzella, A., and Puxeddu, M.: Crustal models of the geothermal areas of southern Tuscany (Italy), Tectonophysics, 281, 221–239, 1997a.

Gianelli, G., Ruggieri, G., and Mussi, M.: Isotopic and fluid inclusion study of hydrothermal and metamorphic carbonates in the Larderello geothermal field and surrounding areas, Italy, Geothermics, 26, 393–417. 1997b

Güleç, N. and Hilton, D. R.: Helium and heat distribution in Western Anatolia, Turkey. Relationship to active extension and volcanism, in: Postcollisional tectonics and magmatism in the Mediterranean region and Asia, edited by: Dilek, Y. and Pavlides, S., Geol. S. Am. S., 409, 305–319, doi:10.1130/2006.2409(16), 2006.

Gürer, Ö.F., Sarica-Filoreau, N., Özburan, M., Sangu, E., and Dogan, B.: Progressive development of the Büyük Menderes Graben based on new data, western Turkey, Geol. Mag., 146, 652–673, doi:10.1017/S0016756809006359, 2009.

Hulen, J. B. and Nielson, L. D.: Interim Report on Geology of The Geysers Felsite, Northwestern California, Geoth. Res. T., 17, 249–258, 1993.

Hulen, J. B. and Nielson, L. D.: The Geysers felsite, Geoth. Res. T., 20, 295–306, 1996.

Hulen, J. B. and Norton, D. L.: Wrench-fault tectonics and emplacement of The Geysers Felsite, Geoth. Res. T., 24, 289–298, 2000.

IGA: Workshop: Developing Best Practice for Geothermal Exploration and Resource/Reserve Classification, 14 November 2013, Essen, Germany, available at: http://www.geothermal-energy.org/reserves_and_resources/workshop_essen.html (last access: September 2015), 2013.

IGA and IFC: Best practices guide for geothermal exploration, IGA Service GmbH, 2014.

Jolivet, L., Faccenna, C., Goffé, B., Mattei, M., Rossetti, F., Brunet, C., Storti, F., Funiciello, R., Cadet, J.P., d'Agostino, N., and Parra, T.: Midcrustal shear zones in postorogenic extension: Example from the northern Tyrrhenian Sea, J. Geophys. Res, 103, 12123–12160, 1998.

Kindap, A., Kaya, T., Haklıdır, F. S. T., and Bükülmez, A. A.: Privatization of Kizildere Geothermal Power Plant and New Approaches for Field and Plant, Proceedings of the World Geothermal Congress, Bali, Indonesia, 25–29 April 2010, 1–4, 2010.

Manzella, A., Ruggieri, G., Gianelli, G., and Puxeddu, M.: Plutonic-Geothermal systems of southern Tuscany: a review of the crustal models, Mem. Soc. Geol. It., 52, 283–294, 1998.

Minissale A.: The Larderello geothermal field: a review, Earth-Sci. Rev., 31, 133–151, 1991.

Minissale, A., Magro, G., Vaselli, O., Verrucchi, C., and Perticone, I.: Geochemistry of water and gas discharges from the Mt. Amiata silicic complex and surrounding areas (central Italy), J. Volcanol. Geoth. Res., 79, 223–251, 1997.

Moeck, I. S.: Catalog of geothermal play types based on geologic controls, Renew. Sust. Energ. Rev., 37, 867–882, doi:10.1016/j.rser.2014.05.032, 2014.

Mongelli, F., Palumbo, F., Puxeddu, M., Villa, I. M., and Zito, G.: Interpretation of the geothermal anomaly of Larderello, Italy, Mem. Soc. Geol. It., 52, 305–318, 1998.

Montone P., Mariucci, M. T., and Pierdominici, S.: The Italian present-day stress map, Geophys. J. Int., 189, 705–716, doi:10.1111/j.1365-246X.2012.05391.x, 2012.

Muffler, L. J. P: Assessment of geothermal resources of the United States – 1978, USGS, Circular 790, 163 pp., 1979.

Nielson, D. and Moore, J. N.: The Deeper Parts of The Geysers Thermal System-Implications for Heat Recovery, Geoth. Res. T., 24, 299–302, 2000.

Norton, D and Knight, J.: Transport phenomena in hydrothermal systems: cooling plutons, Am. J. Sci., 277, 937–981, 1977.

Oppenheimer, D. H.: Extensional tectonics at The Geysers geothermal area, J. Geophys. Res., 91, 11463–11476, 1986.

Orlando, A., Conte, A. M., Borrini, D., Perinelli, C., Gianelli, G., and Tassi, F.: Experimental investigation of $CO_2$-rich fluids production in a geothermal area: The Mt. Amiata (Tuscany, Italy) case study, Chem. Geol., 274, 177–186, 2010.

Özgür, N.: Geochemical Signature of the Kizildere Geothermal Field, Western Anatolia, Turkey, Int. Geol. Rev., 44, 153–163, 2002.

Özgür, N. and Karamenderesi, İ. H.: An Update of the Geothermal Potential in the Continental Rift Zone of the Büyük Menderes, Western Anatolia, Turkey, Proceedings, Fortieth Workshop on Geothermal Reservoir Engineering Stanford University, 26–28 January 2015, Stanford, CA, USA, sgp-tr-204, 1–7, 2015.

Pandeli, E., Puxeddu, M., Gianelli, G., Bertini, G., and Castellucci, P.: Paleozoic sequences crossed by deep drillings in the Monte Amiata geothermal region (Italy), Boll. Soc. Geol. Ital., 107, 593–606, 1988.

Ruggieri G., Giolito C., Gianelli G., Manzella A., and Boiron M. C.: Application of fluid inclusions to the study of Bagnore geothermal field (Tuscany, Italy), Geothermics, 33, 675–692, 2004.

Serpen, U. and Ugur, Z.: Reassessment of Geochemistry of the Kizildere Geothermal Field, Geoth. Res. T., 22, 135–140, 1998.

Serpen, U. and Satman, A., Reassessment of the Kizildere geothermal reservoir, Proceedings of World Geothermal Congress, 28 May–10 June 2000, Beppu-Morioka, Japan, 2869–2874, 2000.

Sibson, R. H.: Structural permeability of fluid-driven fault-fracture meshes, J. Struct. Geol., 18, 1031–1042, 1996.

Simsek, S., Hydrogeological and isotopic survey of geothermal fields in the Buyuk Menderes graben, Turkey, Geothermics, 32, 669–678, doi:10.1016/S0375-6505(03)00072-5, 2003.

Stanley, W. D. and Rodriguez, B. D.: A Revised Tectonic Model for The Geysers- Clear Lake Geothermal Region, California, Proceedings of the World Geothermal Congress, Florence, Italy, 18–31 May 1995, 1193–1198, 1995.

Süer, S., Wiersberg, T., Güleç, N., Erzinger, J., and Parlaktuna, M.: Geochemical Monitoring of the Seismic Activities and Noble Gas Characterization of the Geothermal Fields along the Eastern Segment of the Büyük Menderes Graben, Proceedings of World Geothermal Congress, 25–29 April 2010, Bali, Indonesia, 1–8, 2010.

Tezcan, A. K. and Turgay, M. I.: Heat flow and temperature distribution in Turkey, in: Geothermal Atlas of Europe, edited by: Hurtig, E., Cermak, V., Haenal, R., and Zui, V., 84–85, 1991.

Walters, M. and Combs, J.: Heat Flow Regime in The Geysers-Clear Lake Area of Northern California, USA, Geoth. Res. T., 13, 491–502, 1989.

**3**

# Stored-heat assessments: a review in the light of field experience

**M. A. Grant**

MAGAK, 14A Rewi Rd, Auckland 1023, New Zealand

*Correspondence to:* M. A. Grant (malcolm@grant.net.nz)

**Abstract.** Stored-heat or volumetric assessments of geothermal resources are appealingly simple: the resource being exploited is heat. A stored-heat calculation simply computes the amount of heat in the resource, similarly to computing the amount of ore in an ore body. The method has theoretical support in numerical simulations of resource production. While there are significant unknowns in any resource, some of these can be covered by probabilistic approaches, notably a Monte Carlo method. The Australian Geothermal Reporting Code represents one specification of such stored-heat assessments.

However the experience of recent decades, with the development of significant numbers of geothermal resources, has shown that the method is highly unreliable and usually biased high. The tendency to overestimates, in particular, has led to the reduced credibility of the method. An example is quoted where simple application of the apparently simple rules gives a ridiculous result. Much of the problem lies in the "recovery factor", the proportion of the resource that can actually be exploited, where comparison with actual performance shows past values have been in all cases too high, as is the current version of the Australian code.

There are further problems, usually overlooked, in the way that the reservoir volume and "cutoff temperature" are defined. Differing approaches mean that results between different reports are not comparable. The different approaches also imply unrecognised assumptions about the physical processes controlling reservoir depletion. The failure of Monte Carlo methods is similarly due to unrecognised violation of logical consistency in the use of probabilities.

The net effect of these problems is that the method is not a simple means to generate a rough resource estimate, and it often generates faulty results. Usually, such results are overestimates. Monte Carlo methods do not provide a protection against these errors.

The Australian Geothermal Reporting Code should be used for hydrothermal systems with an average recovery factor of 10 %. With this average, results are subject to an error of ±70 %. For enhanced geothermal systems (EGS), the recovery factor should be a few percent.

## 1 Stored-heat definitions

The amount of heat stored in a geothermal reservoir is given, in simple form, by

$$Q = \int \rho_f C_f (T - T_{\mathrm{ref}}) \mathrm{d}V, \tag{1}$$

where $\rho_f C_f$ is the heat capacity of the rock (including pore fluid), $T$ is the rock temperature and $T_{\mathrm{ref}}$ is a cutoff or base temperature. This formulation omits dependence on porosity to which the result is only slightly sensitive. The integral is taken over the entire volume of the reservoir which is being assessed. The practicalities of geological complexity mean that only a fraction ($\eta$) of this heat will in fact be recoverable, and the amount is reduced by this "recovery factor". Finally, the produced heat will be converted to electricity by plant with thermal efficiency ($\eta'$). This gives the amount of electrical energy recoverable from the resource as

$$E = \eta\eta' Q = \eta\eta' \int \rho_f C_f (T - T_{\mathrm{ref}}) \mathrm{d}V. \tag{2}$$

The method is appealingly simple: draw isotherms, evaluate the integral over the volume and multiply by the recovery factor and thermal efficiency to get a total quantum of electricity. Then dividing by the planned lifetime gives a power capacity in megawatts.

There are some parameters in the formula. The base temperature ($T_{\text{ref}}$) is normally taken as the reject temperature ($T_{\text{r}}$) of the turbine, on the basis that it is the heat supplied to the plant, above this temperature, that is being used. The thermal efficiency ($\eta'$) is determined by the laws of thermodynamics and plant design. For the binary plant that is commonly used, $T_{\text{ref}}$ is usually around 80–90 °C, and $\eta'$ is frequently around 10 %. For the recovery factor ($\eta$), a value must be assumed.

The stored-heat method was first developed by the U.S. Geological Survey (USGS; White and Williams, 1975), for a national assessment of geothermal reserves. Simple models of flow in fractured porous medium were used to estimate an average value of the recovery factor of 25 % (Nathenson, 1975). This value of 25 % has been used in most applications of the method since then. Sometimes the inevitable uncertainty about the resource is reflected by probabilistic results – usually a Monte Carlo approach, with a range of values being assumed for various parameters, sometimes including the recovery factor. It is hoped that by doing this a more robust final answer will be obtained.

More recently the stored-heat method has been codified in the Australian Geothermal Reporting Code (AGEA AGEG 2010). The method is not changed from the USGS original.

In all reserve assessments there is an assumption about the technology and economics – proven reserves use current technology and current economics. Reserves dependent upon an anticipated change in either technology or economics form a contingent reserve.

## 2   Stored heat in practice

The stored-heat method is purely theoretical, lacking any support from observation. Ideally such a method would be validated by an estimate being made, and then later compared with actual results when the field is subsequently developed. There were many geothermal developments of hydrothermal systems through the late-20th century, and although there were no explicit comparisons it became apparent that past estimates had been too high. This was first reported by Grant (2000), and later by Sanyal et al. (2002, 2004) and Stefansson (2005). It was observed that past estimates had in some cases been several-fold overestimates. Because it is an heuristic fudge factor, effort has tended to focus on the recovery factor, with these observations generally being regarded as showing that past values were too high.

Sanyal et al. (2002) compared past stored-heat assessments against actual performance and numerical simulation, and found that factors of 5–10 % were "a more reasonable range of values". Sanyal et al. (2004) similarly reviewed the

USGS assessments (Muffler, 1978) and found that the total resource was one-third of the original estimate, and that recovery factors should lie in the range 3–17 % with a mean of 11 %. Williams (2004) similarly reviewed performance in three US fields and found recovery factors closer to 10 than 25 %, and that recovery varied strongly between fields. Note that in all three studies the recovery factor is in every case under 25 %: the implication is that every field reviewed was initially overestimated. These results are supported by the observation that US geothermal generation operates at around 60 % of capacity (Lund et al., 2010), arguing for much past oversizing. The same proportion applies to the state of Nevada, so this result is not over-weighted by The Geysers.

These results represent the only published evaluation of the stored-heat method compared against observation. They clearly establish that 25 % is too high a recovery factor and that an average value of around 10 % corresponds to observed results. Beyond establishing the correct average recovery factor, there are a wide range of recovery factors: 3–17 % covers the entire range of observed results. This indicates that any result is subject to an error of at least a factor of 2, or alternatively ±70 %.

No method was proposed for discriminating between fields with low and high recovery. In a different context, Wilmarth and Stimac (2014) indicate differences in power density (which imply similar differences in recovery factor) depending on tectonic environment, with fields in extensional tectonic environments having higher recovery than those in compressive environments.

The Australian code, in its first draft, recommended 25 %. In the second edition, in the light of Williams' work, this was revised to 14–17.5 %. It is difficult to see any justification for this value, which is still too high, and in consequence the code, on average, still overstates reserves by up to 75 %.

## 3   Other parameters

### 3.1   Base temperature

The focus on the recovery factor has diverted attention from other uncertainty within the stored-heat model. For example, Sarmiento and Björnsson (2007) reviewed past reserve estimates in the Philippines and found that they were roughly in line with subsequent performance. The author can also confirm this latter observation from experience in the Philippines. However, it is critical to note that the stored-heat method in use in the Philippines, derived from Icelandic practice (Pálmason et al., 1985), is different from the USGS method. The base temperature, $T_{\text{ref}}$, normally used is not the reject temperature of the turbine, but the temperature $T_{P_{\text{min}}}$ at which wells cease to operate. This is typically 180–200 °C. The practical consequence is that stored heat, computed using the Philippine method, for typical Philippine resources, is 35–40 % of the value that would be computed using the

USGS method. The author has seen simultaneous evaluations of the same resource, using the Philippine and the USGS method. The latter gave a result 2.5 times larger. Thus in fact the results of Sarmiento and Björnsson, while supporting the Philippine results with a recovery factor of 25 %, also indicate that had the USGS method been used, the recovery factor should have been 10 % rather than 25 %.

Which choice for the base temperature is appropriate depends upon an assumption about the state of the reservoir at abandonment. The heat that is extracted from the reservoir is the difference between the initial and abandonment states, integrated over the entire reservoir. Consider the contrasting examples of a reservoir produced by cold sweep and a reservoir produced by boiling down or mixing. Assume that the production and injection wells are all ideally placed so that production continues from all wells, until all fail simultaneously as their temperature falls too low.

If there is a homogeneous reservoir containing liquid, with a smooth inward sweep of water from the injection wells, the production area remains near original temperature until the thermal front arrives. This will be at the injection temperature, which is the turbine reject temperature. So in this case all the heat above the reject temperature is produced, and the reject temperature is the proper choice for the base temperature.

Conversely, consider a high-temperature reservoir containing two-phase fluid which boils continuously under production, or a reservoir subject to rapid mixing of injectate into the production area. In both cases the reservoir cools steadily under production, and production ceases when temperature falls to the minimum producible temperature. That temperature is the temperature of the reservoir at abandonment, and so only the heat above this temperature has been produced.

The difference can be shown clearly by considering a hypothetical liquid reservoir at 180 °C. The minimum production temperature ($T_{Pmin}$) is also 180 °C. A binary turbine using the produced fluid has a reject temperature ($T_r$) of 80 °C. If the latter is used as the cutoff temperature, there is a substantial resource. If $T_{Pmin}$ is used as the cutoff, there is no resource at all. Now consider actual exploitation of this resource. If waste water is injected at a distant site, from which it returns, sweeping water at reservoir temperature ahead of it, the wells will continue to produce for an extended period and there is a resource to exploit. On the other hand, if injectate returns immediately into the production area, mixing with and cooling the reservoir, the reservoir temperature quickly drops below 180 °C, and the wells and the project cease operation. In the first case $T_r$ is the correct choice for cutoff; in the second case $T_{Pmin}$ is the correct choice and there is very little exploitable resource present.

Of course in the latter case the use of downhole pumps would make the resource exploitable. This is a change in the assumed technology, as the presumed $T_{Pmin}$ would no longer apply. Such a technological change does not modify the prior assessment; it introduces a new assessment dependent upon different assumptions.

Actual field behaviour normally lies between these extremes, so there is some argument for either choice. It is important that the choice made be clear, and consistently applied. In particular, if some field experience is being used to justify the choice of recovery factor, it is important that the same method be used throughout.

Garg and Combs (2011) give further elaboration on reference temperature, conversion efficiency and how these apparently simple parameters depend on the specific case.

## 3.2 Volume

Further assumptions arise in the definition of the reservoir volume. The problem is that many geothermal fields contain regions of very poor permeability, such that they do not constitute a reservoir at all. For a volume of hot rock to contribute to production, and hence to the field's reserves, there must be sufficient permeability within it for fluid flow, and pressure changes, to propagate through the rock during the production lifetime. Often part of the field fails to reach this minimal level of permeability.

One approach has been to ignore this problem and assume this is one of the many complications swept into the recovery factor: there is within the recovery factor an implicit allowance for part of the reservoir being impermeable. A contrasting approach has been to consider that proven reserves derive only from regions known to be permeable. At whatever the state of exploration, contours are drawn around the productive wells and this region only is counted. This second approach tends to err on the low side, because of the probable reserves that may exist in regions not yet drilled.

Either approach can be used provided it is used consistently. For undrilled fields where the only information on area is geophysical, there is little alternative to using the entire area outlined by geophysical data. It would seem that the recovery factor for such an approach should be lower than when the area considered is restricted to that drilled successfully.

A further issue arises with respect to the depth of the reservoir. Consider a hypothetical field in which permeability decreases markedly with increasing depth (as is often the case). An array of wells have already been drilled to mid-depths, supporting some level of production. A new well is drilled to significantly greater depth. It finds increasing temperature but no significant permeability below that already found. Has it added to the field reserves by adding more volume at the reservoir bottom? There is at this point no known productive capacity at greater depth. The hot rock and fluid at this greater depth only increase the field's long-term production if, over field life, fluid circulates to these depths – the most likely scenario would be injection at depth. If the permeability is too low, the additional hot rock is worthless for pro-

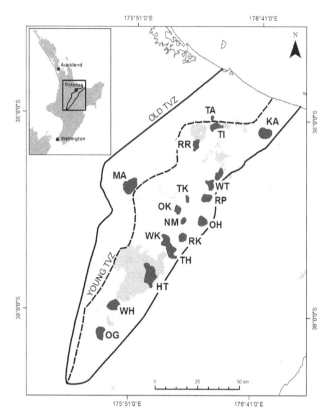

**Figure 1.** Geothermal fields in the Taupo Volcanic Zone of the North Island, New Zealand. MA = Mangakino, OH = Ohaaki (GNS Science, personal communication, 2014).

duction. The same issues apply to the definition of reservoir depth as to the area.

## 4  Probability

Monte Carlo methods are frequently used to provide some accounting for the inevitable uncertainties in a resource at best only partly explored. Table 1 reproduces two results from a national assessment of New Zealand by Lawless and Lovelock (2002), using the method of the Australian code. In the data columns (area, thickness, mean temperature) the data are given as minimum/mode/maximum, and the capacity is given as 10th percentile/median/90th percentile. The percent void space is also included because the calculation uses a recovery factor proportional to void space (2.5 × void). Figure 1 shows the location of the fields discussed below.

There were two versions of the paper, with different estimates for Mangakino. Mangakino was drilled after this assessment. After four deep wells, the field was abandoned. In a result unique in New Zealand geothermal exploration, the thermal anomaly was present but proved to be conductively heated in rock of very low permeability. There is no ability to produce this resource and so the actual capacity is 0 MWe. This result lies far outside the range of the assessed proba-

bility distribution. In any undrilled field, there is some risk of failure. The second Mangakino estimate extends the area range down to zero, but this still produces a vanishingly small probability of a zero outcome, and so this modification still does not cover the actual realisation – it would be necessary to add a discrete quantum at 0 MWe. Rather than so modifying the probability distribution, it is probably better to assign a prior probability of failure (~5–10 %) and then consider the current distribution as applying if this first contingency does not eventuate.

Ohaaki Power Station came online in 1988, with installed capacity of 112 MWe. There was significant surplus production capacity at startup. The station ran at full load for 5 years, then ran down rapidly and production since has been about half the installed level, despite significant drilling of deeper wells (Clearwater et al., 2011). Again, the actual result lies well outside the range of possible outcomes from the Monte Carlo assessment. In this particular case the error is clear. Ohaaki has a permeable area of only 3 km$^2$, despite the much larger field area. It is an extreme case of the issue discussed above, of hot but impermeable regions within the geothermal anomaly.

A more detailed examination of the risks involved in Monte Carlo assessments is given by Garg and Combs (2010), from which Table 2 is reproduced. The tabulated rows give the capacity in megawatts at the 90th, 50th and 10th percentile, and the ratio of the 10th and 90th percentile values.

Case 1 is the field evaluation previously published by GeothermEx (2004). The other four cases are simply variations on the original assessment, making changes that appear to lie within the range of possibility. The upper (10 %) capacity varies somewhat, but the median varies more and the 90th percentile varies by a factor of 4. As this percentile is normally taken as the proven capacity, it is clear that there is far greater uncertainty in the proven capacity than is represented in the original assessment (Case 1). Garg and Combs also give a detailed discussion of the sensitivity to reject temperature and the effect of changes in this parameter on other assumptions, reinforcing that the stored-heat method is not nearly as simple as it first appears.

The problem in all these cases is that the concept of the Monte Carlo method has been misapplied, by using overly restrictive estimates of parameter ranges. "Reasonable" assumptions are made about the possible ranges of parameters. But a probability distribution must include not just reasonable outcomes. It must include all outcomes that are possible in the given state of knowledge. If there is information unknown, which may produce a particular outcome, that outcome must be included within the probability distribution. The parameter ranges must not be reasonable, but inclusive of all possibilities. This is a very common fault in the use of the Monte Carlo method in geothermal resources.

In the case of Mangakino above, the assessment was made before deep drilling (there had been one shallow well ear-

**Table 1.** New Zealand national assessment (Lawless and Lovelock, 2002), selected fields.

| Field | Area | Thickness | Mean temp | Void % | Capacity, MWe |
|---|---|---|---|---|---|
| Mangakino (1) | 8/9.5/17 | 1500/1700/2200 | 220/230/250 | 8/10/12 | 65/85/120 |
| Mangakino (2) | 0/8/10 | 1500/1700/2200 | 220/230/250 | 8/10/12 | 20/47/70 |
| Ohaaki | 5/11/12 | 1800/2100/2500 | 260/275/280 | 6/8/10 | 100/135/175 |

**Table 2.** Megawatt capacity for Silver Peak, Nevada (Garg and Combs, 2010).

| | Case 1 | Case 2 | Case 3 | Case 4 | Case 5 |
|---|---|---|---|---|---|
| 90 % | 41 | 41 | 13 | 9 | 8 |
| 50 % | 82 | 82 | 64 | 44 | 44 |
| 10 % | 146 | 150 | 140 | 102 | 137 |
| 10 % MW/90 % MW | 3.6 | 3.7 | 10.8 | 11.3 | 17.1 |

lier). It is an observational fact that some fields are drilled and abandoned. Usually the reason is adverse chemistry, but lack of permeability is another reason sometimes seen, and realised here. Before drilling, there is some possibility that the field will be abandoned, and hence some possibility of a capacity of 0. This possibility must be contained within the range of probability. The revision to produce the second estimate for Mangakino also shows the failure to include all possibilities. In the light of additional information, the area range was extended outside the first range. If the first range were properly constructed, additional information would only produce a restriction – anything that could result from additional information should already be included. Also note in Table 1 the quite narrow range of possible reservoir thicknesses. Geothermal fields vary greatly. Some have essentially only one permeable interval, and so are simply hosted within this one aquifer. There are other fields where drilling beyond 3km depth has continued to find permeability and production. Thus, the range of thicknesses actually realised ranges from a few hundred metres to a few kilometres. Absent any more knowledge, it is very surprising to so restrict the range of possible thicknesses.

In the case of Ohaaki, Lawless and Lovelock argue, "Some of resource has demonstrated low permeability, but only to a certain depth, and so is all included in the higher areal estimate." That is, knowing that the resource as then drilled has only a very limited permeable area, because there might be permeability at depth over a wider area, the greater area is included. This argument deliberately excludes from the range of possible areas the area as then known and consequently incorrectly modifies the probability distribution. In the absence of knowledge about permeability at greater depth, the possible outcomes must include both good and bad results of deeper drilling, and the argument advanced by Lawless and Lovelock is an argument for extending the range of possible areas, not for excluding part of that range.

The Silver Peak case similarly shows too-restrictive assumptions about possible outcomes of further exploration. All the cases presented by Garg and Combs can be supported by argument, and so, in the present state of knowledge, a probability distribution describing this knowledge must include them all.

It can be further observed that the use of the Monte Carlo method does not provide protection against over-estimation. If the parameters are assumed with values too high, the resultant probability distribution will also average too high. Ohaaki provides a clear example – actual performance lies well below the 10th percentile. Having a distribution on the inputs does not correct for, in this case, the area and recovery factors being assumed too high over their respective entire assumed ranges. When the mean values of the parameter distributions assumed are biased high, the result will be similarly biased.

## 5 Enhanced geothermal systems (EGS)

The conclusions above apply only to hydrothermal systems, as they are the source of all data used. There are very few data for any other systems. Initial estimates of heat recovery from enhanced geothermal systems (EGS) use fractured medium models, similar to the reasoning of Nathenson (1975). Sanyal and Butler (2005), using such modelling, found a recovery factor to be typically 40–50 %, which was used as a starting point for MIT (2006).

Williams (2010) concluded that "field observations and modelling studies indicate that values for the recovery factor" less than 10 % are likely to be representative. There is one published analysis for EGS based upon field data (Grant and Garg 2012), which indicates even lower recovery factors: 2 %. It would be expected that EGS would have lower recovery than naturally fractured systems, as the fractures will be less pervasive. Although microseismicity indicates significant volumes stimulated by fracturing, the reservoir that developed, through which there is significant fluid flow, appears to be markedly smaller (Williams, 2010; Grant and Garg, 2012). As with hydrothermal systems, it appears that modelling studies have erred on the optimistic side, and for EGS by an even greater amount. Recovery factors for EGS should be at best a few percent.

## 6  Conclusions

One conclusion is immediate: past recovery factors have been too high, and comparison with actual performance shows that an average value of 10 % should be used. For this reason the Australian code, even in its second edition, is biased high. With the recovery factor corrected, the results are subject to an error of $\pm 70$ %. For EGS, smaller recovery factors – only a few percent – are indicated and the Australian code is again biased high.

There are significant implicit assumptions about the underlying physical processes in the use of the stored-heat method, which means in practice that there is great uncertainty in the calculation – this may or may not relate to the variation in outcomes just noted. Monte Carlo methods do not provide protection against these uncertainties.

**Acknowledgements.** Thanks are due to the reviewers, and in particular to E. Juliusson for drawing attention to the "Icelandic method".

## References

AGEA AGEG: The geothermal reporting code Second edition, 2010a.

AGEA AGEG: Geothermal lexicon for resources and reserves definition and reporting, 2010b.

Clearwater, E. K., O'Sullivan, M. J., and Brockbank, K.: An update on modelling the Ohaaki geothermal system, NZ Geothermal Workshop, 2011.

Garg, S. K. and Combs, J.: Appropriate use of USGS volumetric heat in place method and Monte Carlo calculations, Proc, 35th Workshop on geothermal reservoir engineering, Stanford University, 2010.

Garg, S. K. and Combs, J.: A re-examination of USGS volumetric heat in place, method, Proc, 36th Workshop on geothermal reservoir engineering, Stanford University, 2011.

GeothermEx: New geothermal site identification and qualification, Consultant report for California Energy Commission PIER report no. P500-04-051, 2004.

Grant, M. A. and Garg, S. K.: Recovery factor for EGS, Proc, 37th Workshop on Geothermal Reservoir Engineering, Stanford University, 2012.

Grant, M. A.: Geothermal resource proving criteria, Proceedings, World Geothermal Congress, 2581–2584, 2000.

Lawless, J. and Lovelock, B.: New Zealand's geothermal resource, paper presented at NZ Geothermal Association, 2002.

Lund, J. W., Gawell, K., Boyd, T. L., and Jennejohn, D.: The United States of America country update 2010, World Geothermal Congress paper 0102, 2010.

MIT: The future of geothermal energy, 2006.

Muffler, L. P. J.: Assessment of geothermal resources of the United States – 1978, U.S. Geological Survey, Circular 790, 163p., 1978.

Nathenson, M.: Physical factors determining the fraction of stored energy recoverable from hydrothermal convection systems and conduction-dominated areas, USGS Open-file report 75–525, 1975.

Pálmason, G., Johnsen, G. V., Torfason, H., Sæmundsson, Ragnars, K., Haraldsson, G. I., and Halldórsson, G. K.: Mat á Jarðvarma Íslands, (in Icelandic) Orkustofnun report OS-85076/JHD-10, 1985.

Sanyal, S. K., Henneberger, R. C., Klein, C. W., and Decker, R. W.: A methodology for the assessment of geothermal energy reserves associated with volcanic systems, Transactions, Geothermal Resources Council 26, 59–64, 2002.

Sanyal, S. K., Klein, C. W., Lovekin, J. W., and Henneberger, R. C.: National assessment of U.S. geothermal resources – a perspective, Transactions, Geothermal Resources Council 28, 355–362, 2004.

Sanyal, S. K. and Butler, S. J.: An analysis of power generation prospects from Enhanced geothermal systems Transactions, Geothermal Resources Council, 29, 131–137, 2005.

Sarmiento, Z. F. and Björnsson, G.: Reliability of early modelling studies for high-temperature reservoirs in Iceland and The Philippines, Proceedings, 32nd Workshop on Geothermal Reservoir Engineering, Stanford University, 2007.

Stefansson, V.: World geothermal assessment, Proceedings, World Geothermal Congress paper 0001, 2005.

White, D. E. and Williams, D. L.: Assessment of geothermal resources of the United States – 1975, Geological Survey Circular 726, USGS, 1975.

Williams, C. F.: Development of revised techniques for assessing geothermal resources, Proceedings, 29th Workshop on Geothermal Reservoir Engineering, Stanford University, 276–280, 2004.

Williams, C. F.: Updated methods for estimating recovery factors for geothermal resources, Proceedings, 32nd Workshop on geothermal reservoir engineering, Stanford University, 2007.

Williams, C. F.: Thermal energy recovery from enhanced geothermal systems – evaluating the potential from deep, high-temperature resources Proceedings, 35th Workshop on geothermal reservoir engineering, Stanford University, 2010.

Wilmarth, M. and Stimac, J.: Worldwide power density review Proc., 39th Workshop on geothermal reservoir engineering, Stanford University, 2014.

# 4

# Proposal of a consistent framework to integrate geothermal potential classification with energy extraction

**G. Falcone**

Department of Geothermal Engineering and Integrated Energy Systems, Institute of Petroleum Engineering,
Clausthal University of Technology, Clausthal-Zellerfeld, Germany

*Correspondence to:* G. Falcone (gioia.falcone@tu-clausthal.de)

**Abstract.** The classification of geothermal resources is dependent on the estimate of their corresponding geothermal potential, so adopting a common assessment methodology would greatly benefit operators, investors, government regulators and consumers.

Several geothermal classification schemes have been proposed, but, to date, no universally recognised standard exists. This is due to the difficulty in standardising fundamentally different geothermal source and product types. The situation is not helped by the accepted use of inconsistent jargon among the geothermal community. In fact, the term "geothermal potential" is often interpreted differently by different geothermal practitioners.

This paper highlights the importance of integrating the classification of geothermal potential with that of geothermal energy extraction from well-defined development projects. A structured progression, from estimates of in situ quantities for a given prospect to actual production, is needed. Employing a unique, unambiguous framework would ensure that the same resource cannot exist simultaneously under different levels of maturity of the estimate (as in double bookings of resources), which would let stakeholders better assess the level of risk involved and the steps needed for a geothermal potential to achieve commercial extraction.

## 1 Introduction

There is a real challenge to reconcile nomenclature for in situ or in-place quantities of geothermal energy, potential (technical or economic or sustainable or developable), resources (inferred or indicated or measured) and reserves (probable or proven), to name but a few.

According to Rybach (2010), the theoretical potential describes the physically usable energy supply (heat in place). Due to technical, structural and administrative limitations only small fractions of the theoretical potential can actually be used. The technical potential describes the fraction of the theoretical potential that can be used under the existing technical restrictions (currently available technology). The economic potential describes the time- and location-dependent fraction of the technical potential that can be economically utilised within the energy system under consideration. The sustainable potential is a fraction of the economic potential; it describes the fraction that can be utilised by applying sustainable production levels. The developable potential describes the fraction of the sustainable potential that can be developed under realistic conditions (regulations, environmental restrictions).

Often, the generic term "potential" is used in the public domain, without clear indication of what particular type of potential is being referred to (theoretical, technical, economic, sustainable or developable). Inevitably, this generates confusion among the stakeholders as to the actual expectations from a given geothermal prospect or development.

Classifying geothermal resources on the basis of the theoretical potential (or heat in place) leads to large figures that can be misleading and may be wrongly interpreted as recoverable energy. However, when the process to be implemented in order to recover a given resource is uncertain, the theoretical potential may represent the only estimate on hand.

Different methods are used to estimate geothermal potential, including power density (usually expressed in terms of $MW\,km^{-2}$), stored heat (or volume method, independent of the method or rate of heat extraction) and numerical reservoir

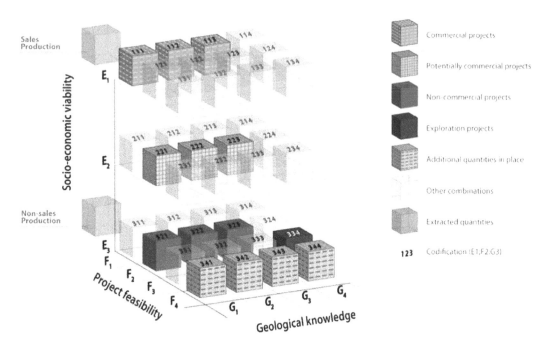

**Figure 1.** UNFC-2009 categories and examples of classes (ECE, 2013).

modelling (generally not suitable in the exploration and early development stage of a project). The resulting estimates carry different levels of confidence depending on whether they are based on indirect or direct evidence (e.g. geological studies and surface exploration vs. well logging and flow testing) and on the level of field maturity (exploration, appraisal or development). Inevitably, broad estimates of geothermal potential at country, continent or world scale rely on the assumption of parameters that are unknown.

Given the current global trend of increasing energy demand and the constant dilemma of whether renewable energy resources will be able to supplement or even replace conventional energy resources, it is crucial to assess future geothermal supplies at a large scale. This translates into a focus today on estimating and classifying geothermal theoretical potential, i.e. additional quantities in place that could be converted into actual production in the foreseeable future.

Unconventional, undiscovered and/or currently unrecoverable geothermal resources therefore need to be captured vis-à-vis the conventional, discovered and/or recoverable resources, in a consistent and structured way, so as to help stakeholders understand the level of risk involved and the steps needed for a geothermal potential to achieve commercial extraction.

## 2   The UNFC-2009

Several geothermal resources classification schemes have been proposed to date, yet no universally recognised standard exists. Falcone et al. (2013) provided a critical review

of key approaches proposed to date for geothermal resources classification.

The United Nations Framework Classification for Fossil Energy and Mineral Reserves and Resources 2009 (UNFC-2009) differs from other systems as it classifies estimated resource quantities using three axes: socio-economic variability ($E$), project feasibility ($F$) and geological knowledge ($G$). The first set of categories (the $E$ axis) designates the degree of favourability of social and economic conditions in establishing the commercial viability of the project, including consideration of market prices and relevant legal, regulatory, environmental and contractual conditions. The second set (the $F$ axis) designates the maturity of studies and commitments necessary to implement mining plans or development projects. These extend from early exploration efforts before a deposit or accumulation has been confirmed to exist through to a project that is extracting and selling a commodity; they reflect standard value chain management principles. The third set of categories (the $G$ axis) designates the level of confidence in the geological knowledge and potential recoverability of the quantities (ECE, 2013). Combinations of these criteria create a three-dimensional system as shown in Fig. 1.

The UNFC-2009 can already be used to normalise the classification of hydrocarbon and mineral resources. It also ensures alignment with widely used systems such as the Committee for Mineral Reserves International Reporting Standards (CRIRSCO) Intenational Reporting Template (IRT) and the Society of Petroleum Engineers (SPE)/World Petroleum Council (WPC)/American Association of Petroleum Geologists (AAPG)/Society of Petroleum

| UNFC Classes Defined by Categories and Sub-categories | | | | | |
|---|---|---|---|---|---|
| **Extracted** | Sales Production | | | | |
| | Non-sales Production | | | | |
| | Class | Sub-class | Categories | | |
| | | | E | F | G |
| **Known Deposit** | Commercial Projects | On Production | 1 | 1.1 | 1, 2, 3 |
| | | Approved for Development | 1 | 1.2 | 1, 2, 3 |
| | | Justified for Development | 1 | 1.3 | 1, 2, 3 |
| | Potentially Commercial Projects | Development Pending | 2[b] | 2.1 | 1, 2, 3 |
| | | Development On Hold | 2 | 2.2 | 1, 2, 3 |
| | Non-Commercial Projects | Development Unclarified | 3.2 | 2.2 | 1, 2, 3 |
| | | Development Not Viable | 3.3 | 2.3 | 1, 2, 3 |
| | Additional Quantities in Place | | 3.3 | 4 | 1, 2, 3 |
| **Potential Deposit** | Exploration Projects | [No sub-classes defined][c] | 3.2 | 3 | 4 |
| | Additional Quantities in Place | | 3.3 | 4 | 4 |

[a]  Refer also to the notes for Figure 2.

[b]  Development Pending Projects may satisfy the requirements for E1.

[c]  Generic sub-classes have not been defined here, but it is noted that in petroleum the terms Prospect, Lead and Play are commonly adopted.

**Figure 2.** UNFC-2009 classes and sub-classes defined by sub-categories[a] (ECE, 2013). For the definition of categories and sub-categories and supporting explanations, see Annex I and Annex II of ECE (2013).

Evaluation Engineers (SPEE) Petroleum Resource Management System (PRMS).

The United Nations Economic Commission for Europe (UNECE, 2012) called upon its Expert Group on Resource Classification (EGRC) to develop ideas on how the UNFC-2009 could apply to and integrate renewable energy resources. Following agreement at the fourth session of the UNECE EGRC held in Geneva, April 2013, a Task Force on the Application of UNFC-2009 to Renewable Energy Resources was established. The Task Force (2014) recently released specifications for the application to renewable energy resources for public comment. The aim is that of complementing these specifications with commodity-specific specifications for each type of renewable energy source, including geothermal.

In September 2014, the International Geothermal Association (IGA) and the UNECE signed a Memorandum of Understanding (MoU) to develop a globally applicable harmonised standard for reporting geothermal resources. Such a standard will ensure greater consistency and transparency in financial reporting and enhance management of geothermal resources. Under this MoU, the IGA will work towards providing technology-specific rules ("specifications") for the application of the UNFC-2009 to geothermal resources. This work will be overseen by the UNECE EGRC. The geothermal specifications will provide the foundation and guidelines

for consistent application of UNFC-2009 for geothermal resources, as well as a meaningful comparison of geothermal resource estimates with other energy resources.

Below there is a discussion on how UNFC-2009 could apply to geothermal energy resources, with a focus on the classification of potential future recovery by successful exploration activities and additional quantities in place associated with known and potential deposits, using the same framework that also allows classifying economic projects.

## 2.1  Potential vs. known deposit in UNFC-2009

With reference to Fig. 2, the UNFC-2009 allows differentiating between a known deposit and a potential deposit. A known deposit is a deposit that has been demonstrated to exist by direct evidence; a potential deposit is a deposit that has not yet been demonstrated to exist by direct evidence (e.g. drilling and/or sampling) but is assessed as potentially existing based primarily on indirect evidence (e.g. surface or airborne geophysical measurements) (ECE, 2013). Within UNFC-2009, the geologic uncertainty for discovered quantities is described using categories G1 to G3, while the geologic uncertainty for undiscovered quantities is described using category G4. Thus, a potential deposit includes quantities classified on the $G$ axis as G4: "estimated quantities associated with a potential deposit, based primarily on indirect evidence" (ECE, 2013).

## 2.2  Additional quantities in place in UNFC-2009

With reference to Fig. 2, the UNFC-2009 allows reporting additional quantities in place as both quantities associated with a known deposit that will not be recovered by any currently defined development project or mining operation and quantities associated with a potential deposit that would not be expected to be recovered even if the deposit is confirmed (ECE, 2013). Additional quantities are classified on the $F$ axis as F4 when "no development project or mining operation has been identified/in situ (in-place) quantities that will not be extracted by any currently defined development project or mining operation" (ECE, 2013). F4 means that no recovery factor is superimposed to estimates of in situ quantities. Additional quantities in place can be sub-classified as follows on the basis of the current state of technological developments (ECE, 2013):

a. "F4.1: the technology necessary to recover some or all of the these quantities is currently under active development, following successful pilot studies on other deposits, but has yet to be demonstrated to be technically feasible for the style and nature of deposit in which that commodity or product type is located;

b. F4.2: the technology necessary to recover some or all of the these quantities is currently being researched, but no successful pilot studies have yet been completed;

| | | Low Estimate | Best Estimate | High Estimate |
|---|---|---|---|---|
| Prospective Resources | Prospect | E3.2,F3.1,G4.1 | E3.2,F3.1,G4.1+G4.2 | E3.2,F3.1,G4.1+G4.2+G4.3 |
| | Lead | E3.2,F3.2,G4.1 | E3.2,F3.2,G4.1+G4.2 | E3.2,F3.2,G4.1+G4.2+G4.3 |
| | Play | E3.2,F3.3,G4.1 | E3.2,F3.3,G4.1+G4.2 | E3.2,F3.3,G4.1+G4.2+G4.3 |

**Figure 3.** Mapping of UNFC-2009 exploration projects to PRMS prospective resources (ECE, 2013). For the definitions of prospect, lead and play, see PRMS (SPE et al., 2007). For the definition of categories and sub-categories and supporting explanations, see Annex I and Annex II of ECE (2013). "Low", "best" and "high" reflect the degree of uncertainty associated with the estimates. A low-estimate scenario is directly equivalent to a high confidence estimate, whereas a best-estimate scenario is equivalent to the combination of the high confidence and moderate confidence estimates. A high-estimate scenario is equivalent to the combination of high, moderate and low confidence estimates. Quantities may be estimated using deterministic or probabilistic methods.

    c. F4.3: the technology necessary to recover some or all of these quantities is not currently under research or development."

## 2.3   Exploration projects in UNFC-2009

With reference to Fig. 2, in the UNFC-2009 exploration projects are classified on the *G* axis as G4 and on the *F* axis as F3: "feasibility of extraction by a defined development project or mining operation cannot be evaluated due to limited technical data", and "very preliminary studies (e.g. during the exploration phase), which may be based on a defined (at least in conceptual terms) development project or mining operation, indicate the need for further data acquisition in order to confirm the existence of a deposit in such form, quality and quantity that the feasibility of extraction can be evaluated" (ECE, 2013).

## 2.4   Combining *E*, *F* and *G* for exploration projects and additional quantities in place in UNFC-2009

Exploration projects and additional quantities in place are classified on the *E* axis as E3.

    When combining the *E*, *F* and *G* categories, an exploration project in UNFC-2009 would fall under E3 F3 G4, and additional quantities in place would fall under E3 F4 G1, 2, 3 (for known deposits) or E3 F4 G4 (for unknown deposits). E3 implies "extraction and sale is not expected to become economically viable in the foreseeable future or evaluation is at too early a stage to determine economic viability" (ECE, 2013).

    Considering that geothermal systems, like petroleum systems, deal with continuous fluid flow, the current bridging document between the PRMS (SPE et al., 2007) and UNFC-2009 could be used to map geothermal exploration projects and additional quantities in place to UNFC-2009.

| | | Low Estimate | Best Estimate | High Estimate |
|---|---|---|---|---|
| Unrecoverable | Discovered | E3.3,F4,G1 | E3.3,F4,G1+G2 | E3.3,F4,G1+G2+G3 |
| | Undiscovered | E3.3,F4,G4.1 | E3.3,F4,G4.1+G4.2 | E3.3,F4,G4.1+G4.2+G4.3 |

**Figure 4.** Mapping of UNFC-2009 additional quantities in place to PRMS unrecoverable quantities (ECE, 2013). "Unrecoverable" implies that the technology has not been demonstrated to be commercially viable and is not currently under active development, and/or there is not yet any direct evidence to indicate that it may reasonably be expected to be available for commercial application within 5 years. For the definition of categories and sub-categories and supporting explanations, see Annex I and Annex II of ECE (2013). "Low", "best" and "high" reflect the degree of uncertainty associated with the estimates. A low-estimate scenario is directly equivalent to a high confidence estimate, whereas a best-estimate scenario is equivalent to the combination of the high confidence and moderate confidence estimates. A high-estimate scenario is equivalent to the combination of high, moderate and low confidence estimates. Quantities may be estimated using deterministic or probabilistic methods.

There are four cells within the *E–F* matrix that map directly and uniquely to corresponding PRMS project maturity classes; these cells relate to exploration projects (prospective resources in PRMS) and additional quantities in place (unrecoverable in PRMS) (ECE, 2013). Accordingly, geothermal exploration projects could be further defined via subcategories as E3.2 F3.1,2,3 G4, where the F3 sub-categories allow specifying the level of maturity of the project (Fig. 3). Similarly, additional quantities in place could be defined as E3.3 F4.1, 2, 3 G4 (Fig. 4), with the F4 sub-categories indicating the current state of technological developments.

    E3.2 indicates that "economic viability of extraction cannot yet be determined due to insufficient information (e.g. during the exploration phase)" (ECE, 2013), while E3.3 suggests that, "on the basis of realistic assumptions of future market conditions, it is currently considered that there are not reasonable prospects for economic extraction and sale in the foreseeable future" (ECE, 2013).

    Considering that, for recoverable estimates of energy resources that are extracted as fluids, their mobile nature generally precludes assigning recoverable quantities to discrete parts of an accumulation, recoverable quantities should be evaluated on the basis of the impact of the development scheme on the accumulation as a whole and are usually categorised on the basis of three scenarios or outcomes that are equivalent to G1, G1+G2 and G1+G2+G3 (ECE, 2013). G1, G1+G2 and G1+G2+G3 represent single specific scenarios that are representative of the extent of the range of uncertainty of the recoverable/potentially recoverable quantities and they correspond to low estimates (high confidence), best estimates (high and moderate confidence) and high estimates (high, moderate and low confidence), respectively.

## 3 Conclusions

The threefold UNFC-2009 framework allows one to capture the nuances of each geothermal quantity, from in situ potential to confirmed economic production. Its flexibility and granularity permits the classification of quantities that have been estimated primarily on the basis of indirect evidence (e.g. country-based resource mapping studies), as well as quantities at a more specific project level.

UNFC-2009 deliberately avoids the use of ambiguous jargon, having benefited from valuable lessons learnt from the fossil fuels and solid mineral sectors. Nevertheless, the development of dedicated geothermal specifications is needed to make UNFC-2009 fully functional for geothermal resources classification, as it was originally developed for non-renewable energy.

## References

ECE – Economic Commission for Europe: United Nations Framework Classification for Fossil Energy and Mineral Reserves and Resources 2009 incorporating Specifications for its Application, United Nations Publication, ECE Energy Series No.42, ISBN 978-92-1-117073-3, 2013.

Falcone, G., Gnoni, A., Harrison, B., and Alimonti, C.: Classification and Reporting Requirements for Geothermal Resources, European Geothermal Congress 2013, Pisa, Italy, 3–7 June 2013.

Rybach, L.: The Future of Geothermal Energy, Proceedings of World Geothermal Congress, Bali, Indonesia, 25–30 April 2010.

Society of Petroleum Engineers (SPE), American Association of Petroleum Geologists (AAPG), World Petroleum Council (WPC), Society of Petroleum Evaluation Engineers (SPEE): Petroleum Resources Management System, available at: http://www.spe.org/industry/docs/Petroleum_Resources_Management_System_2007.pdf, downloaded on 26 May 2014.

Task Force on the Application of UNFC-2009: Draft Specifications for Application of UNFC-2009 to Renewable Energy Resources, presented for public comment on 12 June 2014, available at: http://www.unece.org/fileadmin/DAM/energy/se/pdfs/UNFC/unfc2009_RE_Specs_publcomm_14/RE_Specs_12.06.2014.pdf, downloaded on 16 June 2014.

UNECE: United Nations Economic Commission for Europe Energy industry calls for UNFC to embrace renewables, press release, 5 November 2012.

# 5

# Reservoir characterization of the Upper Jurassic geothermal target formations (Molasse Basin, Germany): role of thermofacies as exploration tool

**S. Homuth[1], A. E. Götz[2], and I. Sass[3]**

[1]Züblin Spezialtiefbau GmbH, Ground Engineering, Europa Allee 50, 60327 Frankfurt a. M., Germany
[2]University of Pretoria, Department of Geology, Private Bag X20, Hatfield, 0028 Pretoria, South Africa
[3]Technische Universität Darmstadt, Geothermal Science and Technology, Schnittspahnstraße 9, 64287 Darmstadt, Germany

*Correspondence to:* S. Homuth (sebastian.homuth@zueblin.de)

**Abstract.** The Upper Jurassic carbonates of the southern German Molasse Basin are the target of numerous geothermal combined heat and power production projects since the year 2000. A production-oriented reservoir characterization is therefore of high economic interest. Outcrop analogue studies enable reservoir property prediction by determination and correlation of lithofacies-related thermo- and petrophysical parameters. A thermofacies classification of the carbonate formations serves to identify heterogeneities and production zones. The hydraulic conductivity is mainly controlled by tectonic structures and karstification, whilst the type and grade of karstification is facies related. The rock permeability has only a minor effect on the reservoir's sustainability. Physical parameters determined on oven-dried samples have to be corrected, applying reservoir transfer models to water-saturated reservoir conditions. To validate these calculated parameters, a Thermo-Triaxial-Cell simulating the temperature and pressure conditions of the reservoir is used and calorimetric and thermal conductivity measurements under elevated temperature conditions are performed. Additionally, core and cutting material from a 1600 m deep research drilling and a 4850 m (total vertical depth, measured depth: 6020 m) deep well is used to validate the reservoir property predictions. Under reservoir conditions a decrease in permeability of 2–3 magnitudes is observed due to the thermal expansion of the rock matrix. For tight carbonates the matrix permeability is temperature-controlled; the thermophysical matrix parameters are density-controlled. Density increases typically with depth and especially with higher dolomite content. Therefore, thermal conductivity increases; however the dominant factor temperature also decreases the thermal conductivity. Specific heat capacity typically increases with increasing depth and temperature. The lithofacies-related characterization and prediction of reservoir properties based on outcrop and drilling data demonstrates that this approach is a powerful tool for exploration and operation of geothermal reservoirs.

## 1 Introduction

To assess the potential and productivity of a hydrothermal or petrothermal reservoir, detailed knowledge of the thermo- and petrophysical as well as mechanical rock and formation properties is mandatory. In general, the determination of these reservoir properties is limited to costly and time-consuming exploration drillings, which give only a limited insight into the entire reservoir system. Especially in terms of carbonates it is difficult to evaluate the heterogeneity of different facies zones in seismic sections (Chilingarian et al., 1992), which applies to the Upper Jurassic (Malm) target formation of numerous planned geothermal power plant projects in the southern German Molasse Basin. These carbonates are characterized by a karst-fractured aquifer system (Schulz et al., 2012) located 3500–5500 m below the surface in the southern part of the Molasse Basin. This basin represents a typical example of a conduction-dominated geother-

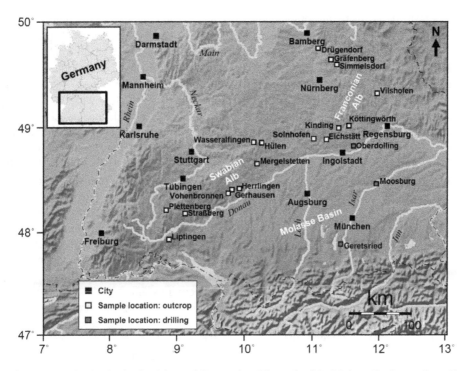

**Figure 1.** Investigated outcrop analogues in the Swabian and Franconian Alb north of the Molasse Basin, southern Germany.

mal play type where the sedimentary sequences of the fore-land basin (Molasse Basin) are influenced by significant crustal subsidence towards the orogenic belt (Alps) due to the weight of the thickened crust of the orogenic belt and loading of erosional products from the mountain belt on the non-thickened crust (Moeck and Beardsmore, 2014).

Outcrop analogue studies enable the determination and correlation of thermo- and petrophysical parameters as well as structural geology data with regional facies patterns. The outcrop analogues of the Swabian and Franconian Alb (Fig. 1) represent the reservoir formations of the Molasse Basin and can be used for detailed facies and thermo- and petrophysical investigations on a low-cost basis. The integrated analysis of lithology, facies, and corresponding thermo- and petrophysical rock properties as well as the application of relevant reservoir transfer models lead to an improved prognosis of the reservoir properties. An outcrop analogue study of the target formation Malm, which is the most prospective formation for deep geothermal projects in the German Molasse Basin, has to include facies studies following a thermofacies concept (Sass and Götz, 2012).

The investigations are carried out on three different scales: (1) the macroscale, including an outcrop mapping to detect the lithotypes, structural elements and facies patterns in the outcrop; (2) the mesoscale, selecting representative rock samples to determine thermo- and petrophysical properties of different lithotypes in the laboratory; and (3) the microscale, to analyse microstructures, cements, porosities, etc. in thin sections.

## 2 Geology

During the Mesozoic the study area was part of an epicontinental sea north of the Tethys Ocean. To the north the shelf region was confined by an island archipelago of changing dimensions. In the southern, deeper part of this epeiric sea, an extensive siliceous sponge-microbial reef belt developed. With the burial of the Vindelician Ridge a direct connection of the southern German Jurassic Sea with the Tethys Ocean was established (Meyer and Schmidt-Kaler, 1989). During the entire Upper Jurassic a high carbonate production on the shallow shelf resulted in thick limestone series (Selg and Wagenplast, 1990). In the southern, deeper shelf area, a reefal facies a reefal facies developed in the Middle Oxfordian and was part of a facies belt characterized by silicious sponge reefs spanning the northern Tethys shelf (Pieńkowski et al., 2008). Most reefs are microbial crusts; however, coral reefs are also present and become increasingly important towards the end of the Upper Jurassic, mirroring the overall mirroring the overall sea-level fall (Pieńkowski et al., 2008). In addition, clay-rich sediments from the Mid-German High were transported into the shelf area. During times of low carbonate production, the clay content of the sediments increased, which resulted in the sedimentation of marl (Meyer and Schmidt-Kaler, 1990). According to Meyer and Schmidt-Kaler (1989, 1990), the Swabian facies as the central part of the reef belt formed a deeper-water area between the shallower Franconian–southern-Bavarian platform in the east and the Swiss platform in the west. In the southwest, the Swabian shelf facies deepens gradually towards the pelagic facies of the Helvetian Basin. The Helvetic facies is characterized by dense, bituminous limestones with, in places, intercalated oolithic layers. This facies describes the transition of the Germanic facies into the Helvetic facies, which is considered as sediments of a deeper shelf area of bedded limestones with very low permeabilities. Karstification is not observed either; thus the northern boundary of the Helvetic facies is considered as the southern boundary of the Malm aquifer of the Molasse Basin (Villinger, 1988).

In general 400–600 m of carbonate rocks were deposited during the Upper Jurassic, and two major facies can be distinguished (Geyer and Gwinner, 1979; Pawellek and Aigner, 2003):

1. the basin facies, consisting of well-bedded limestones and calcareous marls (mud-/wackestones);

2. the reefal or massive facies when bedding is absent, indistinct or very irregular (rud-/float-/grainstones).

The massive limestones are built by microbial crusts (stromatolites and thrombolites) and siliceous sponges that have been interpreted by various authors as relatively deep and quiet water "reefs", mounds or bioherms (Gwinner, 1976; Leinfelder et al., 1994, 1996; Pawellek and Aigner, 2003). The basin facies may either interfinger with the reefs or onlap onto the reefs (Gwinner, 1976; Pawellek, 2001). In the upper parts of the Upper Jurassic, a coral facies developed locally upon the microbial crust-sponge reefs. The abundance of reef facies differs regularly through time. Reef expansion phases correlate with an increase in the carbonate content within the basin facies, while phases of reef retreat correlate with increasing abundance of marls within the basin facies (Meyer and Schmidt-Kaler, 1989, 1990; Pawellek 2001).

Variations in hydraulic conductivity, particularly within the Upper Jurassic aquifer, are related to lateral changes in lithofacies and degree of karstification (Birner et al., 2012). Thus, it can be assumed that the geothermal potential of the Jurassic aquifer shows a distinct facies-related regional pattern: in the western part of the Molasse Basin (Baden-Wuerttemberg) the potential is significantly lower than in the eastern part (Bavaria), where the Upper Jurassic aquifer is the major producer of geothermal energy in the area around Munich (Stober, 2013).

## 3  Materials and methods

Reservoir characterization based on thermo- and petrophysical parameters – including permeability, porosity, density, specific heat capacity, thermal diffusivity and thermal conductivity data measured at the same sample – was rarely performed in previous works (e.g. Clauser et al., 2002; Stober et al., 2013). In this study, for direct correlation all parameters are determined at the same sample. More than 350 rock samples from 19 outcrop locations as well as shallow and deep boreholes in Baden-Wuerttemberg and Bavaria (Fig. 1) were

collected and analysed. For statistical purposes 3–10 single measurements of different rock properties were conducted on each rock sample; i.e. in total over 1150 measurements for distinct parameters were collected. According to the Dunham (1962) and Embry and Klovan (1971) classification of carbonate rocks the following lithofacies types are detected in the study area: mudstones, wackestones, grain-/packstones and float-/rudstones. The rock classification is also based on previous studies of the Malm formations in southern Germany by Schauer (1998) and Pawellek (2001).

To determine the thermophysical properties of the sampled formations and to generate reproducible results, the samples were dried at 105 °C to mass constancy and afterwards cooled down to 20 °C in an exsiccator. A thermal conductivity scanner (optical scanning method after Popov et al., 1985), a gas pressure permeameter (Jaritz, 1999; Hornung and Aigner, 2004) and porosimeter were used. The thermal conductivity scanner is also equipped to determine the thermal diffusivity. The measurement is based on a contact-free temperature measurement with infrared temperature sensors (Bär et al., 2011). The measurement accuracy is about 3 %. The determination of the grain and bulk density as well as the porosity was done by measuring the grain and bulk volume of the samples, using a helium pycnometer and a powder pycnometer. The specific heat capacity $c_p$ was measured with a C-80 calorimeter in a temperature range from 20 to 200 °C for selected samples. Additionally $c_p$ was calculated with the measured thermal conductivity $\lambda$, density $\rho$ and thermal diffusivity $\alpha$ (converted Debye equation):

$$c_p = \frac{\lambda}{(\rho \cdot \alpha)}. \tag{1}$$

For the determination of the rock permeability a combined column and mini permeameter was used. The method offers either the measurement of the apparent gas permeability which afterwards is converted in permeability or the direct measurement of the intrinsic permeability. The basis for the gas driven permeameter is the Darcy law, which is enhanced by the terms of compressibility and viscosity of gases. To simulate geothermal reservoir conditions, temperature and pressure-dependent parameters must be considered. It is possible to calculate these values for water-saturated rocks under reservoir pressure and temperature conditions for relevant depths (Vosteen and Schellschmidt, 2003; Popov et al., 2003). These parameters can be validated in a Thermo-Triaxial-Cell simulating the existing temperature and pressure conditions of the target horizon of a distinct geothermal reservoir and furthermore induces a pore pressure on the rock sample (Pei et al., 2014). The unique design of the Thermo-Triaxial-Cell allows experiments with tempered rocks and fluids up to about 170 °C by applying up to 500 MPa lithostatic pressure and 70 MPa confining pressure. Built of V4A premium steel, the cell can be operated with highly aggressive (corrosive) fluids. Both fluid and rock can be individu-

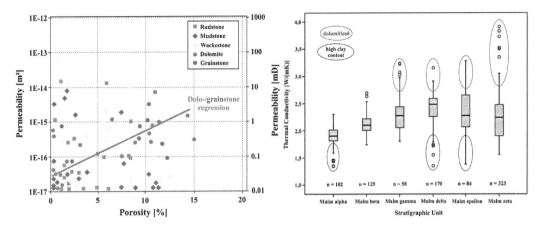

**Figure 2.** Left: porosity–permeability relationship of different lithotypes (mean values); right: a stratigraphic trend of increasing thermal conductivity is detected from Malm $\alpha$ to Malm $\zeta$ depending on clay and dolomite content (Homuth et al., 2014).

ally tempered, thus allowing a wide range of testing setups to simulate reservoir conditions.

## 4   Results

### 4.1   Permeability and porosity

The matrix permeability of all measured carbonate core samples is quite low except for some grain- and dolostones with higher permeabilities and porosities (Fig. 2). Permeabilities range from $10^{-18}$ to $10^{-13}$ m$^2$ (0.001–10 mD) (K in m$^2$ = K in D $\cdot$ 9.8692 $\times$ $10^{-13}$). The grain density of the outcrop samples ranges between 2.59 and 2.80 g cm$^{-3}$; the bulk density is between 2.31 and 2.75 g cm$^{-3}$. The porosity calculation based on these values results accordingly in less than 15 %. The massive limestones have porosities below 8 %, while grainstones and dolomitized zones show increased porosities up to 18 %. The permeability measurements state in general very low matrix permeabilities. Only grainstones, reef/coral debris limestones and dolomitized zones show higher permeability ranges up to $10^{-14}$ m$^2$ (10 mD). A comparison of permeability and porosity indicates that high porosities occur in grain- and dolostones and also cause higher permeability. For all other lithofacies types no correlation between porosity and permeability in regard to interconnected porosity can be inferred (Fig. 2). Diagenetic processes caused dolomitization and de-dolomitization of reef structures and their adjacent transition zones to the basin facies, resulting in an increase of inter-crystalline porosity and therefore increased matrix permeability. On the other hand, if de-dolomitization led to the formation of saccharoidal limestone, permeability decreases due to reduced crystalline porosity. With increasing dolomite content an increase of thermal conductivity is observed due to the higher thermal conductivity of the dolomite crystal structure. The dolomitized areas, related to the geometry of the massive reefal limestone complexes, can span over several stratigraphic units of the Malm, predominantly in the

vertical direction (Stober and Villinger, 1997; Schauer, 1998; Koch, 2011; Birner et al., 2012). The dolomitization and de-dolomitization processes can have a significant influence on rock permeability, either increasing or reducing the average rock permeability. Including fracture network, dolomitization and karstification, a positive shift of the permeability-porosity relationship across several magnitudes can be observed. Jodocy and Stober (2011) and Stober et al. (2013) published permeability data obtained from drill core measurements and pump tests, as well as data inferred from geophysical logs and drilling documentation showing hydraulic conductivities of core samples within the same range of values as presented in this study. They also determined an average increase of permeability over 3 magnitudes from core data to pump test data. This shift indicates a high hydrothermal potential of the deep Malm aquifer system in the Molasse Basin. The assumption of a positive permeability correction within the range of 2–3 magnitudes is also based on pump test data and comparisons of matrix and formation productivity from different deep drilling locations in the Molasse Basin (Böhm et al., 2013; Schulz et al., 2012).

### 4.2   Thermal conductivity

Thick-bedded and platy limestones have thermal conductivities of about 2 W (m $\cdot$ K)$^{-1}$, characteristic of limestones. Marly limestones have lower thermal conductivities than thick-bedded and platy limestones, showing the same range of permeabilities as the thick-bedded limestones. It seems that the higher clay content of the marly limestones decreases the thermal conductivity by insulating the heat conduction and at the same time showing only minor effects on permeability, which shows the same range for mud- and wackestones. The thermal conductivities of different reefal limestones have values of 1.8–3.9 W (m $\cdot$ K)$^{-1}$, related to the higher content of secondarily silicified reef bodies and due to dolomitization of reefal structures. The layers with increased silica content are identified as layers with silicified sponge build-ups (Leinfelder et al., 1994, 1996). The dolomitized carbonates show the highest values of thermal conductivity of all investigated carbonates in this study. For some stratigraphic units trends are detectable (Fig. 2): increasing thermal conductivity from Malm $\alpha$ to Malm $\zeta$, due to decreasing clay content and increasing dolomitization (the maximum of the dolomitization is also found within the Malm $\delta$ by Schauer (1996, 1998)). A peak of thermal conductivity observed in the Malm $\delta$ also correlates with an increased silica content of silicified sponge build-ups.

## 4.3 Comparison of results from shallow and deep drill cores and cuttings

The presented data of outcrop analogue studies are based on rock measurements on oven-dried cores, which are conducted under laboratory conditions with atmospheric pressure and room temperature of 20 °C. This approach guarantees a very good reproducibility of the results but also requires a correction of the measured data for reservoir conditions. It is assumed that the reservoir is completely saturated. For the following analyses, the temperature and pressure conditions of a 5000 m deep (= lithostatic pressure: 130 MPa) and 150 °C hot reservoir, which are realistic values for the Molasse Basin, are estimated.

The thermal conductivity of water-saturated rocks can be calculated following the model of Lichtenecker and numerous other authors (Clauser and Huenges, 1995; Popov et al., 2003; Hartmann et al., 2005). Temperature dependency models of thermophysical properties of different rock types can be found in Somerton (1992), Vosteen and Schellschmidt (2003) and Abdulagatova et al. (2009). In general, the thermal conductivity decreases with increasing temperature and increases with increasing pressure (Clauser and Huenges, 1995). The fundamental effects are the reduction of pore space and the increasing temperature with increasing depth (Clauser et al., 2002). Both parameters control the fluid and matrix conditions, although in terms of tight carbonates the temperature-dependent porosity reduction is the dominant factor. Also for tight carbonates the lithostatic pressure has only a minor influence on the porosity–permeability relationship (Bjørkum et al., 1998).

Table 1 shows the range of measured values and calculated transfer values of different thermophysical rock properties for different facies types of the Malm carbonates. For comparison also the calculated reservoir rock properties with respect to the distinct reservoir conditions (150 °C, 5000 m depth) and accordingly applied correction functions are listed. In terms of matrix porosity and permeability it is concluded that the low rock porosity measured on the outcrop samples will not change significantly with increasing depth in regards to effective hydraulic conductivity. In terms of the mean reservoir porosity the temperature of the carbonate systems is the dominant factor with regard to the thermal

**Table 1.** Facies types and associated thermophysical properties at laboratory (measured values) and reservoir conditions (temperature: 150 °C, lithostatic pressure: 130 MPa; calculated values) according to different transfer models; matrix dominated: mud-/wackestone; grain-/component dominated: rud-/floatstone, grain- and dolostone.

| Rock property | Facies type | Value range, dry, 20 °C | Value range, water-saturated, 20 °C | Value range, water-saturated, 150 °C | Temperature transfer model | Pressure correction[a] |
|---|---|---|---|---|---|---|
| Thermal conductivity [W (m · K)$^{-1}$] | matrix dominated ($n = 478$) | 1.35–2.62 | 1.60–2.79 | 1.80–2.46 | Zoth and Hänel (1988), modified after Clauser (2003) | 1.88–2.71 |
| | grain/comp. dominated ($n = 418$) | 1.72–4.87 | 1.85–4.94 | 1.92–3.53 | | 2.03–4.16 |
| | matrix dominated ($n = 478$) | 1.35–2.62 | 1.60–2.79 | 1.83–2.51 | Sass et al. (1971) | 1.92–2.78 |
| | grain/comp. dominated ($n = 418$) | 1.72–4.87 | 1.85–4.94 | 1.92–3.92 | | 2.03–4.72 |
| | matrix dominated ($n = 478$) | 1.35–2.62 | 1.60–2.79 | 1.41–2.25 | Vosteen and Schellschmidt (2003), Clauser et al. (2002) | 1.41–2.44 |
| | grain/comp. dominated ($n = 418$) | 1.72–4.87 | 1.85–4.94 | 1.56–3.71 | | 1.59–4.42 |
| | matrix dominated ($n = 478$) | 1.35–2.62 | 1.60–2.79 | 1.53–2.52 | Somerton (1992) | 1.55–2.79 |
| | grain/comp. dominated ($n = 418$) | 1.72–4.87 | 1.85–4.94 | 1.70–3.58 | | 1.76–4.23 |
| Specific heat capacity [J (kg · K)$^{-1}$] | matrix dominated ($n = 410$) | 613–1045 | 645–1081 | 806–1227 | Vosteen and Schellschmidt (2003) | n/a |
| | grain/comp. dominated ($n = 386$) | 547–1167 | 565–1184 | 711–1330 | | |
| | Limestone, matrix dominated ($n = 410$) | 613–1045 | 645–1081 | 976 (mean) | derived from calorimeter measurements | |
| | Dolostone, grain-/comp. dominated ($n = 386$) | 547–1167 | 565–1184 | 1007 (mean) | | |

[a] after Fuchs and Förster (2013)

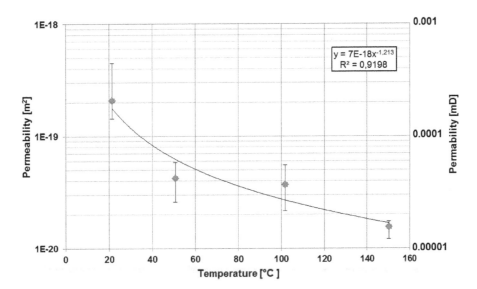

**Figure 3.** Decrease of permeability of four different lithotype outcrop samples under reservoir pressure and increasing temperature regime starting at 20 °C up to 150 °C (reservoir temperature). Points show the average permeability; bars indicate the value range of the four tested samples.

expansion and carbonate chemistry, and not the depth of the reservoir (Bjørkum and Nadeau, 1998). A comparison with other carbonate reservoir data (Ehrenberg and Nadeau, 2005) and a thermo-triaxial test series (Fig. 3) on outcrop samples confirms this approach.

The thermo-triaxial apparatus of the TU Darmstadt Geothermal Laboratory has been developed to facilitate research on petrophysical properties of rock samples under simulated geothermal reservoir conditions. The test device consists of control systems for vertical stress and horizontal confining pressure, a pair of independent pore pressure controllers for applying different upstream and downstream pore pressures at the base and top of rock specimens, an external heater and a data logging system. The permeability of rocks is measured using steady-state and transient flow methods (Pei et al., 2014). Different lithotype samples tested with the thermo-triaxial cell showed initial permeabilities, measured under laboratory conditions with an air-driven permeameter, of about $3.5 \times 10^{-16}\,\mathrm{m}^2$. After complete water saturation of the samples an average decrease of permeability of about 2 magnitudes is observed ($4.3 \times 10^{-18}\,\mathrm{m}^2$). When applying reservoir pressure (vertical stress: 130 MPa = 5000 m depth, confining stress: 30 MPa (due to experimental cell setup); pore pressure: 1 MPa) and temperature (150 °C), a total shift of permeability of about 2–3 magnitudes compared to the samples origin permeability measured under laboratory conditions is measured (Fig. 3). Based on these experiments the following matrix permeability–temperature relationship for the Malm carbonates is inferred:

$$K_{\mathrm{temp}} = K_{(0,\mathrm{sat})} \cdot T^{(-1,213)}, \qquad (2)$$

where $K_{\mathrm{temp}}$ is the temperature-dependent permeability in $\mathrm{m}^2$, $K_{(0,\mathrm{sat})}$ is the water-saturated permeability in $\mathrm{m}^2$ at 20 °C and $T$ is temperature in °C.

The measurements of shallow (Solnhofen-Maxberg, Oberdolling) and deep drill cores from a 1600 m deep research core drilling (Moosburg SC4) and a 4850 m deep production well (GEN-1, cuttings only) confirm the above-stated assumptions and correction functions applied on the outcrop values. In terms of permeability a significant change is only inferable for greater depth and higher temperature. The permeability values obtained from cores in depth of 1600 m show typical values of permeability comparable to the value range of outcrop samples. The thermal conductivity shows only minor to negligible differences compared with the outcrop results, except for the depth range of 230, 1300–1500, 4400 and 4700–4850 m where dolomitized zones of massive facies in the Malm $\zeta 1$ and $\zeta 2$ are encountered. The dolomitization process results here in a significant increase (up to $2.5\,\mathrm{W\,(m \cdot K)^{-1}}$) of thermal conductivity. The temperature influence at greater depth (4850 m) shows a decrease of thermal conductivity and an increase of specific heat capacity (Fig. 4).

Thermophysical correlations between different reservoir properties are controlled by lithofacies. Based on the Debye equation (Eq. 1) and the analyses of measurements (Fig. 4), it can be inferred that the thermophysical properties for tight carbonate rocks are density-controlled. Density itself is strongly dependent on the lithofacies of the carbonate rock; i.e. the massive and basin facies have direct influence on the formations' hydraulic conductivity. In the transition zone of basin to massive (reef) facies, sub-vertical fractures caused by differential compaction between massive facies and ad-

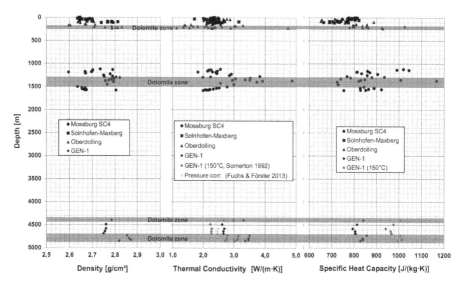

**Figure 4.** Comparison of density, thermal conductivity and specific heat capacity of drilling locations in the outcrop analogue area (Solnhofen-Maxberg) and the Molasse Basin (Oberdolling, Moosburg SC4 and Geretsried GEN-1).

jacent basin facies can also be observed in the studied outcrops. Due to the increased fracture density in this zone, the karstification process is favoured, which results in dissolution of carbonate. The increased hydraulic conductivity results either in the disintegration into dolomite sand or in the process of de-dolomitization (re-calcification). In this context it is important to consider that dolomitized zones, due to their primary facies and genesis, even on a small scale are laterally variable and developed across fractures and porous zones into adjacent facies (Koch, 2011). Therefore, the identification and location of such zones is of special interest for the geothermal reservoir prognosis in terms of hydraulically prospective reservoir formations.

The thermal conductivity, porosity and permeability values presented herein are in good accordance with results of recent studies on a limited number of rock samples from the Molasse Basin (Koch et al., 2007, 2009; Böhm et al. 2013).

## 5   Conclusions

The studied rocks of the Upper Jurassic are not a homogenous formation of limestones. Even on a small scale, different facies types and their interfingering – which can be differentiated in geometry, structure, fabric and composition – can be identified. These differences affect the thermophysical properties of the rocks and show facies-related trends. The hydraulic parameters vary on the order of 4 magnitudes within a stratigraphic unit or facies zone, but in general they show a range of poor to very poor matrix hydraulic conductivity (cf. Stober et al., 2013). From outcrop studies it can be inferred that hydraulic active pathways are bound to fracture networks, faults and adjacent karstification and/or dolomitized zones. The secondary reservoir permeability is strongly

related to the tectonic setting and facies-controlled diagenesis. Additionally, reservoir permeability depends on the hydrochemical conditions of the carbonate reservoir to maintain open flow paths. Based on the investigation of the matrix parameters, the sustainable heat transport into the utilized geothermal reservoir can be assessed. Thus, the long-term capacities for different utilization scenarios can be calculated more precisely. With the help of 3-D seismic surveys the investigation of lateral extension and related facies heterogeneity will give valuable information on the transmissibility of different target horizons/facies. The thermofacies characterization and prediction of geothermal reservoir parameters enables the identification of prospective exploration areas. However, the structural hydraulic conductivity of fault zones has to be addressed as a first step in exploration, followed by lithofacies studies to ensure a successful exploration strategy for the Upper Jurassic aquifer exploitation.

The data from the Upper Jurassic limestones of southern Germany show that the prognosis of reservoir properties applying facies models can be implemented as an additional exploration tool. The determination of geothermal reservoir properties serves in general to distinguish between petrothermal and hydrothermal systems (Sass and Götz, 2012) and can also be used to optimize the drilling and stimulation design. Outcrop analogue studies are an effective tool to create a database in an early project phase. Ultimately, the assessment of production capacities of geothermal reservoirs becomes more reliable; applying reservoir transfer models to the database predicted reservoir properties at greater depths and higher temperatures. Furthermore, these studies provide a sufficient database to determine thermophysical reservoir characteristics of the rock matrix which can be used for optimized temperature distribution modelling of geothermal reservoir formations. Facies concepts are applied as an exploration tool producing conservative results. Adding information on secondary porosities, karstification, dolomitization and stress field into a reservoir model will enable estimating realistic reservoir capacities. The key to a reliable reservoir prognosis, reservoir stimulation and sustainable reservoir utilization for the Malm in the Molasse Basin is to integrate statistically tested databases of tectonic, hydraulic and thermofacies features into 3-D reservoir models.

**Acknowledgements.** The authors thank the Bayerisches Landesamt für Umwelt (LfU) for the permission to sample shallow and deep research drillings. We also thank the ENEX Power Germany GmbH for support and access to cutting material from the drill site GEN-1 in Geretsried. Furthermore, we acknowledge the contribution of numerous students who performed measurements of geothermal rock parameters in the framework of their theses at the Institute of Applied Geosciences, Technische Universität Darmstadt. The constructive comments of Ingrid Stober and an anonymous reviewer greatly improved the manuscript.

# References

Abdulagatova, Z., Abdulagatov, I. M., and Emirov, V. N.: Effect of temperature and pressure on the thermal conductivity of sandstone, Int. J. Rock Mech. Min., 46, 1055–1071, 2009.

Bär, K., Arndt, D., Fritsche, J.-G., Götz, A. E., Kracht, M., Hoppe, A., and Sass, I.: 3-D-Modellierung der tiefengeothermischen Potenziale von Hessen: Eingangsdaten und Potenzialausweisung, Z. Dt. Ges. Geowiss., 162, 371–388, 2011.

Birner, J., Fritzer, T., Jodocy, M., Savvatis, A., Schneider, M., and Stober, I.: Hydraulische Eigenschaften des Malmaquifers im Süddeutschen Molassebecken und ihre Bedeutung für die geothermische Erschließung, Z. Geol. Wiss., 40, 133–156, 2012.

Bjørkum, P. A. and Nadeau, P. H.: Temperature controlled porosity/permeability reduction, fluid migration, and petroleum exploration in sedimentary basins, Australian Pet. Prod. Expl. Assoc. J., 38, 453–464, 1998.

Böhm, F., Savvatis, A., Steiner, U., Schneider, M., and Koch, R.: Lithofazielle Reservoircharakterisierung zur geothermischen Nutzung des Malm im Großraum München, Grundwasser, 18, 3–13, 2013.

Chilingarian, G. V., Mazzullo, S. J., and Rieke, H. H.: Carbonate Reservoir Characterization: A Geologic-Engineering Analysis, Elsevier Sci. Publs. B. V., Amsterdam, the Netherlands, 639 pp., 1992.

Clauser, C. (Ed.): Numerical Simulation of Reactive Flow in Hot Aquifers – SHEMAT and Processing SHEMAT, Springer-Verlag, Berlin Heidelberg, Germany, doi:10.1007/978-3-642-55684-5, 2003.

Clauser, C. and Huenges, E.: Thermal Conductivity of Rocks and Minerals, Rock Physics and Phase Relations, A Handbook of Physical Constants, AGU Reference Shelf, 3, 105–126, 1995.

Clauser, C., Deetjen, H., Höhne, F., Rühaak, W., Hartmann, A., Schellschmidt, R., Rath, V., and Zschocke, A.: Erkennen und Quantifizieren von Strömung: Eine geothermische Rasteranalyse zur Klassifizierung des tiefen Untergrundes in Deutschland hinsichtlich seiner Eignung zur Endlagerung radioaktiver Stoffe, Endbericht zum Auftrag 9X0009-8390-0 des Bundesamtes für Strahlenschutz (BfS), Applied Geophysics and Geothermal Energy E.ON Energy Research Center, RWTH Aachen, Germany, 159 pp., 2002.

Dunham, R. J.: Classification of carbonate rocks according to depositional texture, in: Classification of carbonate rocks, edited by: Ham, W. E., AAPG Memoir, 1, 108–171, 1962.

Ehrenberg, S. N. and Nadeau, P. H.: Sandstone versus carbonate petroleum reservoirs: a global perspective on porosity-depth and porosity-permeability relationships, AAPG Bulletin, 89, 435–445, 2005.

Embry, A. F. and Klovan, J. E.: A Late Devonian reef tract on Northeastern Banks Island, NWT: Canadian Petroleum Geology Bulletin, 19, 730–781, 1971.

Fuchs, S. and Förster, A.: Well-log based prediction of thermal conductivity of sedimentary successions: a case study from the North German Basin, Geophys. J. Int., 196, 291–311 doi:10.1093/gji/ggt382, 2013.

Geyer, O. F. and Gwinner, M. P.: Die Schwäbische Alb und ihr Vorland, Slg. Geol. Führer, 67, 271 pp., 1979.

Gwinner, M. P.: Origin of the Upper Jurassic of the Swabian Alb, Contrib. Sedimentol., 5, 1–75, 1976.

Hartmann, A., Rath, V., and Clauser, C.: Thermal conductivity from core and well log data, Int. J. Rock Mech. Min., 42, 1042–1055, 2005.

Homuth, S., Götz, A. E., and Sass, I.: Facies relation and depth dependency of thermo- and petrophysical rock parameters of the Upper Jurassic geothermal carbonate reservoirs of the Molasse Basin, Z. Dt. Ges. Geowiss., 165, 469–486, 2014.

Hornung, J. and Aigner, T.: Sedimentäre Architektur und Poropermanalyse fluviatiler Sandsteine: Fallbeispiel Coburger Sandstein, Franken, Hallesches Jahrb. Geowiss., Reihe B, 18, 121–138, 2004.

Jaritz, R.: Quantifizierung der Heterogenität einer Sandsteinmatrix am Beispiel des Stubensandstein (Mittlerer Keuper, Württemberg), Tübinger Geol. Abhandlungen, Reihe C, 48, 1–104, 1999.

Jodocy, M. and Stober, I.: Porosities and Permeabilities in the Upper Rhine Graben and in the SW Molasse Basin (Germany), Erdöl Erdgas Kohle, 127, 20–27, 2011.

Koch, R.: Dolomit und Dolomit-Zerfall im Malm Süddeutschlands – Verbreitung, Bildungsmodelle, Dolomit-Karst, Laichinger Höhlenfreund, 46, 75–92, 2011.

Koch, A., Hartmann, A., Jorand, R., Mottaghy, D., Pechnig, R., Rath, V., Wolf, A., and Clauser, C.: Erstellung statistisch abgesicherter thermischer und hydraulischer Gesteinseigenschaften für den flachen und tiefen Untergrund in Deutschland (Phase 1 – Westliche Molasse und nördlich angrenzendes Süddeutsches Schichtstufenland), Schlussbericht zum BMU-Projekt FKZ 0329985, RWTH Aachen, Germany, 220 pp., 2007.

Koch, A., Jorand, R., Arnold, J., Pechnig, R., Mottaghy, D., Vogt, C., and Clauser, C.: Erstellung statistisch abgesicherter thermischer und hydraulischer Gesteinseigenschaften für den flachen und tiefen Untergrund in Deutschland (Phase 2 – Westliches Nordrhein-Westfalen und bayerisches Molassebecken), Abschlussbericht zum BMU-Projekt FKZ 0329985, Aachen (RWTH), Germany, 174 pp., 2009.

Leinfelder, R. R., Krautter, M., Laternser, R., Nose, M., Schmid, D. U., Schweigert, G., Werner, W., Keupp, H., Brugger, H., Hermmann, R., Rehfeld-Kiefer, U., Schroeder, J. H., Reinhold, C., Koch, R., Zeiss, A., Schweizer, V., Christmann, H., Menges, G., and Luterbacher, H.: The origin of Jurassic reefs: current research developments and results, Facies, 31, 1–56, 1994.

Leinfelder, R. R., Werner, W., Nose, M., Schmid, D. U., Krautter, M., Laternser, R., Takacs, M., and Hartmann, D.: Paleoecology, growth parameters and dynamics of coral, sponge and microbolite reefs from the late Jurassic, Gött. Arb. Geol. Paläontol., 2, 227–248, 1996.

Meyer, R. K. F. and Schmidt-Kaler, H.: Paläogeographischer Atlas des süddeutschen Oberjura (Malm), Geol. Jb., A/115, 3–77, 1989.

Meyer, R. K. F. and Schmidt-Kaler, H.: Paläogeographie und Schwammriffentwicklung des süddeutschen Malm – ein Überblick, Facies, 23, 175–184, 1990.

Moeck, I. and Beardsmore, G.: A new "geothermal play type" catalog: Streamlining exploration decision making, Proceedings, Thirty-Ninth Workshop on Geothermal Reservoir Engineering Stanford University, Stanford, California, 2014.

Pawellek, T.: Fazies-, Sequenz-, und Gamma-Ray-Analyse im höheren Malm der Schwäbischen Alb (SW-Deutschland) mit Bemerkungen zur Rohstoffgeologie (hochreine Kalke), Tübinger Geol. Arb., Reihe A, 61, 1–246, 2001.

Pawellek, T. and Aigner, T.: Apparently homogenous "reef"-limestones built by high-frequency cycles Upper Jurassic, SW-Germany, Sediment. Geol., 160, 259–284, 2003.

Pei, L., Rühaak, W., Stegner, J., Bär, K., Homuth, S., Mielke, P., and Sass, I.: Thermo-Triax: An Apparatus for Testing Petrophysical Properties of Rocks Under Simulated Geothermal Reservoir Conditions, Geotech. Testing J., 38, 1–20, 2014.

Pieńkowski, G., Schudack, M. E., Bosák, P., Enay, R., Feldman-Olszewska, A., Golonka, J., Gutowski, J., Herngreen, G. F. W., Jordan, P., Krobicki, M., Lathuilière, B., Leinfelder, R. R., Michalík, J., Mönnig, E., Noe-Nygaard, N., Pálfy, J., Pint, A., Rasser, M. W., Reisdorf, A. G., Schmid, D. U., Schweigert, G., Surlyk, F., Wetzel, A., and Wong, T. E.: Jurassic, in: The Geology of Central Europe, edited by: McCann, T., The Geological Society, London, UK, 2: Mesozoic and Cenozoic: 823–922, 2008.

Popov, Y. A., Berezin, V. V., Semenov, V. G., and Korostelev, V. M.: Complex detailed investigations of the thermal properties of rocks on the basis of a moving point source, Phys. Solid Earth, 21, 64–70, 1985.

Popov, Y. A., Tertychnyi, V., Romushkevich, R., Korobkov, D., and Pohl, J.: Interrelations between thermal conductivity and other physical properties of rocks: experimental data, Pure App. Geophys., 160, 1137–1161, 2003.

Sass, I. and Götz, A. E.: Geothermal reservoir characterization: a thermofacies concept, Terra Nova, 24, 142–147, 2012.

Sass, J. H., Lachenbruch, A. H., Munroe, R. J., Greene, G. W., and Moses Jr., T. H.,: Heat Flow in the Western United States, J. Geophys. Res., 76, 6376–6413, 1971.

Schauer, M.: Untersuchungen an Dolomitsteinen und zuckerkörnigen Kalksteinen auf der Mittleren Schwäbischen Alb, Laichinger Höhlenfreund, 31, 5–18, 1996.

Schauer, M.: Dynamische Stratigraphie, Diagenese und Rohstoffpotenzial des Oberjura (Kimmeridge 1–5 der mittleren Schwäbischen Alb, Tübinger geowissenschaftliche Arbeiten, Tübingen Universität Tübingen, Germany, Reihe A, 36, 1998.

Schulz, R. (Ed.), Thomas, R., Dussel, M., Lüschen, E., Wenderoth, F., Fritzer, T., Birner, J., Schneider, M., Wolfgramm, M., Bartels, J., Hiber, B., Megies, T., and Wassermann, J.: Geothermische Charakterisierung von karstig-klüftigen Aquiferen im Großraum München, Final report, Hannover (LIAG), Germany, Förderkennzeichen 0325013A, 2012.

Selg, M. and Wagenplast, P.: Beckenarchitektur im süddeutschen Weißen Jura und die Bildung der Schwammriffe, Jh. Geol. Landesamt Baden-Württemberg, 32, 171–206, 1990.

Somerton, W. H.: Thermal properties and temperature-related behavior of rock-fluid systems, Dev. Petr. Sci., 37, 257 pp., 1992.

Stober, I.: Die thermalen Karbonat-Aquifere Oberjura und Oberer Muschelkalk im Südwestdeutschen Alpenvorland, Grundwasser, 18, 259–269, 2013.

Stober, I. and Villinger, E.: Hydraulisches Potential und Durchlässigkeit des höheren Oberjuras und des Oberen Muschelkalks unter dem Baden-Württembergischen Molassebecken, Jh. geol. L.-Amt Baden-Württemberg, 37, 7–24, 1997.

Stober, I., Jodocy, M., and Hintersberger, B.: Gegenüberstellung von Durchlässigkeiten aus verschiedenen Verfahren im tief liegenden Oberjura des südwestdeutschen Molassebeckens, Z. Dt. Ges. Geowiss (German J. Geosci.), 164, 663–679, 2013.

Villinger, E.: Hydrogeologische Ergebnisse, in: Ergebnisse der Hydrogeothermiebohrungen in Baden-Württemberg, edited by: Bertleff, B. W., Joachim, H., Koziorowski, G., Leiber, J., Ohmert, W., Prestel, R., Stober, J., Strayle, G., Villinger, E., and Werner, W., Jh. Geol. Landesamtes Baden-Württemberg, 30, 27–116, 1988.

Vosteen, H.-D. and Schellschmidt, R.: Influence of temperature on thermal conductivity, thermal capacity and thermal diffusivity for different types of rock, Phys. Chem. Earth, 28,, 499–509, 2003.

Zoth, G. and Hänel, R.: Thermal conductivity, in: Handbook of Terrestrial Heat Flow Density Determination, edited by: Hänel, R., Rybach, L., and Stegena, L., Kluwer Academic Publishers, 449–453, 1988.

# 6

# Empirical relations of rock properties of outcrop and core samples from the Northwest German Basin for geothermal drilling

**D. Reyer**[1,*] **and S. L. Philipp**[1]

[1]Georg August University of Göttingen, Geoscience Centre, Department of Structural Geology and
Geodynamics, Germany
[*]now at: State Authority of Mining, Energy and Geology – Zentrum für TiefenGeothermie, Celle, Germany

*Correspondence to:* D. Reyer (dorothea.reyer@geo.uni-goettingen.de)

**Abstract.** Information about geomechanical and physical rock properties, particularly uniaxial compressive strength (UCS), are needed for geomechanical model development and updating with logging-while-drilling methods to minimise costs and risks of the drilling process. The following parameters with importance at different stages of geothermal exploitation and drilling are presented for typical sedimentary and volcanic rocks of the Northwest German Basin (NWGB): physical ($P$ wave velocities, porosity, and bulk and grain density) and geomechanical parameters (UCS, static Young's modulus, destruction work and indirect tensile strength both perpendicular and parallel to bedding) for 35 rock samples from quarries and 14 core samples of sandstones and carbonate rocks.

With regression analyses (linear- and non-linear) empirical relations are developed to predict UCS values from all other parameters. Analyses focus on sedimentary rocks and were repeated separately for clastic rock samples or carbonate rock samples as well as for outcrop samples or core samples. Empirical relations have high statistical significance for Young's modulus, tensile strength and destruction work; for physical properties, there is a wider scatter of data and prediction of UCS is less precise. For most relations, properties of core samples plot within the scatter of outcrop samples and lie within the 90 % prediction bands of developed regression functions. The results indicate the applicability of empirical relations that are based on outcrop data on questions related to drilling operations when the database contains a sufficient number of samples with varying rock properties. The presented equations may help to predict UCS values for sedimentary rocks at depth, and thus develop suitable geomechanical models for the adaptation of the drilling strategy on rock mechanical conditions in the NWGB.

## 1  Introduction

In Germany, the North German Basin (NGB) is one region with considerable geothermal low-enthalpy potential (Paschen et al., 2003). To utilise this potential, deep wellbores have to be drilled to reach prospective geothermal reservoir rocks at depths of 3000–6000 m. Well construction is therefore the main expense factor of geothermal projects in this region. In sedimentary successions such as the NGB, one of the major problems and expenditures may be related to wellbore stability issues (e.g. Dusseault, 2011; Zeynali, 2012). Such wellbore instabilities are recognised as a drilling

challenge that may considerably increase drilling costs and safety risks (Proehl, 2002; York et al., 2009; Li et al., 2012). The profit margin of geothermal projects, however, is rather small compared with hydrocarbon projects. Therefore, a substantial reduction of costs for well construction and completion is desirable (cf. www.gebo-nds.de).

Evaluation of in situ rock mechanical behaviour requires different information. Important input data include estimates of mechanical conditions, pore pressures, and stress state. According to Zeynali (2012), two of the most important mechanical factors affecting wellbore stability are the mechanical properties of rock – including anisotropy of strengths and elastic moduli (e.g. Heap et al., 2010) – and in situ stresses existing in different layers of rock. Development of a geomechanical model before starting the drilling operation is a powerful tool to prevent wellbore instabilities and minimise drilling costs of geothermal wells (Khaksar et al., 2009). For drilling through a rock mass, such model captures the initial equilibrium state that describes the stresses, pore pressure, and geomechanical properties. With logging-while-drilling data the initially computed geomechanical model can be continuously adapted to the conditions at depth.

For such geomechanical modelling, the uniaxial compressive strength (UCS) is the most important geomechanical input parameter (Chang et al., 2006; Nabaei and Shahbazi, 2012; Vogt et al., 2012). There already exist several software approaches for building and updating geomechanical models (Settari and Walters, 2001; geomechanics software, e.g. GMI – http://www.baker-hughes.com). Generally, such geomechanical modelling software uses empirical relationships that were developed for hydrocarbon reservoirs. To date there do not exist such relationships for geothermal reservoirs of the NGB. Here, the geological setting may be completely different leading to other rock mechanical conditions. Therefore, existing methods for geomechanical modelling have to be reviewed carefully and adapted where needed.

There are several relevant parameters with importance given to different stages of geothermal exploitation and drilling. Physical properties such as density, $\rho$, and $P$ and $S$ wave velocities, $v_p$ and $v_s$ (compressional and shear wave velocities), are parameters that can be measured directly in wellbores; the porosity, $\Phi$, is derivable from such well logs (Edlmann et al., 1998). The dynamic Young modulus is derived from velocity and density logs (Fricke and Schön, 1999; Zoback, 2007; Rider and Kennedy, 2011). Geomechanical parameters are important for reservoir exploitation and drilling operations. The static Young modulus, $E_s$, is interesting in terms of predictions of fracture propagation (Jaeger et al., 2007; Gudmundsson, 2011). The indirect tensile strength, $T_0$, gives information about the rock's resistance to tensile fractures. These parameters are of interest in terms of dimensioning of hydraulic fracturing operations, wellbore stability and drilling mud selection (e.g. Zoback, 2007). The destruction work, $W$, is one parameter providing information on the amount of energy needed to destroy the rock while drilling. It is known to correlate with the drilling efficiency which is a term used to describe the effects of a number of geological and machine parameters on the drilling velocity (Thuro, 1997). Therefore, it is desirable to make reasonable assumptions about these parameters for drilling through the rock units. To do so, we need empirical relations between UCS and parameters which are either knowable before drilling or determinable with logging-while-drilling tools. With well logs from existing adjacent boreholes, a geomechanical model can be built using empirical relations between rock-strength values and physical parameters. Empirical relations can then be used for validation

of the geomechanical model while- and after-drilling by updating the model continuously with logging data.

Determining geomechanical and physical parameters directly from core material, however, is expensive and time-consuming because a large number of core samples are needed, and core material is rare (e.g. Khaksar et al., 2009). Therefore, in this study we aim to improve predictions of mechanical properties for rocks at depth. First, we present data on geomechanical and physical properties of representative rock types of the NGB. We sampled 35 mainly sedimentary rocks of the western sub-basin of the NGB, the Northwest German Basin (NWGB), from Lower Permian to Upper Cretaceous, exposed in outcrop analogues, i.e. quarries. In addition to these outcrop samples, we analysed 14 core samples from two wellbores with the same stratigraphic units, comparable lithologies and facies as equivalent samples to analyse mechanical property changes due to uplift and alteration. Secondly, we used the data of sedimentary rocks to perform regression analyses, together with calculation of coefficients of determination ($R^2$), between UCS and the described parameters, separately for outcrop samples only and including core samples. To analyse the statistical significance of the developed regression functions, 90 % confidence and prediction bands are added. The rock properties of core samples are compared with the results of outcrop samples from the developed equations of outcrop samples to examine the relevance of outcrop samples for predicting rock properties at depths. The regression functions may help predict UCS values for sedimentary rocks at depth, and thus develop a suitable geomechanical model for the adaptation of the drilling strategy on rock mechanical conditions.

## 2 Geologic setting and sample locations

The study area is part of the NWGB, the western part of the NGB, located in northwestern Germany (Walter, 2007). The NGB initiated in the Late Carboniferous–Permian due to rifting processes subsequent to the Variscan orogenesis (Betz et al., 1987; Ziegler, 1990). From marine to continental conditions, the sedimentary succession is characterised by changing sedimentation environments. Therefore, the NWGB is comprised of mainly carbonate and clastic rocks with some intercalated evaporates leading to very heterogeneous rock mechanical conditions.

The study area is located at the southern and western margins of the western region of the North German Basin (Fig. 1; cf. Reyer et al., 2012). Sedimentary rocks that occur at geothermally relevant depths in the centre and north of the NWGB crop out at the basin margins and can be sampled in quarries. In such outcrop analogues, listed in Table 1, we took samples of two main rock types: carbonate rocks (Triassic, Jurassic, and Cretaceous age) and sandstones (Permian, Triassic, Jurassic, and Cretaceous age; Table 1). Three Rotliegend volcanic rock (Permian) samples are included to

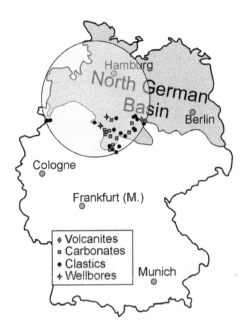

**Figure 1.** North German Basin (modified after http://www.geotis. de) with the locations of sampled wellbores and quarries and the exposed rock types (see key) in the NWGB (rough location marked).

obtain rock property data over a wide range of lithologies present in the NWGB (Fig. 1). For four carbonate rock units and three sandstone units, the equivalent core samples were identified and sampled from two wellbores: Groß Buchholz (Gt1) and Eulenflucht 1 (EF1; Table 1).

## 3    Methods

### 3.1   Density and porosity

The bulk density, $\rho_d$ [g cm$^{-3}$], was determined from dry cylindrical specimens with a GeoPyc 1360 (Micromeritics), setting measured volume and mass in relation. For the same specimens, we measured the grain density, $\rho_0$ [g cm$^{-3}$], with an Ultrapycnometer 1000 (Quantachrome) at room temperature using 99.9 % helium, previously measured $\rho_d$ and specimen's mass.

The total porosity, $\Phi$, given in [%], was calculated from $\rho_0$ and $\rho_d$. Samples are separated in low- (0–10 %), medium- (10–20 %), and high-porosity (> 20 %) rocks for further interpretation of rock properties.

### 3.2   Rock testing

Uniaxial compression tests were performed stress-controlled at a constant rate of 0.5 MPa s$^{-1}$ on specimens with length–diameter ratios of 2–2.5 to determine UCS and $E_s$ (ISRM, 2007). For each outcrop sample, six specimens with diameters of 40 mm were measured, both parallel and perpendicular to sedimentary bedding or, for volcanic rocks, with respect to surface orientation. Core samples were tested only

perpendicular to bedding due to limited core material. $v_p$ is measured (Tektronix TDS 5034B; 1 MHz rectangular pulse) to eliminate defective specimens. $E_s$ is determined at the linear–elastic deformation path of the stress–strain curve. For rock samples showing brittle failure, we calculated $W$ (Thuro, 1997) as the area below the stress-strain curve given in kilojoules per cubic metre.

$T_0$ is measured both parallel and perpendicular to sedimentary bedding on specimens with diameters of 40 mm and lengths of 15–20 mm with Brazilian tests (ISRM, 2007). Both parallel and perpendicular to bedding, a minimum of nine (outcrop samples) and four specimens (core samples), respectively, were tested.

### 3.3   Statistical analyses

For each sample, both parallel and perpendicular to bedding, mean values and standard deviations of the tested specimens were calculated for geomechanical parameters and $v_p$. We performed regression analyses (linear and non-linear) of mean values for UCS with $\Phi$, $\rho_d$, $v_p$, and $E_s$ and for $W$ and $T_0$ with UCS, respectively. Different regression analyses were made for each pair of parameters: (1) all samples to obtain a good overview, (2) sandstone samples only, and (3) carbonate samples only. In each case, regressions were made both for outcrop samples only and for all samples including core samples. For outcrop sample equations, 90 % confidence and prediction bands are included. Confidence bands represent the 90 % certainty of regression curve estimation based on limited sample data (Wooldridge, 2009; Brink, 2010). Prediction bands cover the range in which the values of future measurements of associated samples lie with a probability of 90 %. Based on these bands core sample results are compared with outcrop results.

## 4    Experimental results

### 4.1   Physical properties

In Tables 2 and 3, mean values of dry bulk density, grain density, calculated porosities, and $P$ wave velocities of all rock samples are listed. The approximate lithology is given to better appraise the following data analyses.

For sandstones and carbonates, we have sample data over a wide range of porosities; the lowest porosities occur in core samples. Accordingly, the dry bulk density values show a wide range. Grain densities of carbonates are highest due to a higher mineral density of the carbonates' main component calcite as compared with quartz. The grain densities strongly depend on the amount of heavy minerals: (1) hematite-rich Triassic sandstones have high $\rho_0$ values (> 2.7 g cm$^{-3}$); (2) carbonate samples with increased grain densities contain large amounts of ferrous carbonates.

$v_p$ values clearly depend on lithology. Carbonate samples show mean values of $v_p$ from 3277 m s$^{-1}$ (porous chalk marl:

**Table 1.** All samples from outcrops and wellbores with sample ID, local name, lithology, stratigraphical units, and total vertical depths of core samples.

| Sample ID | Lithology | System | Local name | |
|---|---|---|---|---|
| KrCa | Chalk marl | | Kreidemergel | |
| GoSa | Sandstone | | Sudmerberg F. | |
| HoT | Marl | | Rotpläner | |
| BrCe | Limestone | Cretaceous | Cenoman-Kalk | |
| OLH | Sandstone | | Hils Sst. | |
| GiUK | Sandstone | | Gildehaus Sst. | |
| FrUK | Sandstone | | Bentheimer Sst. | |
| OK | Sandstone | | Wealden Sst. | |
| ThüJ | Limestone | | Serpulit | |
| GVa | Limestone | | Gigas Schichten | |
| OKDa | Limestone | Jurassic | Oberer Kimmeridge | |
| ShJk | Limestone | | Korallenoolith | |
| HSDi, HSDi2 | Limestones | | Heersumer Schichten | |
| AlWo | Sandstone | | Aalen Sst. | |
| koQ | Sandstone | | Rhät Sst. | |
| koVe | Sandstone | | Rhät Sst. | |
| kuWe | Siltstone | | Lettenkohlen Sst. | |
| EM | Limestone | | Trochitenkalk | |
| H | Limestone | | Schaumkalk | |
| EL1, EL2, EL3 | Limestones | Triassic | Wellenkalk | |
| soWa | Shale-Gypsum | | Röt 1 | |
| smHN | Sandstone | | Hardegsen-Folge | |
| smD | Sandstone | | Detfurth-Folge | |
| smVG, smVG2 | Sandstones | | Volpriehausen-Folge | |
| suHe | Limestone | | Rogenstein | |
| BiSu | Sandstone | | Bernburg-Folge | |
| BeRo, BeRoK | Sandstones | | Rotliegend Sst. | |
| DöRo | Andesite | Permian | Rotliegend-Vulkanit | |
| FL2, FL6 | Rhyolites | | Rotliegend-Vulkanit | |
| | Wellbore 1: Eulenflucht 1 (EF1) | | | |
| | Wellbore 2: Groß Buchholz (Gt1) | | | TVD [m] |
| Gt1WS1 | Sandstone | | Wealden Sst. | 1.221 |
| Gt1WS2 | Sandstone | Cretaceous | Wealden Sst. | 1.211 |
| EF1WS | Sandstone | | Wealden Sst. | 135 |
| EF1GS | Limestone | | Gigas Schichten | 210 |
| EF1OK | Limestone | | Oberer Kimmeridge | 243 |
| EF1UKK | Limestone | Jurassic | Korallenoolith | 282 |
| EF1KO | Limestone | | Korallenoolith | 286 |
| EF1HS | Limestone | | Heersumer Schichten | 325 |
| Gt1DU1 | Sandstone | | Detfurth-Folge | ~3535.8 |
| Gt1DU2 | Sandstone | | Detfurth-Folge | ~3534.3 |
| Gt1DU3 | Sandstone | Triassic | Detfurth-Folge | ~3534.7 |
| Gt1DW | Siltstone | | Detfurth-Folge | ~3537.2 |
| Gt1VS1 | Sandstone | | Volpriehausen-Folge | ~3655.6 |
| Gt1VS2 | Sandstone | | Volpriehausen-Folge | ~3657.8 |

Sst.: sandstone, F.: formation; TVD: total vertical depth

**Table 2.** Lithology, dry bulk density, grain density, porosity and $P$ wave velocity for outcrop samples.

| Sample ID | Specified lithology | $\rho_d$ [g cm$^{-3}$] | $\rho_0$ [g cm$^{-3}$] | $\Phi$ [%] | $v_p$ [m s$^{-1}$] + SD |
|---|---|---|---|---|---|
| KrCa | Porous chalk marl | 2.18 | 2.86 | 23.9 | $3277 \pm 84$ |
| GoSa | Medium-grained sandstone | 2.53 | 2.69 | 6 | $3772 \pm 70$ |
| HoT | Marl | 2.59 | 2.73 | 5.2 | $5116 \pm 199$ |
| BrCe | Bioclast-bearing matrix LS | 2.66 | 2.77 | 3.8 | $4674 \pm 258$ |
| OLH | Medium-grained sandstone | 2.09 | 2.77 | 24.6 | $2291 \pm 63$ |
| GiUK | Medium-grained sandstone | 2.11 | 2.68 | 21.6 | $2576 \pm 130$ |
| FrUK | Fine-grained sandstone | 2.36 | 2.68 | 12.1 | $2172 \pm 87$ |
| OK | Medium-grained sandstone | 2.29 | 2.80 | 18.3 | $2942 \pm 120$ |
| ThüJ | Bioclast-rich matrix LS | 2.07 | 2.83 | 26.7 | $4262 \pm 215$ |
| GVa | Porous sparry LS | 2.29 | 2.96 | 22.8 | $3967 \pm 106$ |
| OKDa | Bioclast-rich matrix LS | 2.63 | 2.83 | 7.2 | $5134 \pm 100$ |
| ShJk | Bioclast-bearing oolite | 2.61 | 2.74 | 4.6 | $5171 \pm 154$ |
| HSDi | Micro-sparry LS | 2.53 | 2.78 | 9.1 | $5084 \pm 350$ |
| HSDi2 | Bioclast-rich sparry LS | 2.40 | 2.78 | 13.7 | $4787 \pm 236$ |
| AlWo | Medium-grained sandstone | 2.09 | 2.69 | 22.5 | $3000 \pm 184$ |
| koQ | Medium-grained sandstone | 2.27 | 2.84 | 20.1 | $3222 \pm 36$ |
| koVe | Fine-grained sandstone | 2.34 | 2.77 | 15.6 | $2980 \pm 38$ |
| kuWe | Siltstone | 2.59 | 2.68 | 3.4 | $3951 \pm 126$ |
| EM | Bioclast-rich sparry LS | 2.71 | 2.79 | 2.9 | $5607 \pm 164$ |
| H | Porous sparry LS | 2.40 | 2.77 | 13.2 | $4888 \pm 73$ |
| EL1 | Dolomitic LS | 2.53 | 2.98 | 15.1 | $4683 \pm 133$ |
| EL2 | Massy matrix LS | 2.74 | 2.75 | 0.3 | $6158 \pm 8$ |
| EL3 | Dolomitic LS | 2.66 | 2.94 | 9.4 | $4526 \pm 23$ |
| soWa | Shale-gypsum alternation | 2.33 | 2.39 | 2.5 | $3690 \pm 120$ |
| smHN | Medium-grained sandstone | 2.26 | 2.71 | 16.6 | $2574 \pm 64$ |
| smD | Medium-grained sandstone | 2.38 | 2.76 | 13.7 | $2986 \pm 22$ |
| smVG | Medium-grained sandstone | 2.32 | 2.72 | 14.4 | $2948 \pm 78$ |
| smVG2 | Medium-grained sandstone | 2.17 | 2.74 | 20.6 | $2074 \pm 89$ |
| suHe | Sparry oolite | 2.71 | 2.75 | 1.5 | $5368 \pm 136$ |
| BiSu | Medium-grained sandstone | 2.15 | 2.79 | 22.9 | $2110 \pm 6$ |
| BeRoK | Conglomeratic sandstone | 2.58 | 2.67 | 3.2 | $3564 \pm 78$ |
| BeRo | Medium-grained sandstone | 2.52 | 2.69 | 6.6 | $3426 \pm 29$ |
| DöRo | Andesite | 2.72 | 2.72 | 0.1 | $5449 \pm 23$ |
| FL2 | Rhyolite | 2.63 | 2.64 | 0.1 | $5260 \pm 44$ |
| FL6 | Rhyolite | 2.69 | 2.69 | 0.1 | $5342 \pm 64$ |

LS, limestone; $\rho_d$, dry bulk density; $\rho_0$, grain density; $\Phi$, porosity; $v_p$, $P$ wave velocity; SD, standard deviation

KrCa) to 6158 m s$^{-1}$ (massy matrix limestone: EL2). Mostly, the standard deviations of carbonate samples are high. This is pronounced in carbonates with either a high presence of lithoclasts or due to rock heterogeneities. $v_p$ in sandstones are considerably slower than in carbonate rocks. The lowest values relate to high porosities. In volcanic rock samples, $v_p$ is rather high (about 5300 m s$^{-1}$) with small variation and standard deviations.

### 4.2 Rock mechanical properties

In Table 4, mean values of the geomechanical parameters of all samples are listed. The standard deviations of all measurements for every sample are given. Measured parameter values of the eight clastic core samples are higher than those of the 14 outcrop samples. The differences between outcrop and core samples of carbonate rocks are, in contrast, rather small. Parameter values of the three volcanic rock samples are considerably higher than of sedimentary outcrop samples.

### 5 Empirical relations of rock properties with UCS

The rock property data, presented in Tables 2, 3, and 4, may be used directly to calibrate an existing geomechanical model by attaching UCS values to log profiles and deducing equivalent values of tensile strength and destruction work using empirical relations. In situ rocks and core samples, however, may have completely different rock properties. Thus, we compare properties of core samples and outcrop samples to analyse if properties of in situ rocks can be predicted based on data from outcrop samples from the same geologic setting.

**Table 3.** Lithology, dry bulk density, grain density, porosity and $P$ wave velocity for core samples.

| Core samples: | | | | | |
|---|---|---|---|---|---|
| Sample ID | Specified lithology | $\rho_d$ [g cm$^{-3}$] | $\rho_0$ [g cm$^{-3}$] | $\Phi$ [%] | $v_p$ [m s$^{-1}$] + SD |
| Gt1WS1 | Coarse-grained sandstone | 2.40 | 2.84 | 15.5 | $4854 \pm 38$ |
| Gt1WS2 | Medium-grained sandstone | 2.58 | 2.79 | 7.6 | $2950 \pm 267$ |
| EF1WS | Medium-grained sandstone | 2.25 | 2.79 | 19.4 | $2638 \pm 28$ |
| EF1MM | Shale gypsum | 2.88 | 2.95 | 2.4 | $5808 \pm 110$ |
| EF1GS | Sparry LS | 2.48 | 2.78 | 10.8 | $5832 \pm 65$ |
| EF1OK | Bioclast-rich matrix LS | 2.72 | 2.78 | 2.1 | $5732 \pm 50$ |
| EF1UKK | Bioclast-rich sparry LS | 2.76 | 2.77 | 0.2 | $5412 \pm 53$ |
| EF1KO | Sparry oolite | 2.72 | 2.79 | 2.6 | $6053 \pm 59$ |
| EF1HS | Bioclast-rich sparry LS | 2.18 | 2.82 | 22.8 | $3831 \pm 87$ |
| Gt1DU1 | Medium-grained sandstone | 2.69 | 2.70 | 0.4 | $4981 \pm 33$ |
| Gt1DU2 | Coarse-grained sandstone | 2.73 | 2.75 | 1.0 | $3410 \pm 78$ |
| Gt1DU3 | Medium-grained sandstone | 2.67 | 2.77 | 3.6 | $4906 \pm 96$ |
| Gt1DW | Siltstone | 2.83 | 2.87 | 1.1 | $5166 \pm 123$ |
| Gt1VS1 | Medium-grained sandstone | 2.71 | 2.72 | 0.1 | $4745 \pm 62$ |
| Gt1VS2 | Coarse-grained sandstone | 2.69 | 2.77 | 2.8 | $4539 \pm 54$ |

LS, limestone; $\rho_d$, dry bulk density; $\rho_0$, grain density; $\Phi$, porosity; $v_p$, $P$ wave velocity; SD, standard deviation

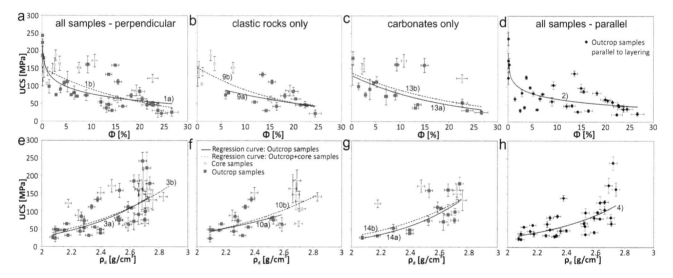

**Figure 2.** UCS measured perpendicular to bedding vs. (**a–c**) $\Phi$ and (**e–g**) $\rho_d$, respectively, separately for all samples (**a, e**; $n = 49$), only clastic rock samples (**b, f**; $n = 24$) and only carbonate samples (**c, g**; $n = 20$) for outcrop and core samples; regression curves shown for both outcrop and core samples and outcrop samples only. UCS measured parallel to bedding vs. (**d**) $\Phi$ and (**h**) $\rho_d$ ($n = 33$); for regression equations see Table 5. For UCS, error bars stand for standard deviations of all measurements of every sample (Table 4). For density and porosity, error bars represent measuring accurancies of 1 % and 5 %, respectively.

In Table 5, the results of regression analyses for the different parameters, presented in following sections, are summarised.

## 5.1 Empirical relations for UCS prediction

### 5.1.1 Density and porosity

Porosity and bulk density are two parameters that can be determined with geophysical logs. Many previous studies showed that there are strong correlations between UCS and

both parameters (e.g. Lama and Vutukuri, 1978; Jizba, 1991; Wong et al., 1997; Palchik, 1999).

In Fig. 2, both porosity and bulk density are plotted against UCS measured perpendicular and parallel to bedding. It is obvious that there is a wide scatter of data resulting in rather poor statistical significance of the empirical relations (Table 5). However, the prediction of in situ properties based on outcrop sample results is one of the main questions of this study. It is conspicuous that in all cases, and especially for carbonates, outcrop and core samples show a similar range

Empirical relations of rock properties of outcrop and core samples from the Northwest German Basin...

47

**Table 4.** Mean values of the geomechanical parameters UCS, Young's modulus, destruction work, and indirect tensile strength, measured perpendicular and parallel to bedding, for all samples including standard deviations.

| Sample | UCS ± SD [MPa] | | $E_s$ ± SD [GPa] | | $W$ ± SD [kJ m$^{-3}$] | | $T_0$ ± SD [MPa] | |
|---|---|---|---|---|---|---|---|---|
| ID | par. | perp. | par. | perp. | par. | perp. | par. | perp. |
| KrCa | 36 ± 7 | 31 ± 4 | 12.7 ± 1.2 | 13.5 ± 1.1 | 106 ± 20 | 133 ± 30 | 2.0 ± 0.6 | 2.7 ± 0.8 |
| GoSa | 35 ± 2 | 75 ± 11 | 7.9 ± 0.5 | 30.3 ± 11.1 | 215 ± 26 | 213 ± 41 | 2.6 ± 0.4 | 5.9 ± 0.3 |
| HoT | 81 ± 7 | 112 ± 15 | 40.6 ± 4.2 | 34.1 ± 6.8 | 263 ± 60 | 332 ± 67 | 5.2 ± 1.0 | 8.0 ± 1.7 |
| BrCe | 126 ± 5 | 91 ± 29 | 44.1 ± 4.0 | 26.8 ± 6.1 | 282 ± 17 | 301 ± 15 | 6.7 ± 0.9 | 7.6 ± 0.6 |
| OLH | 37 ± 7 | 23 ± 10 | 15.1 ± 5.5 | 10.9 ± 6.5 | 118 ± 19 | 50 ± 6 | 3.1 ± 0.3 | 3.1 ± 0.3 |
| GiUK | 56 ± 2 | 47 ± 4 | 19.7 ± 4.6 | 15.7 ± 2.1 | 175 ± 25 | 198 ± 41 | 4.1 ± 0.4 | 3.1 ± 0.2 |
| FrUK | 45 ± 7 | 55 ± 4 | 12.0 ± 3.5 | 13.7 ± 4.5 | 245 ± 21 | 248 ± 38 | 2.4 ± 0.3 | 2.8 ± 0.4 |
| OK | 82 ± 6 | 73 ± 7 | 19.3 ± 1.5 | 18.0 ± 2.6 | 394 ± 34 | 404 ± 17 | 3.8 ± 0.9 | 4.2 ± 0.8 |
| ThüJ | 23 ± 5 | 26 ± 5 | 14.7 ± 1.3 | 13.3 ± 3.9 | 69 ± 10 | 77 ± 9 | 2.3 ± 0.4 | 2.7 ± 0.4 |
| GVa | 48 ± 1 | 53 ± 11 | 14.5 ± 2.4 | 14.9 ± 4.7 | 151 ± 13 | 174 ± 36 | 3.5 ± 0.6 | 4.8 ± 0.4 |
| OKDa | 79 ± 1 | 71 ± 24 | 35.6 ± 6.8 | 30.2 ± 1.8 | 207 ± 24 | 137 ± 20 | 6.3 ± 0.6 | 4.5 ± 0.6 |
| ShJk | 97 ± 3 | 109 ± 3 | 46.4 ± 3.3 | 43.4 ± 8.0 | 499 ± 67 | 558 ± 91 | 5.2 ± 1.1 | 6.5 ± 1.0 |
| HSDi | 58 ± 12 | 74 ± 13 | 37.3 ± 6.6 | 27.7 ± 0.9 | 203 ± 31 | 317 ± 21 | 5.5 ± 1.5 | 6.9 ± 0.7 |
| HSDi2 | – | 48 ± 4 | – | 36.5 ± 3.8 | – | 215 ± 78 | 4.3 ± 1.7 | 6.6 ± 1.0 |
| AlWo | 21 ± 3 | 48 ± 9 | 6.1 ± 0.9 | 15.8 ± 2.5 | 90 ± 11 | 154 ± 17 | 1.3 ± 0.2 | 4.1 ± 0.6 |
| koQ | 64 ± 8 | 85 ± 12 | 16.1 ± 1.5 | 20.1 ± 1.2 | 395 ± 36 | 378 ± 60 | 2.9 ± 0.5 | 3.6 ± 0.7 |
| koVe | 86 ± 5 | 112 ± 6 | 20.6 ± 2.4 | 24.1 ± 2.3 | 638 ± 53 | 699 ± 91 | 4.4 ± 0.7 | 4.9 ± 1.0 |
| kuWe | 41 ± 4 | 63 ± 19 | 17.3 ± 1.7 | 20.7 ± 2.0 | 202 ± 5 | 142 ± 13 | 5.9 ± 1.0 | 6.0 ± 0.7 |
| EM | 82 ± 10 | 75 ± 7 | 47.0 ± 4.2 | 36.9 ± 4.1 | 299 ± 64 | 339 ± 54 | 7.0 ± 1.8 | 6.1 ± 1.2 |
| H | 46 ± 4 | 38 ± 1 | 16.5 ± 2.6 | 24.5 ± 6.5 | 154 ± 43 | 89 ± 18 | 4.2 ± 0.8 | 4.0 ± 1.1 |
| EL1 | 96 ± 12 | 159 ± 20 | 30.8 ± 2.5 | 31.3 ± 1.2 | 410 ± 68 | 542 ± 76 | 5.5 ± 1.2 | 10.2 ± 2.6 |
| EL2 | 162 ± 19 | 179 ± 19 | 81.6 ± 6.5 | 49.2 ± 1.4 | 546 ± 43 | 479 ± 37 | 7.5 ± 1.3 | 9.0 ± 2.2 |
| EL3 | – | 104 ± 11 | – | 28.6 ± 2.3 | – | 424 ± 53 | 6.2 ± 1.5 | 10.3 ± 1.9 |
| soWa | 32 ± 5 | – | 20.6 ± 5.6 | – | 66 ± 10 | – | 2.2 ± 0.4 | 4.2 ± 1.1 |
| smHN | 32 ± 4 | 43 ± 11 | 13.4 ± 2.2 | 12.5 ± 4.9 | 141 ± 8 | 181 ± 26 | 1.8 ± 0.5 | 2.6 ± 0.7 |
| smD | 137 ± 8 | 133 ± 7 | 27.5 ± 2.4 | 27.9 ± 2.5 | 521 ± 86 | 561 ± 57 | 5.5 ± 0.9 | 7.7 ± 0.9 |
| smVG | 61 ± 1 | 64 ± 4 | 13.4 ± 2.4 | 12.4 ± 0.7 | 282 ± 2 | 366 ± 22 | 2.4 ± 0.7 | 2.8 ± 0.4 |
| smVG2 | 29 ± 3 | 31 ± 4 | 6.1 ± 0.7 | 6.8 ± 1.2 | 111 ± 28 | 245 ± 21 | 2.3 ± 0.3 | 2.6 ± 0.3 |
| suHe | 65 ± 6 | 99 ± 10 | 71.5 ± 6.8 | 41.8 ± 4.6 | 426 ± 45 | 476 ± 100 | 6.0 ± 1.4 | 7.7 ± 2.3 |
| BiSu | 44 ± 2 | 46 ± 1 | 8.9 ± 2.3 | 10.3 ± 1.8 | 186 ± 24 | 228 ± 14 | 2.0 ± 0.3 | 2.2 ± 0.3 |
| BeRo | 66 ± 7 | 81 ± 2 | 17.0 ± 4.0 | 19.5 ± 2.5 | 281 ± 42 | 333 ± 73 | 3.1 ± 1.0 | 4.0 ± 0.7 |
| DöRo | 236 ± 19 | 223 ± 25 | 41.8 ± 5.8 | 41.1 ± 4.3 | 873 ± 75 | 1052 ± 116 | 13.9 ± 2.0 | 18.8 ± 2.9 |
| FL2 | 124 ± 18 | 186 ± 18 | 44.1 ± 4.6 | 39.0 ± 5.2 | 738 ± 85 | 744 ± 68 | 9.8 ± 2.3 | 10.6 ± 1.8 |
| FL6 | 173 ± 21 | 243 ± 24 | 46.0 ± 4.8 | 46.2 ± 4.6 | 1004 ± 96 | 986 ± 29 | 12.9 ± 2.5 | 15.7 ± 2.0 |
| Gt1WS1 | – | 152 ± 15 | – | 43.6 ± 4.4 | – | 677 ± 34 | 3.6 ± 1.0 | 7.2 ± 0.2 |
| Gt1WS2 | – | 65 ± 7 | – | 19.8 ± 2.0 | – | 636 ± 32 | 1.3 ± 0.1 | 2.8 ± 1.0 |
| EF1WS | – | 88 ± 15 | – | 28.2 ± 5.4 | – | 333 ± 5 | 3.9 ± 0.5 | 4.1 ± 0.5 |
| EF1MM | – | 16 ± 4 | – | 51.5 ± 5.3 | – | 28 ± 1 | 2.7 ± 1.4 | 5.0 ± 1.3 |
| EF1GS | – | 172 ± 18 | – | 41.3 ± 4.2 | – | 549 ± 27 | 8.2 ± 1.4 | 8.7 ± 2.6 |
| EF1OK | – | 149 ± 15 | – | 35.2 ± 4.0 | – | 334 ± 17 | 6.0 ± 1.7 | 7.7 ± 2.0 |
| EF1UKK | – | 132 ± 19 | – | 49.7 ± 1.4 | – | 320 ± 16 | 6.2 ± 2.1 | 7.4 ± 1.8 |
| EF1KO | – | 160 ± 19 | – | 55.8 ± 1.8 | – | 584 ± 65 | 7.0 ± 1.6 | 7.4 ± 1.9 |
| EF1HS | – | 122 ± 11 | – | 30.2 ± 3.1 | – | 235 ± 12 | 7.3 ± 1.5 | 6.3 ± 1.7 |
| Gt1DU1 | – | 147 ± 23 | – | 33.0 ± 0.3 | – | 635 ± 19 | 4.5 ± 1.6 | 12.1 ± 3.4 |
| Gt1DU2 | – | 107 ± 4 | – | 25.4 ± 3.2 | – | 541 ± 27 | 5.7 ± 0.4 | 6.5 ± 1.5 |
| Gt1DU3 | – | 164 ± 20 | – | 37.0 ± 5.9 | – | 907 ± 45 | 9.3 ± 2.7 | 10.2 ± 0.7 |
| Gt1DW | – | 141 ± 12 | – | 35.9 ± 3.5 | – | 304 ± 49 | 6.1 ± 2.1 | 11.2 ± 3.0 |
| Gt1VS1 | – | 128 ± 28 | – | 37.2 ± 9.6 | – | 342 ± 32 | 3.6 ± 1.5 | 7.9 ± 1.6 |
| Gt1VS2 | – | 187 ± 15 | – | 35.1 ± 2.8 | – | 707 ± 35 | 7.0 ± 1.7 | 8.3 ± 1.2 |

SD, Standard deviation; UCS, uniaxial compressive strength; $E_s$, static Young's modulus; $W$, destruction work; $T_0$, indirect tensile strength.

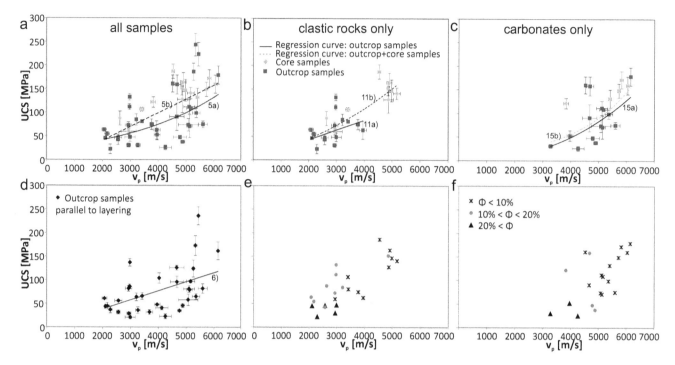

**Figure 3.** $v_p$ vs. UCS for specimens taken perpendicular to bedding for outcrop and core samples separately for **(a)** all samples ($n = 49$), **(b)** only clastic rock samples ($n = 24$) and **(c)** only carbonate samples ($n = 20$); **(d)** $v_p$ vs. UCS for all specimens taken parallel to bedding ($n = 33$); regression curves shown for both outcrop and core samples and outcrop samples only; for regression equations see Table 5. Error bars stand for standard deviations of all measurements of every sample (Tables 2–4). **(e–f)** $v_p$ vs. UCS for low-, medium- and high-porosity samples of clastic rocks **(e)** and carbonates **(f)**.

of both $\Phi$ and $\rho$ values. Though clastic core samples plot far above the regression curve of UCS–$\Phi$ of outcrop samples this is mainly based on the lack of outcrop samples with low porosities (Fig. 2b). For UCS–$\rho$, however, core samples plot along an extension of the regression curve for outcrop data (Fig. 2f). Data therefore show that, if core samples are included, the best fit regression curve is similar to the one with outcrop data only.

### 5.1.2  *P* wave velocity

Many studies show that UCS correlates positively with $v_p$ and travel time, respectively (Freyburg, 1972; McNally, 1987; Kahraman, 2001; Sharma and Singh, 2008). $v_p$ is one parameter determined easily with borehole acoustic logs (e.g. Fricke and Schön, 1999; Rider and Kennedy, 2011) and it may be relevant for the geomechanical model validation and logging-while-drilling.

UCS–$v_p$ data show a wide scatter for all samples both perpendicular and parallel to bedding (Fig. 3a, d). The coefficients of determination are rather poor for both outcrop samples only and core samples included (Table 5). However, there are only small differences between best fit curves for outcrop samples only and core samples included. Especially for carbonates, the regression curve differs only slightly when core samples are included (Fig. 3c). There is some de-

viation for clastic rocks due to lacking low-porosity outcrop samples (Fig. 3b). The coefficient of determination is yet considerably higher if core samples are included (Eq. 11b).

There are conspicuous interdependencies between UCS, $v_p$ and porosity for both clastic rocks and carbonates. High-porosity clastic and carbonate rocks have the lowest UCS and $v_p$ values, and low-porosity samples have the highest values (Fig. 3e, f). If porosities are plotted vs. *P* wave velocities (Fig. 4) there is a clear linear relationship for carbonates at higher $v_p$ values. The mineralogical composition of clastic rock samples is more heterogeneous compared with carbonate samples reflected in a wider scatter of $v_p$ values at lower UCS values. $v_p$ values strongly depend on mineral composition due to the minerals' different elastic wave velocities (e.g. Gebrande et al., 1982). Sandstones' main component quartz has a considerably lower $v_p$ than calcite, the main component of carbonates. $v_p$ of dolomite is lower, too. Consequently, two samples with dolomitic composition (EL1, EL3) plot above the regression curve of carbonates (Fig. 3c).

### 5.1.3  Young's modulus

Former studies showed that, in most cases, there is a strong correlation between $E_s$ and UCS (Sachpazis, 1990; Aggistalis et al., 1996; Palchik, 1999; Dinçer et al., 2004). Our data, shown in Fig. 5, are in good agreement with

**Table 5.** Summarised results of statistical analyses for the correlation of UCS with different parameters of both outcrop and core samples and outcrop samples only with coefficients of determination $R^2$.

| | | Outcrop samples only | | | Outcrop and core samples | |
|---|---|---|---|---|---|---|
| | Eq. | UCS [MPa] | $R^2$ | Eq. | UCS [MPa] | $R^2$ |
| All samples | (1a) | $-28.6\ln(\Phi)+144.2$ | 0.675 | (1b) | $151.95\,e^{-0.051\Phi}$ | 0.526 |
| | (2)* | $-22.2\ln(\Phi)+115.9$ | 0.558 | | | |
| | (3a) | $0.775\,\rho^{5.16}$ | 0.571 | (3b) | $1.285\,\rho^{4.66}$ | 0.520 |
| | (4)* | $0.568\,e^{1.943\rho}$ | 0.498 | | | |
| | (5a) | $23.763\,e^{0.0003\,v_\mathrm{p}}$ | 0.314 | (5b) | $0.029\,v_\mathrm{p}-19.09$ | 0.405 |
| | (6)* | $0.019\,v_\mathrm{p}$ | 0.269 | | | |
| | (7a) | $2.474\,E_\mathrm{s}^{1.102}$ | 0.590 | (7b) | $3.335\,E_\mathrm{s}^{1.008}$ | 0.686 |
| | (8)* | $7.538\,E_\mathrm{s}^{0.698}$ | 0.639 | | | |
| Sandstones | (9a) | $110.73\,e^{-0.037\Phi}$ | 0.206 | (9b) | $152.6\,e^{-0.053\Phi}$ | 0.608 |
| | (10a) | $3.453\,\rho^{3.427}$ | 0.266 | (10b) | $2.245\,\rho^{4.0132}$ | 0.493 |
| | (11a) | $0.025\,v_\mathrm{p}^{0.980}$ | 0.185 | (11b) | $4\times10^{-6}\,v_\mathrm{p}^2+0.009\,v_\mathrm{p}+11.5$ | 0.651 |
| | (12a) | $4.319\,E_\mathrm{s}^{0.944}$ | 0.682 | (12b) | $3.364\,E_\mathrm{s}^{1.035}$ | 0.822 |
| Carbonates | (13a) | $129.95\,e^{-0.051\Phi}$ | 0.517 | (13b) | $137.08\,e^{-0.043\Phi}$ | 0.390 |
| | (14a) | $0.319\,\rho^{5.953}$ | 0.708 | (14b) | $1.116\,\rho^{4.741}$ | 0.476 |
| | (15a) | $2\times10^{-7}\,v_\mathrm{p}^{2.351}$ | 0.351 | (15b) | $8.535\,e^{0.0005\,v_\mathrm{p}}$ | 0.360 |
| | (16a) | $1.928\,E_\mathrm{s}^{1.098}$ | 0.576 | (16b) | $1.783\,E_\mathrm{s}^{1.138}$ | 0.616 |

| | | Outcrop samples only | | | Outcrop and core samples | |
|---|---|---|---|---|---|---|
| | Eq. | Regression function | $R^2$ | Eq. | Regression function | $R^2$ |
| All samples | (17a) | $W=3.953\,\mathrm{UCS}$ | 0.824 | (17b) | $W=5.954\,\mathrm{UCS}^{0.9023}$ | 0.678 |
| | (18)* | $W=3.026\,\mathrm{UCS}^{1.07}$ | 0.816 | | | |
| | (19a) | $T_0=0.0002\mathrm{UCS}^2+0.023\,\mathrm{UCS}+2.30$ | 0.861 | (19b) | $T_0=0.0002\,\mathrm{UCS}^2+0.02\,\mathrm{UCS}+2.35$ | 0.787 |
| | (20)* | $T_0=3\times10^{-5}\,\mathrm{UCS}^2+0.047\,\mathrm{UCS}+1.01$ | 0.797 | | | |
| Sandstones | (21a) | $W=2.867\,\mathrm{UCS}^{1.102}$ | 0.729 | (21b) | $W=7.164\,\mathrm{UCS}^{0.889}$ | 0.611 |
| | (22a) | $T_0=0.0002\,\mathrm{UCS}^2+0.0065\,\mathrm{UCS}+2.46$ | 0.581 | (22b) | $T_0=1.9125e^{0.01\mathrm{UCS}}$ | 0.758 |
| Carbonates | (23a) | $W=3.714\,\mathrm{UCS}^{0.98}$ | 0.804 | (23b) | $W=4.851\,\mathrm{UCS}^{0.906}$ | 0.769 |
| | (24a) | $T_0=3.79\ln(\mathrm{UCS})-9.997$ | 0.862 | (24b) | $T_0=0.407\,\mathrm{UCS}^{0.609}$ | 0.817 |

\* Parallel to bedding.

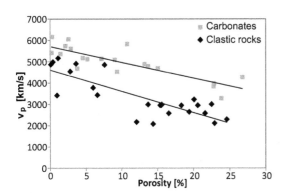

**Figure 4.** Porosity vs. $v_\mathrm{p}$ for carbonates and clastic rocks (see key) with linear regression lines.

these studies, especially for the lithologically separated plots (Fig. 5b, c). Coefficients of determination are in most cases high. To better analyse the statistical significance of the developed regression functions for outcrop samples, 90 % confidence and prediction bands are added.

If all lithologies are plotted together, there is a certain scatter of data both perpendicular and parallel to bedding reflected in wide 90 % prediction bands (Fig. 5a, d). Parallel to bedding the $E_\mathrm{s}$ values tend to be slightly higher than if perpendicular. For small UCS and $E_\mathrm{s}$ values the relationship between the parameters is excellent, and with higher values the scatter increases considerably. The core samples comply with the data of outcrop samples. When core results are included, the quality of regression analysis fit is even improved and is demonstrated by a higher coefficient of determination (Fig. 5a; Table 5).

If sandstone samples are plotted separately, the coefficient of determination is high and confidence and prediction bands, respectively, are narrow (Table 5; Fig. 5b). It has

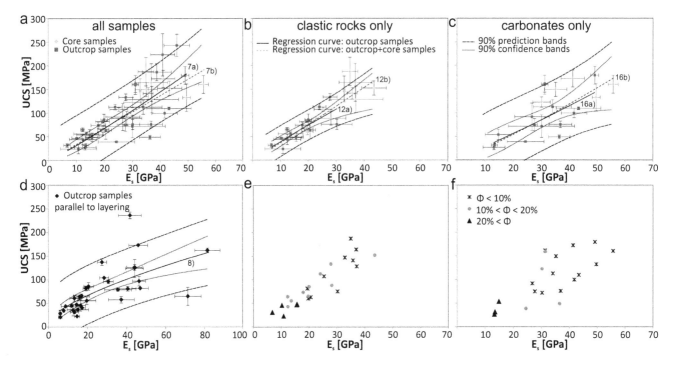

**Figure 5.** $E_S$ vs. UCS for specimens taken perpendicular to bedding for outcrop and core samples separately for (**a**) all samples ($n = 49$), (**b**) only clastic rock samples ($n = 24$) and (**c**) only carbonate samples ($n = 19$); (**d**) $E_S$ vs. UCS for all specimens taken parallel to bedding ($n = 33$). Regression curves shown for both outcrop and core samples and outcrop samples only; 90 % prediction and confidence bands are included; for regression equations see Table 5. Error bars stand for standard deviations of all measurements of every sample (Table 4). (**e, f**) $E_S$ vs. UCS for low-, medium- and high-porosity samples of clastic rocks (**e**) and carbonates (**f**).

to be considered that the sampled carbonates are both matrix and sparry limestones with varying amount of bioclasts (cf. Tables 2, 3). These more-heterogeneous compositions of carbonate samples are reflected in statistically less satisfactory results ($R^2 = 0.576$; Table 5) with wider prediction and confidence bands (Fig. 5c). The increase of the regression curve is lower than for sandstone samples; that is, a carbonate sample is expected to have a higher $E_S$ value than a sandstone sample of similar UCS. For both, sandstones and carbonates, equivalent core samples match the scatter of outcrop data well and lie within the 90 % prediction bands. There are only minor changes of regression curves if core samples are included (Fig. 5b, c).

There is a known relationship between porosity and Young's modulus of rocks (e.g. Rajabzadeh et al., 2012). Therefore, we redraw the UCS–$E_S$ data of sandstones and carbonates with different marks for low-, medium- and high-porosity rocks (Fig. 5e, f). Sandstones and carbonates with high porosities have the lowest UCS and $E_S$ values; the differences between medium- and low-porosity rocks are less pronounced. Both porosity classes include medium UCS and $E_S$ values as well as high values.

## 5.2 Deriving rock properties from UCS

### 5.2.1 Destruction work

The destruction work is an important parameter for dimensioning and planning of drilling projects and correlates with drilling efficiency (Thuro, 1997). Rocks which strongly deform while loading have high destruction-work values because for specimen failure more energy is needed. The destruction work, calculated as the area below the stress–strain curve of the uniaxial compression test, is plotted against UCS of the different samples (Fig. 6).

Regression analyses show that power-law functions fit best in most cases, and coefficients of determination are rather high in all cases. To analyse the statistical significance, 90 % confidence and prediction bands are added.

For outcrop samples parallel and perpendicular to bedding, the fit is excellent with narrow bands (Fig. 6a, d; Table 5). There are, however, clear lithological differences of the destruction-work values. For carbonates, core samples show a considerable deviation from the regression function of outcrop data more to lower $W$ values for similar UCS (Fig. 6c). For sandstones, core samples show a wider scatter, in some cases even beyond the 90 % prediction bands of outcrop samples (Fig. 6b). The slope for clastic rock samples is considerably steeper than that of carbonate rocks (Fig. 6c). That is, more energy is needed to destruct a sandstone sample

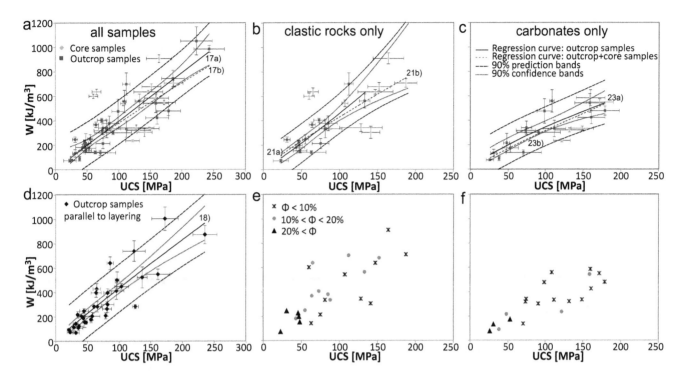

**Figure 6.** UCS vs. $W$ for specimens taken perpendicular to bedding for outcrop and core samples separately for **(a)** all samples ($n = 49$), **(b)** only clastic rock samples ($n = 24$) and **(c)** only carbonate samples ($n = 18$); **(d)** UCS vs. $W$ for all specimens taken parallel to bedding ($n = 33$). Regression curves shown for both outcrop and core samples and outcrop samples only; 90 % prediction and confidence bands are included; for regression equations see Table 5. Error bars stand for standard deviations of all measurements of every sample (Table 4). **(e–f)** UCS vs. $W$ for low-, medium- and high-porosity samples of clastic rocks **(e)** and carbonates **(f)**.

than a carbonate sample of the same UCS value. From that we infer that sandstone samples receive more deformation at the same applied stress than carbonate samples.

In the same way as we did for UCS–$E_s$ values (Fig. 5e, f), UCS–$W$ data of sandstones and carbonates with low-, medium- and high-porosity rocks are plotted separately (Fig. 6e, f). Also in this case, sandstones and carbonates with high porosities have the lowest UCS and $W$ values; the differences between medium- and low-porosity rocks are less clear. For carbonate samples, however, we recognise that low-porosity samples tend to have higher UCS and $W$ values than high-porosity samples (Fig. 6f).

### 5.2.2   Indirect tensile strength

For rocks, there is a known correlation between compressive and tensile strength with a factor of approximately 10 between these two parameters (e.g. Hobbs, 1964; Lockner, 1995). Our results are in good accordance; coefficients of determination are high in all cases with very narrow confidence and prediction bands. Overall, the values of core samples are similar to the values of outcrop samples and plot within the 90 % prediction bands. Both regression functions, developed for clastic rocks, are very similar, and core results fit well within the scatter that is quite similar to outcrop results

(Fig. 7b; Eqs. 22a, b). For carbonates, the equivalent core samples also plot within the 90 % prediction bands (Fig. 7c).

However, there are clear lithological differences in the indirect tensile strength values of the outcrop samples (Fig. 7b, c). For low UCS, $T_0$ values of clastic rock samples are lower than those of carbonates; for high UCS, however, the increase of $T_0$ values is less for carbonates, leading to higher values of clastic rock samples.

We plot UCS–$T_0$ data of sandstones and carbonates with low-, medium- and high-porosity rocks (Fig. 7e, f; see key). This empirical relation also shows that high-porosity samples of clastic rocks and carbonates have the lowest UCS and $T_0$ values; the differences between medium and low-porosity rocks are less clear. In contrast to the UCS–$W$ relation (Fig. 6) where carbonates tend to have higher values, we recognise that in this case low-porosity sandstone samples tend to have higher UCS and $T_0$ values.

## 6   Discussion

### 6.1   Applicability of empirical relations to predict in situ rock properties

A comparison of empirical relations, determined from outcrop samples only, with properties of core samples gives information on parameter changes due to load removal and

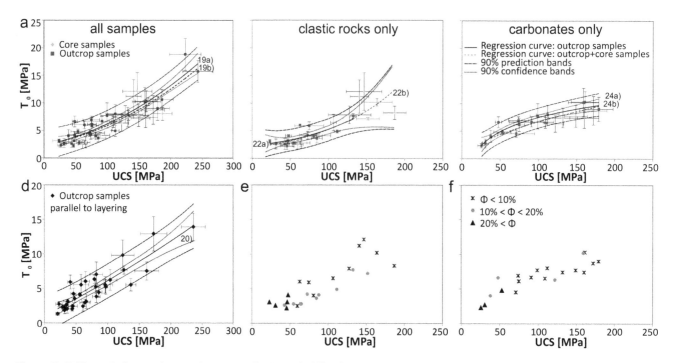

**Figure 7.** UCS vs. $T_0$ for specimens taken perpendicular to bedding for outcrop and core samples separately for (**a**) all samples ($n = 49$), (**b**) only clastic rock samples ($n = 24$) and (**c**) only carbonate samples ($n = 18$); (**d**) UCS vs. $T_0$ for all specimens taken parallel to bedding ($n = 33$). Regression curves shown for both outcrop and core samples and outcrop samples only; 90 % prediction and confidence bands are included; for regression equations see Table 5. Error bars stand for standard deviations of all measurements of every sample (Table 4). (**e–f**) UCS vs. $T_0$ for low-, medium- and high-porosity samples of clastic rocks (**e**) and carbonates (**f**).

beginning of alteration. We found that the developed empirical relations with or without core samples are quite similar for all analysed parameters (cf. Sect. 5, Table 5). Simply, core samples have similar or only slightly higher values than outcrop samples. That is, the ratios of UCS with the considered parameters do not change considerably. Based on these findings it is assumed that these parameter–UCS ratios remain unaffected by unloading. Only the destruction work shows some divergence between outcrop and core samples. For carbonates with high UCS, destruction-work values of core samples tend to be lower than those of outcrop samples with comparable UCS resulting in a steeper regression function for outcrop samples only (Fig. 6c). That is, for the destruction of core samples less energy is needed than for outcrop samples. This may be caused by higher porosities of outcrop samples where more energy can be absorbed by pore-space destruction before brittle failure occurs. The destruction work, measured in laboratory, correlates with the in situ drillability of rocks (Thuro, 1997). Therefore, the destruction work, measured in laboratory, is strongly related to field-work efforts.

The UCS–$E_s$ relationship indicates that clastic and carbonate rocks including their core equivalents show different behaviour. A carbonate rock is expected to have a higher $E_s$ compared with a clastic rock of the same UCS (Fig. 5). The intensity of deformation depends on the rock strength,

the stresses applied and the time over which the stresses are acting and accumulating. It is known that carbonate rocks react differently to stresses than clastic rocks (e.g. Lockner, 1995; Jaeger et al., 2007). On long-term-stress applications clastic rocks may receive more brittle deformation than carbonate rocks due to pressure-solution and slip-folding processes which are typical phenomena in carbonates (Fossen, 2010). These are deformation processes which act on a longer timescale. At drilling operations, however, there is only a short-term-stress application on the rock mass similarly to laboratory experiments. That is, the UCS–$E$ relationship is developed for a similar timescale as the goal of this study, namely drilling applications, and not for long-term-deformation processes.

All data in this study were determined in laboratory measurements of dry rock specimens. Applying the results to in situ conditions is non-trivial for some parameters because rocks at depth are loaded by overburden and confining pressures and are commonly saturated with fluids. Saturation and pressures have strong effects on some of the described parameters.

The compressional wave velocity is one parameter which can be determined easily by using a borehole acoustic log. It has to be taken into account that $v_p$ measurements in boreholes comprise a larger volume which may include fractures and are obtained with different frequencies than laboratory measurements. Therefore, in most cases, saturated samples, measured in laboratory, give higher $v_p$ values than in situ rocks determined from well logs (e.g. Popp and Kern, 1994; Zamora et al., 1994). Laboratory measurements of dry specimens will give lower velocities than those of fully saturated samples (Nur and Simmons, 1969). Kahraman (2007) showed that for sedimentary rocks there is a strong linear correlation between $P$ wave velocities of dry $v_p^d$ and saturated rocks $v_p^w$. Most rocks show significant trends of UCS reduction with increasing degree of saturation (Shakoor and Barefield, 2009; Karakul and Ulusay, 2013). For Miocene limestones, there is a reduction of UCS and $T_0$ values with increasing saturation (Vásárhelyi, 2005). Similarly, Baud et al. (2000) showed that there is a weakening effect of water on sandstone. Triaxial tests have shown that compressive strength and Young's modulus of rocks positively correlate with confining pressure (Nur and Simmons, 1969; You, 2003; Zoback, 2007).

All laboratory measurements have been carried out on high-quality samples where discontinuities such as fractures are absent. In situ rocks, in contrast, typically include fractures. That is, UCS and $E_s$ values measured with laboratory tests tend to be higher than those measured in situ (Priest, 1993; Huang et al., 1995). The presented data of Young's modulus were determined with uniaxial compressive tests, which give static Young's modulus values referring to fracture propagation (cf., Section 1; Jaeger et al., 2007). In boreholes, from acoustic logs, dynamic Young's moduli are obtained (Zoback, 2007; Rider and Kennedy, 2011). The comparison of dynamic and static Young's moduli is complicated. Discontinuities such as fractures have different effects on static measurements of Young's modulus and $P$ wave propagation. Martínez-Martínez et al. (2012), for example, showed that, for carbonate rocks, there is only a poor linear relationship which can be corrected by using $v_p$ and Poisson's ratio.

This shows that transfer to in situ conditions has to be considered carefully for each parameter individually.

For validation purposes, it is advisable to apply the developed equations on logging data of wellbores in the NWGB for UCS calculation. It would then be possible to compare the calculated UCS values with the actual UCS values measured with cores of the same wellbore (cf., Vogt et al., 2012). The estimation of rock strength is not only possible with empirical relations as presented in this study but also with micromechanical methods (e.g, Sammis and Ashby, 1986; Zhu et al., 2011), which are powerful tools to understand failure processes in rock. To build a geomechanical model before starting the drilling operation, such micromechanical methods may be a good supplemental option when using data from adjacent wellbores.

## 6.2  Comparison with previous studies

Many empirical relations between UCS and other parameters were developed. In Table 6, selected equations are presented. None of these relations, however, refer to the NWGB. These functions fit best for the geological situation the analysed samples belong to and are only valid for the defined range of parameter values (cf. Fig. 8). In most cases, the functions relate to a specific lithology.

The presented regression analyses show that coefficients of determination of the regression curves for carbonates have, in most cases, smaller values compared with sandstone samples. Carbonate samples from the NWGB include sparry and matrix limestones, bioclast-rich limestones, oolites, marls, and dense and porous limestones (cf. Tables 2, 3). This means that the lithology of sampled carbonates is much more variable than that of sandstones. This may be one reason for the wider range of mechanical and physical data and the poorer relations of UCS–$E_s$ (Eqs. 12b, 16b), UCS–$v_p$ (Eqs. 11b, 15b), and UCS–$\Phi$ (Eqs. 9b, 13b). In former studies on limestones (e.g. McNally, 1987; Sachpazis, 1990; Bradford et al., 1998; Chang et al., 2006) the lithology, for which the empirical relation was developed, is specified. Accordingly, the presented relationships are more trustworthy if they refer to a specific lithology (cf. Eqs. 9–16, 21–24). If only general assumptions of UCS values are needed (e.g. from well logs of heterogeneous stratifications) or the lithology of the respective wellbore section cannot be defined precisely, it appears to be better to apply the empirical relations generated for all samples (Eqs. 1–8, 17–20).

To compare the regression functions, developed in this study, with the relations of previous studies we use a graphic representation considering the range of parameter values for which the relations were developed (Table 6; Fig. 8). Differences between the functions are depicted. For clastic rocks, there are significant variations for small porosities (Fig. 8a.1). Vernik et al. (1993; Eq. 25) predict much higher UCS for low-porosity sandstones ($\Phi < 15\,\%$) and lower UCS for high-porosity sandstones ($\Phi > 25\,\%$) than Eqs. (9a) and (9b). They, however, determined UCS values from triaxial testing, which gives higher UCS values than uniaxial compressive strength measurements (cf. Zoback, 2007). The effects of small discontinuities on rock strength are smaller when confining pressure is applied.

For carbonate rock samples, however, the calculated regression functions (Eqs. 13a, b) fit perfectly well with previous studies (Fig. 8a.2). Only for high-porosity carbonate rocks ($\Phi > 15\,\%$) are the smallest variations from Eq. (30) in the range of 10 MPa for UCS.

The errors of the empirical relations between UCS–$v_p$ and UCS–$\Delta t$, respectively, are high for all studies (cf. Table 5). The determined regression functions of previous studies are, however, quite similar to Eq. (11b) for clastic rocks (Fig. 8b). The UCS–$v_p$ relation of Freyburg (1972; Eq. 34) is in good accordance with our results. The data relate to sandstones

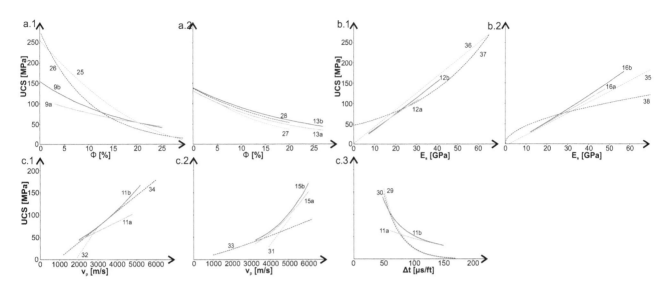

**Figure 8.** Correlations between UCS and the parameters (**a**) porosity, (**b**) $E_s$ and (**c**) $v_p$ separately for clastic rocks and carbonates; correlations from this study and those published by other authors (equation numbers shown) consider the range of parameter values for which the functions are valid.

**Table 6.** Correlations between UCS and the parameters porosity, $P$ wave velocity, travel time and Young's modulus reported by other authors.

| Eq. | Parameter | UCS2 : 1 [MPa] | Rock type | Reference |
|---|---|---|---|---|
| (25) | $\Phi$ | $254\,(1-0.027\Phi)$ [$\Phi$ in %] | Clastic rocks | Vernik et al. (1993) |
| (26) | | $277\,e^{-0.1\Phi}$ [$\Phi$ in %] | Sandstones ($0.2 < \Phi < 33\,\%$) | Chang et al. (2006) |
| (27) | | $143.8\,e^{-0.0695\Phi}$ [$\Phi$ in %] | High UCS limestones ($5 < \Phi < 20\,\%$) | Chang et al. (2006) |
| (28) | | $135.9\,e^{-48\Phi}$ [$\Phi$ in %] | High UCS limestones ($0 < \Phi < 20\,\%$) | Chang et al. (2006) |
| (29) | $v_p\,/\,\Delta t$ | $1277\,e^{-0.036\Delta t}$ [$\Delta t$ in $\mu$s ft$^{-1}$] | Sandstones | McNally (1987) |
| (30) | | $1174\,e^{-0.0358\Delta t}$ [$\Delta t$ in $\mu$s ft$^{-1}$] | Clastic rocks | McNally (1987) |
| (31) | | $56.71\,v_p-192.93$ [$v_p$ in km s$^{-1}$] | Limestones, clastic rocks ($3.9 < v_p < 5.2$ km s$^{-1}$) | Çobanoğlu and Çelik (2008) |
| (32) | | $0.0642\,v_p-117.99$ [$v_p$ in m s$^{-1}$] | Different kinds of rock ($1.8 < v_p < 3.0$ km s$^{-1}$) | Sharma and Singh (2008) |
| (33) | | $9.95\,v_p^{1.21}$ [$v_p$ in km s$^{-1}$] | Different kinds of rock ($1 < v_p < 6.3$ km s$^{-1}$) | Kahraman (2001) |
| (34) | | $0.035\,v_p-31.5$ [$v_p$ in m s$^{-1}$] | Sandstones | Freyburg (1972) |
| (35) | $E_s\,/\,E$ | $2.667\,E_s-4.479$ [$E_s$ in GPa] | Carbonate rocks | Sachpazis (1990) |
| (36) | | $2.28 + 4.1089\,E_s$ [$E_s$ in GPa] | Sandstones | Bradford et al. (1998) |
| (37) | | $46.2\,e^{0.000027E}$ [$E$ in MPa] | Sandstones | Chang et al. (2006) |
| (38) | | $0.4067\,E^{0.51}$ [$E$ in MPa] | Limestones ($10 < $ UCS $< 300$ MPa) | Chang et al. (2006) |

UCS, uniaxial compressive strength; $\Phi$, porosity; $E_s$, static Young's modulus; $\Delta t$, travel time; $v_p$, $P$ wave velocity.

from the Middle Bunter and Lower Bunter (Thuringia, Germany) as well. The comparability of the equations is therefore also based on similar sedimentary conditions of the analysed rocks.

Equations (15a) and (15b) lead to much higher UCS values for high $v_p$ than the relationship published by Kahraman (2001; Eq. 35), who considered not only carbonate samples. There are also bigger differences between our results and other equations (Eqs. 31, 32) which both include different kinds of rock. The regression function obtained by Sharma and Singh (2008), for example, is based on only three sandstone samples together with many other samples of different rock types (volcanic, sedimentary, and metamorphic)

and therefore differs too much from the samples analysed in this study so that they cannot be compared. McNally (1987) published empirical relations of UCS–$\Delta t$ for different stratigraphic units in Australia. The functions relating to clastic units (Eqs. 29, 30) are comparable with Eq. (11b), but only for low $\Delta t$ values ($\Delta t < 75\,\mu$s ft$^{-1}$; Fig. 8b).

In comparison with the two relationships presented above, UCS–$\Phi$ and UCS–$v_p$, it is noteworthy that calculated regression functions for UCS–$E_s$ of both carbonate rock samples and sandstones are in good accordance with previous studies (Table 6; Fig. 8c). Only the limestone function by Chang et al. (2006; Eq. 38) predicts higher UCS for small $E_s$ values and lower UCS for high $E_s$ values than Eq. (8c). The study of Chang et al. (2006), however, is not based on measurements of the static Young's modulus but of the dynamic Young's modulus. As discussed above, the comparability of dynamic and static Young's moduli is complicated because discontinuities have different effects on the measurements of the static Young modulus and acoustic wave propagation.

Overall, the obtained empirical relations are similar to equations developed in previous studies but, in some cases, show considerable differences. These variations mostly relate either to differences in lithologies the study is based on (Eqs. 31, 32), or to different ways of parameter determination (Eqs. 25, 38). The presented data set and empirical relations, however, give new and comprehensive information about mechanical and physical properties of sedimentary and volcanic rocks valid for the NWGB. Nevertheless, they are not only interesting for regional drilling projects or geomechanical modelling. They also supplement and enlarge the existing published results on rock properties, and the new relations may be applied to other sedimentary basins similar to the NWGB.

# 7 Conclusions

Geomechanical and physical parameters with importance in different stages of geothermal exploitation are measured for 35 outcrop samples from quarries and 14 core samples of the Northwest German Basin. Rock properties of these core samples are compared with results of outcrop samples by using regression analyses. The following conclusions can be made:

1. Simple regression analyses for UCS with the parameters porosity, bulk density, and $P$ wave velocity indicate that the statistical significance for these parameters is low. The developed equations yield distinct under- and over-predictions of UCS values. Data show, however, that properties of core samples fit perfectly well within the scatter of outcrop samples. That is, the developed regression functions work well for at least estimating core sample properties with comparatively small deviation. For drilling applications these equations are highly substantial because they allow a continuous update of the original geomechanical model with logging-while-drilling methods for the calculation of optimum mud weights to avoid wellbore instabilities.

2. The developed empirical relations for Young's modulus, destruction work and indirect tensile strength with UCS show high statistical significance. Core samples plot within the 90 % prediction bands. Regression analyses indicate that prediction of destruction work and tensile strength from UCS by outcrop data is possible. The applicability of these equations to rocks from greater depths is therefore assumed. The ratio between UCS and parameter values is the same for both outcrop and core samples. That is, data indicate that parameters

of core sample are predictable from equations developed from an outcrop sample data set.

3. The presented data and regression equations may help to predict UCS values for sedimentary rocks at depth, and thus develop suitable geomechanical models for the adaptation of the drilling strategy on rock mechanical conditions in the Northwest German Basin and similar sedimentary basins.

**Acknowledgements.** The authors appreciate the support of the Niedersächsisches Ministerium für Wissenschaft und Kultur and "Baker Hughes" within the gebo research project (http://www.gebo-nds.de).

We thank the numerous owners for the permission to enter their quarries and to take samples. Special thanks to the LIAG (Leibniz Institute for Applied Geophysics) and BGR (Bundesamt für Geowissenschaften und Rohstoffe) in Hanover, Germany, for permission to sample the drill cores. We also thank the LIAG for the opportunity to perform density and porosity measurements. Review comments by I. Moeck and M. Heap helped in improving the manuscript.

# References

Aggistalis, G., Alivizatos, A., Stamoulis, D., and Stournaras, G.: Correlating uniaxial compressive strength with Schmidt hardness, point load index, Young's modulus, and mineralogy of gabbros and basalts (Northern Greece), Bull. Int. Assoc. Eng. Geol., 54, 3–11, 1996.

Baud, P., Zhu, W., and Wong, T.-f.: Failure mode and weakening effect of water on sandstone, J. Geophys. Res.-Sol. Ea., 105, 16371–16389, 2000.

Betz, D., Führer, F., Greiner, G., and Plein, E.: Evolution of the Lower Saxony Basin, Tectonophysics, 137, 127–170, 1987.

Bradford, I. D. R., Fuller, J., Thompson, J., and Walsgrove, T. R.: Benefits of assessing the solids production risk in a North Sea reservoir using elastoplastic modelling, SPE/ISRM Eurock '98, Trondheim, 261–269, 1998.

Brink, D.: Essentials of Statistics, 2nd Edn., David Brink & Ventus Publishing, available at: http://www.bookboon.com (last access: January 2014), ISBN:978-87-7681-408-3, 2010.

Chang, C., Zoback, M. D., and Khaksar, A.: Empirical relations between rock strength and physical properties in sedimentary rocks, J. Petrol. Sci. Eng., 51, 223–237, 2006.

Çobanoğlu, I. and Çelik, S. B.: Estimation of uniaxial compressive strength from point load strength, Schmidt hardness and $P$ wave velocity, B. Eng. Geol. Environ., 67, 491–498, 2008.

Dinçer, I., Acar, A., Çobanoğlu, I., and Uras, Y.: Correlation between Schmidt hardness, uniaxial compressive strength and Young's modulus for andesites, basalts and tuffs, B. Eng. Geol. Environ., 63, 141–148, 2004.

Dusseault, M. B.: Geomechanical challenges in petroleum reservoir exploitation, KSCE J. Civ. Eng., 15, 669–678, 2011.

Edlmann, K., Somerville, J. M., Smart, B. G. D., Hamilton, S. A., and Crawford, B. R.: Predicting rock mechanical properties from wireline porosities, SPE/ISRM Eurock 47344, 1998.

Fossen, H.: Structural Geology, Cambridge University Press, New York, 2010.

Freyburg, E.: Der Untere und Mittlere Buntsandstein SW-Thüringen in seinen gesteinstechnischen Eigenschaften, Ber. Dtsch. Ges. Geol. Wiss. A Berlin, 17, 911–919, 1972.

Fricke, S. and Schön, J.: Praktische Bohrlochgeophysik, Enke, Stuttgart, 1999.

Gebrande, H., Kern, H., and Rummel, F.: Elasticity and inelasticity, in: Landolt–Bornstein Numerical Data and Functional Relationship in Science and Technology, New Series, edited by: Hellwege, K.-H., Group V. Geophys. Space Res., 1, Physical Properties of Rocks, Subvolume b. Springer, Berlin, 233 pp., 1982.

Gudmundsson, A.: Rock Fractures in Geological Processes, Cambridge University Press, New York, 2011.

Heap, M. J., Faulkner, D. R., Meredith, P. G., and Vinciguerra, S.: Elastic moduli evolution and accompanying stress changes with increasing crack damage: implications for stress changes around fault zones and volcanoes during deformation, Geophys. J. Int., 183, 225–236, 2010.

Hobbs, D. W.: The tensile strength of rocks, Int. J. Rock Mech. Min., 1/3, 385–396, 1964.

Huang, T. H., Chang, C. S., and Yang, Z. Y.: Elastic moduli for fractured rock mass, Rock Mech. Rock Eng., 28, 135–144, 1995.

ISRM: The Complete ISRM Suggested Methods for Rock Characterization, Testing and Monitoring: 1974–2006. Suggested Methods Prepared by the Commission on Testing Methods, International Society for Rock Mechanics, Ulusay, R. and Hudson, J. A., Compilation Arranged by the ISRM Turkish National Group, Ankara, Turkey, 2007.

Jaeger, J. C., Cook, N. G. W., and Zimmerman, R. W.: Fundamentals of Rock Mechanics, Blackwell, Malden USA, 2007.

Jizba, D. L.: Mechanical and Acoustical Properties of Sandstones, Dissertation, Stanford University, 1991.

Kahraman, S.: Evaluation of simple methods for assessing the uniaxial compressive strength of rock, Int. J. Rock Mech. Min., 38, 981–994, 2001.

Kahraman, S.: The correlations between the saturated and dry $P$ wave velocity of rocks, Ultrason, 46, 341–348, 2007.

Karakul, H. and Ulusay, R.: Empirical correlations for predicting strength properties of rocks from $P$ wave velocity under different degrees of saturation, Rock Mech. Rock Eng., 46, 981–999, 2013.

Khaksar, A., Taylor, P. G., Kayes, T., Salazar, A., and Rahman, K.: Rock strength frlom Core and logs: Where we stand and ways to go, SPE EUROPEC/EAGE, 121972, 2009.

Lama, R. D. and Vutukuri, V. S.: Handbook on Mechanical Properties of Rocks – Testing Techniques and Results, Vol. II. Trans Tech Publications, Clausthal, 1978.

Li, S., George, J., and Purdy, C.: Pore-pressure and wellbore-stability prediction to increase drilling efficiency, J. Petrol. Technol., 64, 99–101, 2012.

Lockner, D. A.: Rock Failure, Rock physics and phase relations, American Geophysical Union, Washington D.C., 127–147, 1995.

Martínez-Martínez, J., Benavente, D., and García-del-Cura, M. A.: Comparison of the static and dynamic elastic modulus in carbonate rocks, B. Eng. Geol. Environ., 71, 263–268, 2012.

McNally, G. H.: Estimation of coal measures rock strength using sonic and neutron logs, Geoexploration, 24, 381–395, 1987.

Nabaei, M. and Shahbazi, K.: A new approach for predrilling the unconfined rock compressive strength prediction, Pet. Sci. Technol., 30/4, 350–359, 2012.

Nur, A. and Simmons, G.: The effect of saturation on velocity in low porosity rocks, Earth Planet. Sc. Lett., 7, 183–193, 1969.

Palchik, V.: Influence of porosity and elastic modulus on uniaxial compressive strength in soft brittle porous sandstones, Rock Mech. Rock Eng., 32, 303–309, 1999.

Paschen, H., Oertel, D., and Grünwald, R.: Möglichkeiten geothermischer Stromerzeugung in Deutschland, TAB Arbeitsbericht 84, 2003.

Popp, T. and Kern, H.: The influence of dry and water saturated cracks on seismic velocities of crustal rocks – A comparison of experimental data with theoretical model, Surv. Geophys., 15, 443–465, 1994.

Priest, S. D.: Discontinuity Analysis for Rock Engineering, Chapman and Hall, London, 1993.

Proehl, T. S.: Geomechanical uncertainties and exploratory drilling costs, SPE/ISRM Rock Mechanics Conference, Irving, USA, 2002.

Rajabzadeh, M. A., Moosavinasab, Z., and Rakhshandehroo, G.: Effects of rock classes and porosity on the relation between uniaxial compressive strength and some rock properties for carbonate rocks, Rock Mech. Rock Eng., 45, 113–122, 2012.

Reyer, D., Bauer, J. F., and Philipp, S. L.: Fracture systems in normal fault zones crosscutting sedimentary rocks, Northwest German Basin, J. Struct. Geol., 45, 38–51, 2012.

Rider, M. and Kennedy, M.: The Geological Interpretation of Well Logs, Rider-French Consulting Ltd., 2011.

Sachpazis, C. I.: Correlating Schmidt hammer rebound number with compressive strength and Young's modulus of carbonate rocks, Bull. Int. Assoc. Eng. Geol., 42, 75–83, 1990.

Sammis, C. G. and Ashby, M. F.: The failure of brittle porous solids under compressive stress states, Acta. Metall. Mater., 34, 511–526, doi:10.1016/0001-20656160(86)90087-8, 1986.

Settari, A. and Walters, D. A.: Advances in coupled geomechanical and reservoir modeling with applications to reservoir compaction, SPE 74142, SPEJ (September 2001), 335–342, 2001.

Shakoor, A. and Barefield, E. H.: Relationship between unconfined compressive strength and degree of saturation for selected sandstones, Environ. Eng. Geosci., 15, 29–40, 2009.

Sharma, P. K. and Singh, T. N.: A correlation between $P$ wave velocity, impact strength index, slake durability index and uniaxial compressive strength, B. Eng. Geol. Environ., 67, 17–22, 2008.

Thuro, K.: Drillability prediction: geological influences in hard rock drill and blast tunnelling, Geol. Rundsch., 86, 426–438, 1997.

Vásárhelyi, B.: Statistical analysis of the influence of water content on the strength of the miocene limestone, Rock Mech. Rock Eng., 38, 69–76, 2005.

Vernik, L., Bruno, M., and Bovberg, C.: Empirical relations between compressive strength and porosity of siliciclastic rocks, Int. J. Rock Mech. Min., 30, 677–680, 1993.

Vogt, E., Reyer, D., Schulze, K. C., Bartetzko, A., and Wonik, T.: Modeling of geomechanical parameters required for safe drilling of geothermal wells in the North German Basin, Celle Drilling, Celle, Germany, 2012.

Walter, R.: Geologie von Mitteleuropa. Schweizerbarth, Stuttgart, 2007.

Wong, T.-F., David, C., and Zhu, W.: The transition from brittle faulting to cataclastic flow in porous sandstones: mechanical deformation, J. Geophys. Res., 102, 3009–3025, 1997.

Wooldridge, J. M.: Introductory Econometrics: A Modern Approach, 4th Edn. Mason: South-Western, Cengage Learning, 2009.

York, P., Pritchard, D., Dodson, J. K., Rosenberg, S., Gala, S., and Utama, B.: Eliminating non-productive time associated with drilling trouble zones, Offshore Technology Conference 2009, Houston, USA, 2009.

You, M.: Effect of confining pressure on the Young's modulus of rock specimen, Chin. J. Rock Mech. Eng., 22, 53–60, 2003.

Zamora, M., Sartoris, G., and Chelini, W.: Laboratory measurements of ultrasonic wave velocities in rocks from the Campi Flegrei volcanic system and their relation to other field data, J. Geophys. Res., 99, 13553–13561, 1994.

Zeynali, M. E.: Mechanical and physico-chemical aspects of wellbore stability during drilling operations, J. Petrol. Sci. Eng., 82–83, 120–124, 2012.

Zhu, W., Baud, P., Vinciguerra, S., and Wong, T.-F.: Micromechanics of brittle faulting and cataclastic flow in Alban Hills tuff, J Geophys. Res.-Sol. Ea., 116, B06209, doi:10.1029/2010JB008046, 2011.

Ziegler, P.: Geological Atlas of Western and Central Europe, Geological Society Publishing House/Shell International Petroleum Maatschappij B.V., 1990.

Zoback, M. D.: Reservoir Geomechanics, Cambridge University Press, New York, 2007.

# 7

# Geochemical study on hot-spring water in West New Britain Province, Papua New Guinea

**M. M. Lahan, R. T. Verave, and P. Y. Irarue**

Geological Survey Division, Mineral Resources Authority, P.O. Box 1906, Port Moresby, NCD, Papua New Guinea

*Correspondence to:* M. M. Lahan (mlahan@mra.gov.pg)

**Abstract.** West New Britain Province, which occupies the western part of New Britain Island in Papua New Guinea, is ideally located within an active tectonic region that influences volcanism creating an environment favourable for geothermal activity. Geothermal mapping of surface manifestations reveals high temperature geothermal prospects along the northern coastline of West New Britain Province that are further confirmed by geochemical analysis. The occurrence of geothermal features is confined to the Quaternary Kimbe Volcanics and alluvium in the lowland areas. The features in Talasea appear to be controlled by deep-seated northerly trending faults while structures in Hoskins also appear to be deep seated but have not been identified. The geothermal systems in West New Britain Province have not been drilled, but preliminary reconnaissance geothermal mapping and geochemical analysis reveals four high temperature geothermal prospects suitable for further investigation and development of geothermal energy. These are the Pangalu (Rabili) and Talasea Station geothermal prospects in Talasea and Kasiloli (Magouru) and Silanga (Bakama and Sakalu) geothermal prospects in Hoskins. The calculated reservoir temperatures for these fields are in the range of 245–310 °C. Recommendations are made for further follow-up exploratory investigations.

## 1  Introduction

The volcanic island of New Britain is the most prospective region for the development of geothermal energy resources in Papua New Guinea (PNG). The island is located east of mainland PNG and lies between the Manus Basin to the north and the New Guinea Trench to the southeast (Fig. 1). The geothermal prospects are scattered on the northern coastline from the Willaumez Peninsula in Talasea, West New Britain Province (WNBP), to the Gazelle Peninsula in Rabaul, East New Britain Province (ENBP) (Fig. 1). Volcanically, New Britain Island is comprised of several dormant volcanoes and two currently active volcanoes: Pago in Cape Hoskins area in WNBP and the Rabaul (Tavurvur) volcano in the Gazelle Peninsula in ENBP. WNBP occupies the western and most of the central part of New Britain Island while the eastern portion is occupied by ENBP.

West New Britain Province hosts a large oil palm industry that serves as its main economic activity apart from other business activities in Kimbe (its provincial capital) and its smaller towns, Hoskins and Bialla. There are approximately 128 800 people (NRI Report, 2010) in the Talasea District of WNBP who directly or indirectly depend on oil palm for their livelihood. The main sources of electricity for the province is by the use of four 400–600 kW diesel generators and two mini-hydro power plants supplied by PNG Power Limited (J. Aska, Kimbe generation team leader, personal communication, 2013). The oil palm plantations in the province provide some of their own energy due to insufficient electricity needs and frequent power outages.

Geoscientific information on the geothermal prospects in the country including WNBP is sparse although previous reconnaissance geochemical survey undertaken in Talasea provided temperatures which ranged from 44 to 101 °C (Berhane and Mosusu, 1997). There has never been any geophysical survey for geothermal exploration, nor has there been drilling conducted in the country except on Lihir Island, where a 56 MW geothermal power plant is in operation supporting the gold mine. PNG through Mineral Resources Au-

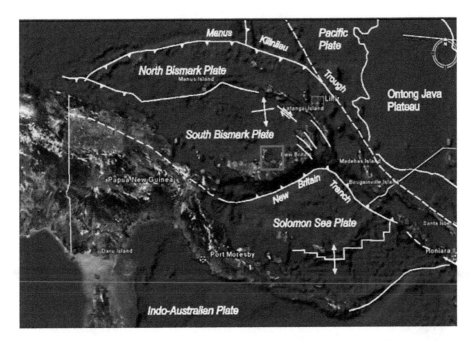

**Figure 1.** Map of PNG showing the tectonic setting and study area (red square) on New Britain Island. Modified from Williamson and Hancock, 2005.

thority (MRA) has now embarked on undertaking geothermal exploration in the country with assistance from the World Bank and the Government of Iceland through training of its staff, as there is a need to diversify its energy source and increase the energy needs of the country.

The reconnaissance field survey of the various geothermal sites in the Talasea District of WNBP was undertaken in November 2012 as part of MRA's ongoing geothermal exploration work. This report provides the results of the geochemical analysis of the water samples and attempts at identifying prospective geothermal sites for further exploratory work.

## 2  Geological setting and geothermal features

### 2.1  Tectonic setting

New Britain Island is part of the New Guinea island arc complex that represents the tectonically active margin of the Australian continental mass. It is situated in a collision zone between three major plates: the northward-moving Indo-Australian, the west-northwest-moving Pacific and the eastward-moving Caroline plates. The boundary between the Pacific and the Australian plates is marked by a number of microplates: the Solomon Plate in the southeast and North and South Bismarck plates to the north of the island (Johnson and Molnar, 1972). These microplates located offshore are bounded by spreading ridges, deep-sea trenches and transform faults while those onshore are represented by thrusts, extensional and strike-slip faults and folds. Earthquakes are located on these spreading ridges (Davies, 2009).

The PNG region experiences very high seismicity due to its location on the "Ring of Fire". The main concentration of seismicity is at the northern and northeastern margins of the Solomon Sea where the Solomon Plate is subducted northward beneath the South Bismarck and the Pacific plates along the New Britain Trench (Fig. 2). Seismicity in this area has been described as the most intense in the world (Ripper and McCue, 1983; Cooper and Taylor, 1989). Seismicity continues from this area towards the southeast through the Solomon Islands, and towards the northwest under the northern part of the New Guinea Island. The volcanoes of New Britain are the result of subduction of the northward-migrating Solomon Sea plate under the South Bismarck Plate. The volcanoes in the Solomon Islands are associated with the Solomon Sea plate as it is subducted beneath the Pacific Plate.

### 2.2  Geology

New Britain Island is mostly comprised of Tertiary to Quaternary volcanic materials. Several dormant and active volcanoes exist along the northern coastline from west to east (Lowder and Carmichael, 1970). Baining Volcanics (*Teb*) which accumulated in an Eocene island arc are the oldest rock type on the island. It is comprised of massive to well-bedded indurated and strongly jointed volcanic breccia, conglomerate, sandstone and siltstone, basic to intermediate lavas and hypabyssal rocks, tuff and minor limestone. There is widespread occurrence of andesitic to basaltic intrusives on the island. In the West New Britain area, deposition of Kapuluk Volcanics (*Tok*) occurred when volcanism resumed in the late Oligocene. The Kapuluk Volcanics are of simi-

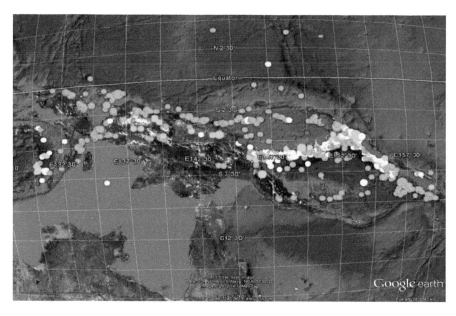

**Figure 2.** Earthquakes of magnitude 6 or greater in the PNG region between 1972 and 2012. Depths: orange – 0–35 km, yellow – 35–70 km, green – 70–150 km, blue – 150–300 km, pink – 300–500 km, and red – 500–800 km. From the USGS/NEI database, accessed at http://earthquakes.usgs.gov/earthquakes/eqarchives/epic.rect.php (after Sheppard and Cranfield, 2012).

lar lithology to the Baining Volcanics but markedly less indurated, jointed and fractured and widely zeolitized. Slow regional subsidence during a period of volcanic quietness in the early Miocene to early Pliocene allowed large thicknesses of the Yalam Limestone (*Tmy*) to accumulate in reefs and inter-reef basins with little or no contaminations from terrestrial sources. It consists of compact or porous, massive to well-bedded bioclastic limestone, chalk, calcareous siltstone and mudstone with minor calcirudite (Ryburn, 1975).

Renewed volcanism in Pliocene formed the Mungu Volcanics (*Tpm*) found southeast of Stettin Bay, which is comprised of dacite, rhyodacite, andesite and pumiceous tuff. They possibly represent the volcanoes that supplied the tuffaceous material in the Kapiura Beds (*Tpk*) found east of Kasiloli thermal area. The Kapiura Beds consist of semi-consolidated, massive to well-bedded acid tuffaceous sandstone, siltstone and conglomerate, tuff, calcareous sediments and limestone. The Talasea Peninsula and Mt Pago area near Hoskins are comprised of Quaternary Kimbe Volcanics, which are basaltic to rhyolitic pyroclastics and lavas as well as hypabyssal intrusives, while the Silanga geothermal area is found in Quaternary alluvium of gravel, sand and silt (Ryburn, 1975). Field observations during this survey noted that the Kimbe Volcanics at geothermal sites are strongly altered to clay due to thermal activity. Structures are difficult to identify; however, observations of the geothermal occurrences show existence of northerly trending faults in the Talasea Peninsula, which is a similar trend to faults mapped by Ryburn (1975). The geology of the survey area by Ryburn (1975) is presented in Fig. 3.

## 2.3 Geothermal prospects and samples

There are at least seven geothermal prospects in WNBP (Grindley and Nairn, 1974) as shown in Fig. 4: Pangalu–Talasea, Bola, Garbuna, Kasiloli–Hoskins, Walo (Silanga), Galloseulo and Bamus. Five of the seven prospects were investigated during this field work: Pangalu–Talasea, Bola and Garbuna (Garu area west of Garbuna) in the Willaumez (Talasea) Peninsula and Silanga and Kasiloli prospects in the Hoskins area. Galloseulo and Bamus prospects located in the east were not visited. In this report, Pangalu and Talasea are presented as separate prospects and Bola is part of Talasea.

A total of 36 geothermal surface features or manifestations (geysers, hot springs, mud pools/pots, fumaroles and/or steaming grounds) were mapped and field observations including basic measurements such as temperature, pH, salinity, conductivity and feature dimension taken. Water and gas samples were collected from selected sites: 13 spring samples, 7 gas samples and a meteoric water sample from Lake Dakataua located at the northern tip of Willaumez Peninsula. Details of sample sites are listed in Table 1 and locations shown in Fig. 3. The samples were sent to the New Zealand Geothermal Analytical Laboratory (NZGAL), GNS Science, Wairakei, for analysis. All gas samples were reportedly contaminated with air and were not analysed. The contamination of the gas samples by air is likely due to samples collected from small gas seeps using improper equipment. The majority hot-spring samples and the meteoric water sample (10) are from Willaumez Peninsula, three from Silanga and one from Kasiloli geothermal prospects. These features are located within oil palm plantations, near settlements and within

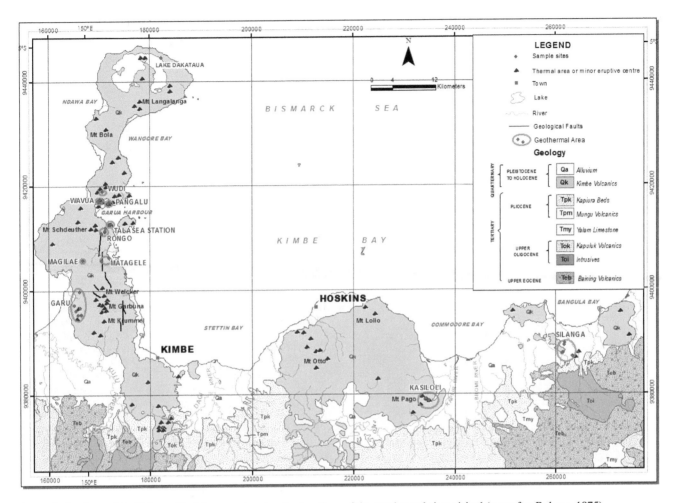

**Figure 3.** Geology of the Talasea–Hoskins area showing the locations of the geothermal sites visited (map after Ryburn, 1975).

close proximity to Kimbe and Hoskins towns, hence easily accessible.

## 3  Fluid chemistry and geothermometry

### 3.1  Fluid chemistry

The chemical compositions of the 14 water samples in terms of major and minor geothermal constituents are presented in Table 2. The local meteoric lake water (i.e. Lake Dakataua) and seawater (Pichler, 2005) are included in the table.

Using the Cl–SO$_4$–HCO$_3$ plot in Fig. 5, different types of thermal waters are distinguished such as steam-heated and volcanic waters based on major anion concentrations (Cl, SO$_4$ and HCO$_3$). As shown in Fig. 5, the Rabili, Talasea Station, Bakama 1 and Magouru springs are classified as matured waters, Galu and Tabero as volcanic waters, Lake Dakataua and Rongo 1 as peripheral waters and the rest of the hot springs as steam heated waters. The matured geothermal waters appear to be clear neutral-chloride boiling springs with very high chloride content compared to the SO$_4$ and HCO$_3$ components. The chloride concentration of

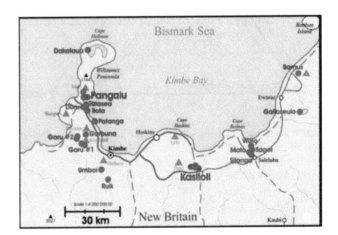

**Figure 4.** Map of central New Britain (see Fig. 2) between Willaumez Peninsula and Banban Island, showing the location of thermal areas (red circles) and active volcanoes (triangles). Map after Grindley and Nairn (1974) taken from McCoy-West et al. (2009).

**Table 1.** List of geothermal features sampled for water and gas analysis and their localities.

| District | Area | Feature name | Water | Gas |
|---|---|---|---|---|
| Talasea (Willaumez Peninsula) | Pangalu | Rabili | 1 | 2 |
| | Wavua | Wavua 1 | 1 | 1 |
| | Wudi | Wudi | 1 | |
| | Talasea Station | Talasea Station | 1 | |
| | | Rongo 1 | 1 | |
| | N. Gabuna | Matagele | 1 | 1 |
| | N. Gabuna | Magilae | 1 | |
| | W. Gabuna | Garu | 1 | 1 |
| | | Tabero | 1 | 1 |
| | Lake Dakataua | Lake Dakataua | 1 | |
| Hoskins | Kasiloli | Magouru | 1 | 1 |
| | Silanga | Bakama 1 | 1 | |
| | | Sakalu/Mato | 1 | |
| | | Taliau | 1 | |

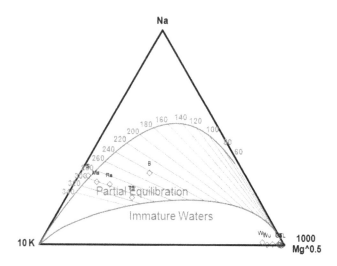

**Figure 6.** Ternary plot showing relative concentrations of Na, K and Mg and the Na–K and K–Mg geothermometers (Giggenbach, 1988). Plot has been generated using a spreadsheet created by Powell and Cumming (2010). Ra – Rabili, Wv – Wavua 1, TS – Talasea Station, Ro – Rongo 1, M – Matagele, MG – Magilae, LD – Lake Dakataua, Wu – Wudi, Ga – Galu, Tb – Tabero, B – Bakama 1, TL – Taliau, S – Sakalu, Ma – Magouru.

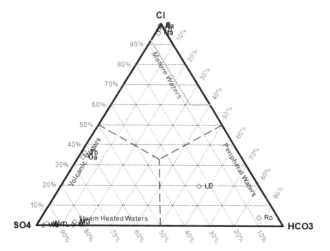

**Figure 5.** Ternary plot showing relative concentrations of the anions chloride (Cl), sulfate ($SO_4$) and bicarbonate ($HCO_3$). Plot has been generated using a spreadsheet created by Powell and Cumming (2010). Ra – Rabili, Wv – Wavua 1, TS – Talasea Station, Ro – Rongo 1, M – Matagele, MG – Magilae, LD – Lake Dakataua, Wu – Wudi, Ga – Galu, Tb – Tabero, B – Bakama 1, TL – Taliau, S – Sakalu, Ma – Magouru.

Rabili hot spring ($18\,993\,mg\,L^{-1}$) is close to that of seawater ($19\,520\,mg\,L^{-1}$); hence there is likely infiltration of seawater due to its location along the shoreline. The volcanic waters discharge sulfuric gas and sulfur deposits as observed in the Garu area as is indicated by their $SO_4$ content. The rest of the steam-heated waters have high $SO_4$ and low Cl and

are very acidic (pH = 1–4). Their acidity is caused by steam heating of shallow groundwater and oxidation of $H_2S$ to $SO_4$ in which there is insignificant Cl discharged.

The Na–K–Mg plot in Fig. 6 further classifies the waters into fully equilibrated, partially equilibrated and immature waters based on the temperature dependence of the full equilibrium assemblage of potassium and sodium minerals that are expected to form after isochemical recrystallization of average crustal rock under conditions of geothermal interest (Giggenbach, 1988). It can be used to predict the equilibrium temperature and also the suitability of thermal waters for ionic geothermometers. As shown in Fig. 6, Sakalu is the only spring that has fully equilibrated with a calculated reservoir temperature of 295 °C while Magouru, Rabili, Talasea Station and Bakama 1 have partially equilibrated with calculated reservoir temperatures of 300, 295, 310 and 245 °C respectively.

## 3.2 Geothermometry

The application of chemical geothermometry to fully and partially equilibrated waters directly estimates the reservoir fluid temperatures. This is based on the principle that specific temperature-dependent mineral-solution equilibria are attained in the geothermal reservoir. As some fluid in this equilibrated reservoir escapes and rises to a hot spring through buoyancy, it will usually cool by conduction and mix faster than it will chemically re-equilibrate. Some chemical species like silica equilibrate faster than others like sodium and potassium; therefore, the chemical composition of hot springs can be used to interpret the temperature and mixing

**Table 2.** Hot-spring compositions in $mg\,L^{-1}$ except where noted.

| Sample name | Lab no. | Date | Temp (°C) | pH | Na | K | Ca | Mg | SiO$_2$ | B | Cl | SO$_4$ | HCO$_3$ | H$_2$S | NH$_3$ | As | Fe | Al | Mn | Cond. $\mu S\,cm^{-1}$ |
|---|---|---|---|---|---|---|---|---|---|---|---|---|---|---|---|---|---|---|---|---|
| Feature-1 RABILI | 2013000037 | 08/11/2012 | 22 | 5.32 | 9697 | 1965 | 961 | 25 | 447 | 24 | 18993 | 153 | <20 | <0.01 | 7.1 | 1.9 | 0.55 | <0.15 | 2.7 | 46830 |
| Feature-5 WAVUA 1 | 2013000038 | 04/11/2012 | 22 | 2.6 | 61 | 35 | 52 | 13.4 | 457 | 0.58 | 5.5 | 543 | <20 | 0.27 | 9.6 | 0.029 | 11.5 | 2.5 | 1.4 | 1956 |
| Feature-6 TALASEA STN | 2013000039 | 05/11/2012 | 22 | 5.91 | 4216 | 976 | 460 | 26 | 209 | 12.4 | 8122 | 194 | <20 | <0.01 | 3.8 | 0.56 | 0.09 | <0.15 | 0.66 | 22715 |
| Feature-11 RONGO 1 | 2013000040 | 05/11/2012 | 22 | 6.99 | 9.7 | 5.3 | 4 | 1.2 | 169 | <0.3 | 2.3 | 4.2 | 47 | <0.01 | 0.009 | <0.015 | <0.02 | <0.15 | <0.005 | 139 |
| Feature-14 MATAGELE | 2013000041 | 06/11/2012 | 22 | 3.51 | 6.2 | 4.1 | 23 | 7.1 | 110 | <0.3 | 0.4 | 117 | <20 | 0.02 | 0.32 | <0.015 | 0.48 | 0.63 | 0.79 | 395 |
| Feature-15 MAGILAE | 2013000042 | 06/11/2012 | 22 | 3.64 | 9.1 | 2.5 | 5.8 | 2.3 | 77 | <0.3 | 2.4 | 117 | <20 | 9.6 | 0.03 | <0.015 | 5.2 | 1.1 | 0.19 | 242 |
| Feature-17 LAKE DAKATAUA | 2013000043 | 07/11/2012 | 22 | 7.33 | 41 | 3.1 | 11.7 | 6.8 | 31 | <0.3 | 24 | 30 | 68 | 0.02 | 0.03 | <0.015 | 0.02 | <0.15 | <0.005 | 284 |
| Feature-18 WUDI | 2013000044 | 08/11/2012 | 22 | 2.01 | 6.1 | 8.3 | 4.1 | 1.2 | 344 | 0.47 | 1.9 | 1075 | 29 | 14.4 | 5.2 | 0.015 | 20 | 43 | 0.22 | 4783 |
| Feature-23 GALU | 2013000045 | 10/11/2012 | 22 | 1.93 | 29 | 10.5 | 58 | 25 | 256 | 2 | 415 | 790 | 21 | 24.3 | 0.39 | 0.07 | 8.8 | 56 | 1.2 | 6109 |
| Feature-24 TABERO | 2013000046 | 10/11/2012 | 22 | 1.82 | 39 | 13 | 75 | 33 | 334 | 3.2 | 595 | 1025 | <20 | 0.4 | 0.51 | 0.09 | 15 | 76 | 1.7 | 8035 |
| Feature-28 BAKAMA 1 | 2013000047 | 11/11/2012 | 22 | 6.19 | 1710 | 193 | 200 | 2 | 131 | 24 | 12888 | 28 | | 27 | <0.01 | 0.46 | 0.84 | <0.08 | <0.15 | 0.1 | 33725 |
| Feature-29 TALIAU | 2013000048 | 11/11/2012 | 22 | 3.2 | 11.4 | 1.8 | 25 | 7.8 | 128 | 1.6 | 1.1 | 235 | <20 | 0.04 | 11.1 | <0.015 | 11.8 | 7.8 | 0.31 | 856 |
| Feature-30 SAKALU | 2013000049 | 11/11/2012 | 22 | 7.43 | 5968 | 1177 | 462 | 0.45 | 205 | 105 | 11357 | 32 | <20 | <0.01 | 0.46 | 7.2 | <0.08 | <0.15 | 0.19 | 30260 |
| Feature-32 MAGOURU | 2013000050 | 12/11/2012 | 21 | 7.2 | 6626 | 1401 | 496 | 3.9 | 420 | 92 | 3097 | 101 | 34 | 0.06 | 16.7 | 9.5 | <0.08 | <0.15 | 4 | 9335 |
| Seawater | | | 8 | | 10450 | 354 | 405 | 1235 | 0.428 | 4.1 | 19520 | 2748 | 154 | | | 3.7* | 15* | | 16* | |

\* in $\mu g\,L^{-1}$.

**Table 3.** Water (solute) geothermometers (temperatures in C) generated using a spreadsheet created by Powell and Cumming (2010), based on Giggenbach (1991). Immature waters are omitted.

| Sample name | Meas. T °C | $T_{AmS}$ Cond. | $T_{Ch}$ Cond. | $T_{QZ}$ Cond. | $T_{Na–K–Ca}$ | $T_{Na–K–Ca}$ Mg corr | $T_{Na/K}$ Fournier (1979)* | $T_{Na/K}$ Giggenbach (1988) | $T_{Na/K}$ Giggenbach (1986)* |
|---|---|---|---|---|---|---|---|---|---|
| Rabili | 100 | 118 | 233 | 246 | 273 | 265 | 286 | 296 | 230 |
| Talasea Station | 100 | 59 | 162 | 184 | 271 | 239 | 301 | 310 | 197 |
| Bakama 1 | 70 | 31 | 128 | 153 | 210 | 210 | 228 | 242 | 182 |
| Sakalu | 100 | 58 | 161 | 182 | 267 | 267 | 283 | 293 | 318 |
| Magouru | 100 | 112 | 226 | 240 | 275 | 275 | 291 | 300 | 262 |

\* These references can be found within the work of Powell and Cumming (2010).

history of a fluid in its path from the reservoir to the surface (Giggenbach, 1988). The most widely used liquid geothermometers involve silica concentration and relative concentrations of the cations Na, K, Mg and Ca. Hence, geothermometry calculations were applied to Sakalu, Magouru, Rabili, Talasea Station and Bakama 1 chemical analyses, which generated the calculated geothermometry temperatures in Table 3.

Amorphous silica temperatures for Talasea Station, Bakama 1 and Sakalu are lower than the measured values (Table 3) while higher for Rabili and Magouru. This low silica geothermometry is a possible indication of dilution with cold water before reaching the surface or precipita-

tion of silica before sample collection. The subsurface reservoir temperatures obtained by Na–K–Ca and Na/K geothermometers compare well with the values obtained from the quartz and chalcedony geothermometer temperatures for Rabili and Magouru but not so for Talasea Station, Bakama 1 and Sakalu. The calculated quartz geothermometer temperatures for the Rabili, Talasea Station, Bakama 1, Sakalu and Magouru hot springs indicate reservoir temperatures of 246, 184, 153, 182 and 240 °C respectively. The Na/K and Na–K–Ca geothermometers indicate reservoir temperatures of 270–295, 270–310, 210–240, 265–290 and 275–300 °C for Rabili, Talasea Station, Bakama 1, Sakalu and Magouru hot springs respectively.

## 4  Conclusions

West New Britain Province is ideally located within an active tectonic region and is a favourable environment for geothermal activity. The occurrence of geothermal surface features is confined to the Quaternary Kimbe Volcanics, which are comprised mainly of basaltic to rhyolitic pyroclastics and lavas and in recent alluvium areas in the lowlands. The geothermal activities in the Willaumez Peninsula appear to be controlled by deep-seated northerly trending faults while those in Hoskins are unknown.

The geochemical analysis of water samples from the area reveals four prospective geothermal prospects:

1. Pangalu (Rabili) – 270–295 °C

2. Talasea Station – 270–310 °C

3. Silanga (Bakama and Sakalu) – 210–290 °C

4. Kasiloli (Magouru) – 275–300 °C.

The Pangalu and Talasea Station geothermal prospects are located near the shoreline and are likely to be influenced by seawater. Isotopic analysis is required in determining the source of the geothermal waters in these prospects as well as determining the likely contamination of seawater. It is noted that there is lack of gas and isotopic data analysis on the prospects investigated. It is therefore recommended that isotopic analysis is performed including the use of proper equipment in collecting gas samples to further analyse and correlate these data with the available geochemistry data.

Further investigations are required in surface geology mapping, geophysics surveys and geochemical surveys of hot springs not sampled yet to delineate the prospectivity of these geothermal prospects for development of geothermal energy.

**Acknowledgements.** The authors wish to thank the PNG Mineral Resources Authority and the World Bank through its Technical Assistance (WBTA2) program for funding this investigation. We thank Lynn Orari and Dorothy Pion of Geological Survey Division, Mineral Resources Authority, for their assistance in reproducing the geology map of the study area. Our sincere thanks and gratitude go to the West New Britain Provincial authorities and the local people for their support and assistance during the field work.

## References

Berhane, D. and Mosusu, N.: A review of the geothermal resources in Papua New Guinea, Proceedings, PNG Geology, Exploration and Mining Conference, Madang, Papua New Guinea, 15–27, 1997.

Cooper, P. and Taylor, B.: Seismicity and focal mechanisms at the New Britain Trench related to deformation of the lithosphere, Technophysics, 164, 25–40, 1989.

Davies, H. L.: New Guinea Geology, Encyclopedia of islands edited by RG Gilespie and DA Clague, University of California Press, Berkeley, CA, USA, 659–665, 2009.

Giggenbach, W. F.: Geothermal solute equilibria, Derivation of Na-K-Mg-Ca geoindicators, Geochim. Cosmochim. Ac., 52, 2749–2765, 1988.

Giggenbach, W. F.: Chemical techniques in geothermal exploration in Application of geochemistry in geothermal reservoir development, UNITAR/UNDP publication, Rome, Italy, 119–142, 1991.

Grindley, G. W. and Nairn, I. A.: Geothermal investigations in Papua New Guinea, New Zealand Geological Survey Unpublished Report M9, Department of Scientific and Industrial Research, New Zealand, 30 pp., 1974.

Johnson, T. and Molnar, P.: Focal mechanisms and plate tectonics of the Southwest Pacific, J. Geophys. Res., 77, 5000–5032, 1972.

Lowder, G. G. and Carmichael, I. S. E.: The volcanoes and caldera of Talasea, New Britain: geology and petrology, Geol. Soc. Am. Bull., 81, 17–38, 1970.

McCoy-West, A. J., Bignall, G., and Harvey, C. C.: Geothermal power potential of selected Pacific Nations, GNS Science Consultancy Report 2009/180, GNS Science, New Zealand, 95 pp., 2009.

NRI Report 2010: Papua New Guinea District and Provincial Profiles, The National Research Institute, 182 pp., March 2010, 2010.

Pichler, T.: Stable and radiogenic isotopes as tracers for the origin, mixing and subsurface history of fluids in shallow-water hydrothermal systems, J. Volcanol. Geoth. Res., 193, 211–226, 2005.

Powell, T. and Cumming, W.: Spreadsheets for geothermal water and gas geochemistry, Proceedings, Thirty-Fifth Workshop on Geothermal Reservoir Engineering, Standford University, Standford, California, USA, SGP-TR-188, 10 pp., 2010.

Ripper, I. D. and McCue, J. F.: The seismic zone of the Papuan Fold Belt: BMR, Journal of Australian Geology and Geophysics, 8, 147–156, 1983.

Ryburn, R. J.: Talasea-Gasmata, New Britain – 1 : 250 000 Geological Series, Bureau of Mineral Resources, Geology and Geophysics, Explanatory Notes, SB/56-5 and BS/56-9, 26 pp., 1975.

Sheppard, S. and Cranfield, L. C.: Geological framework and mineralization of Papua New Guinea, Mineral Resources Authority, Independent State of Papua New Guinea, 65 pp., 2012.

Williamson, A. and Hancock, G.: The geology and mineral potential of Papua New Guinea, Papua New Guinea department of Mining, 152 pp., 2005.

# 8

# Overcoming challenges in the classification of deep geothermal potential

**K. Breede, K. Dzebisashvili, and G. Falcone**

Dept. of Geothermal Engineering and Integrated Energy Systems, Institute of Petroleum Engineering, Clausthal University of Technology, Clausthal, Germany

*Correspondence to:* K. Breede (katrin.breede@tu-clausthal.de)

**Abstract.** The geothermal community lacks a universal definition of deep geothermal systems. A minimum depth of 400 m is often assumed, with a further sub-classification into middle-deep geothermal systems for reservoirs found between 400 and 1000 m. Yet, the simplistic use of a depth cut-off is insufficient to uniquely determine the type of resource and its associated potential. Different definitions and criteria have been proposed in the past to frame deep geothermal systems. However, although they have valid assumptions, these frameworks lack systematic integration of correlated factors. To further complicate matters, new definitions such as hot dry rock (HDR), enhanced or engineered geothermal systems (EGSs) or deep heat mining have been introduced over the years. A clear and transparent approach is needed to estimate the potential of deep geothermal systems and be capable of distinguishing between resources of a different nature. In order to overcome the ambiguity associated with some past definitions such as EGS, this paper proposes the return to a more rigorous petrothermal versus hydrothermal classification. This would be superimposed with numerical criteria for the following: depth and temperature; predominance of conduction, convection or advection; formation type; rock properties; heat source type; requirement for formation stimulation and corresponding efficiency; requirement to provide the carrier fluid; well productivity (or injectivity); production (or circulation) flow rate; and heat recharge mode. Using the results from data mining of past and present deep geothermal projects worldwide, a classification of the same, according to the aforementioned criteria is proposed.

## 1 Review

In the past, definitions such as hydrothermal and petrothermal have been created to categorize deep geothermal systems, i.e. systems with a depth greater than 400 m, into two groups. The first group includes geothermal reservoirs that provide a heat source, a natural reservoir with high enough permeability, and a water recharge. The second group comprises geothermal systems where only a natural heat source exists, while the underground heat exchanger must be created artificially and water must be supplied for water circulation within. Hydrothermal systems (HSs) are clearly dominant in comparison to petrothermal systems (PSs) with regards to number of occurrences worldwide and megawatts of electricity generated.

In 1970, the hot dry rock (HDR) concept was introduced to describe a system which uses hot and dry rock as a heat source and where an artificial underground heat exchanger had to be created (Cummings and Morris; 1979; Tester et al., 1989; Potter et al., 1974). However, during the history of deep drilling, it was found that most rocks are actually not completely dry, but contain at least some naturally occurring water. This finding led to the development of a definition of hot wet rock (Duchane, 1998). In addition, the category of hot fractured rock was created to describe geothermal reservoirs that consist of hot rocks, typically crystalline, that are already naturally fractured due to fault systems or that require artificial fracturing (Genter et al., 2003). Stimulated geothermal systems, deep heat mining (Häring, 2007), and deep earth geothermal were also introduced to describe deep geothermal systems that are typically created in crystalline rocks and are independent from water-bearing structures. All these definitions are actually related to PSs.

Recently, the new definition of enhanced or engineered geothermal systems (EGSs) was introduced for deep geothermal systems, which required technical enhancement such as stimulation to create an artificial reservoir or the supply of water (MIT, 2006a; AGRCC, 2010; Williams et al., 2011; BMU, 2011). This definition is not solely related to PSs, but can also be applied to HSs that require technical enhancement such as stimulation techniques or artificial water supply for water circulation in order to increase the productivity of the system.

On the hydrothermal side of deep geothermal systems, only the recently developed definition of hot sedimentary aquifer (HSA) was additionally introduced to describe HSs as having a heat source that is conduction-dominated, rather than convection-dominated.

However, the creation of so many definitions for deep geothermal systems and the fact that they are not recognized as internationally standards has created some confusion about the actual classifications and which geological setting or geothermal system is being described. An additional complication is that, at a given geothermal site, different systems can exist; e.g. at Soultz-sous-Forêts, where at one depth, an HFR system is present, and at another depth, an HDR system is found.

This paper tries to meet the challenge of the classification of deep geothermal systems by reintroducing the categories of petrothermal, hydrothermal and, additionally, HSA. The term EGS is excluded from our new classification as it carries a vague definition and provides insufficient information about the system, e.g. if natural water is available in the underground heat exchanger and if the permeability is high enough to produce heat or electricity.

## 2 Definition of deep geothermal energy

Deep geothermal energy is defined by its depth, which has to be at least 400 m and a temperature of at least 20 °C. However, some authors recommend using the term deep geothermal energy only for depths of at least 1000 m and temperatures of more than 60 °C. The depth range from 400 to 1000 m is sometimes referred to as middle-deep geothermal. Deep geothermal systems are commonly divided into HSs and PSs, but deep geothermal energy can also be used from mines, caverns, and tunnels. (PK Tiefe Geothermie, 2007; VDI-Richtlinie 4640, 2010)

### 2.1 Definition of enhanced geothermal systems

In recent times, the term EGS has been used more and more. However, as already reported by Breede et al. (2013), the definition of EGS is vague and exists in different forms. For example, MIT (2006a) defines EGSs as "engineered reservoirs that have been created to extract economical amounts of heat from low permeability and/or porosity geothermal resources". Another definition is provided by the Australian

Geothermal Reporting Code Committee, which defines an EGS as "a body of rock containing useful energy, the recoverability of which has been increased by artificial means such as fracturing" (AGRCC, 2010).

### 2.2 Definition of petrothermal systems

The terminology petrothermal was first mentioned by Roberts and Kruger (1982), while the term EGS was first proposed by Grassiani et al. (1999). Petrothermal systems (PSs) are commonly defined as hot (>150 °C) and dry crystalline or dense sedimentary rocks, which do not have high enough natural permeability and therefore require the application of stimulation techniques in order to create an artificial reservoir (Nag, 2008). Hence, these systems are independent from water-bearing structures and it is essential to provide water for both hydraulic fracturing and as a carrier fluid (via water injection for circulation through the underground heat exchanger, and subsequent production). The natural permeability of the production well before stimulation, as opposed to the injection well, defines the term petrothermal (Schulz, 2008); thus, the injection horizon could be an aquifer, which can be used for water disposal. By this definition, Landau in Germany is not a PS, but an HS, as hydraulic fracturing was only required for the injection well in order to increase the injectivity index (Schindler et al., 2010). However, in many geothermal projects, the injection well and production well have the same technical design. Thus, they can be used alternatively as injector or producer, according to the hydraulic schemes. This is the case at Soultz, for example, where some wells were first used as producers and then as injectors. In order to create the artificial heat exchanger and to use the PS, at least two wells, one injection well and one production well, are required.

Schulz (2008) and Kreuter (2011b) state that the following criteria have to be fulfilled simultaneously in the case of a PS:

1. average natural permeability, before stimulation, of less than $10^{-14}$ m$^2$;

2. production well does not allow for an economically relevant production; i.e. the productivity index is less than $10^{-2}$ m$^3$/(MPa s), without the application of stimulation techniques;

3. using hydraulic fracturing, the production of the formation must be increased by at least 50 %.

In his second draft for the renewable energy law in Germany (EEG), Schulz (2009) recommended that the productivity enhancement factor should be 100 % (a factor of 2) instead of only 50 % (a factor of 1.5). How high this factor should be depends on the determined productivity index prior to hydraulic fracturing and is thus site dependent. The idea behind the enhancement factor is that the productivity must be increased in such a way that it is economical to produce geothermal energy at the given site. The productivity index has to be determined using hydraulic tests before any hydraulic fracturing techniques are applied. However, prior application of chemical stimulations is possible.

The values for the permeability threshold, productivity index and the productivity enhancement factor of 50 % are based on field experience, mainly gathered from the European HDR project at Soultz-sous-Forêts in France. However, it is difficult to generalize from this site alone, as different productivity indices have been determined at different depths varying from 1 to more than 100 (Schill et al., 2013). Thus, the complexity of the geological conditions has to be taken into account before determining which productivity enhancement factor is suitable for a given formation.

When considering past nomenclature, PSs could fall into the following categories (GtV, 2014c):

- enhanced geothermal systems (EGSs),

- engineered geothermal systems (EGSs),

- hot dry rock (HDR),

- hot wet rock (HWR),

- deep heat mining (DHM),

- stimulated geothermal systems (SGSs),

- deep geothermal probes.

PSs are used most commonly for electricity generation (Hirschberg et al., 2015a) and combined heat and power (CHP) production due to drilling costs being much higher than for HSs. However, with increasing costs for heating oil, PSs could also become economic for heating in the future. The exception is the deep geothermal probe, which is a closed-loop system that employs a heat transfer medium to recover heat being stored in any rock formation. Geothermal probes are used for heating purposes only.

PSs are always conduction-dominated (Sass and Goetz, 2011); i.e. the heat moves through the material from a hotter zone to a cooler zone.

There exists a transition zone between HSs and PSs, where a project could be classified as either petrothermal or hydrothermal. Thus, at the same geothermal site, different geothermal systems can co-exist at different depths, as it is the case for Soultz and Landau. Experience gained from deep wells showed that the classic definition of the HDR Technology, which refers to a hot and almost completely dry basement rock, is invalid (Schulz, 2008).

## 2.3 Definition of hydrothermal systems

Hydrothermal systems (HSs) are defined by the availability of a water-bearing structure, such as an aquifer, which is used by the production and injection well (Bertani, 2012). To ensure high enough flow rates and thus high productivity of

the wells, high permeabilities are required and the water-bearing structure should be vertically and laterally extensive to guarantee the sustainability of the HS (GtV, 2014d). Looking at the definition of PS proposed by Schulz (2008), the permeability of the productive horizon in HSs should be at least $10^{-14}\,\mathrm{m^2}$ and the productivity index at least $10^{-2}\,\mathrm{m^3/(MPa\,s)}$. Thus, HSs are convection-dominated; i.e. the heat is transported by the movement of hot material (Huenges, 2010a). Volcanic systems are the most representative type of HSs worldwide. Additional common hydrothermal reservoir rocks are sedimentary porous aquifers, such as sandstones or conglomerates, secondary fractured and/or cavernous rocks, such as limestones, or young and deep fault systems, such as those found in the Upper Rhine Valley (Huenges, 2010b; GtV, 2014d). Often major fault zones are targeted for HSs, as they commonly provide much higher permeability values. However, due to the existing pre-stresses, these fault zones might present more risk for induced seismicity than initially estimated (Hirschberg et al., 2015b). Typically, hydrothermal reservoirs in Germany are found in the North German Basin, the Upper Rhine Graben and the Molasse Basin, located in the north, south-west and south of Germany, respectively.

Besides the original exploration well in a HS field, at least one further appraisal well must be drilled. In some cases, an additional third well is drilled to reduce hydro-mechanical shearing in the reservoir, which thereby reduces the risk of induced seismicity (Cuenot, 2013). Although HSs do not require stimulation, Huenges (2010c) states that it might be sensible to use chemical stimulation in order to enhance permeability in the near-wellbore region.

## 2.4 Definition of hot sedimentary aquifers

In recent years, the term HSA has been created for deep and hot sedimentary aquifers that are, in contrast to common HSs, conduction-dominated (Mortimer et al., 2010; Huddlestone-Holmes and Hayward, 2011; Huddlestone-Holmes and Russel, 2012). However, Clean Energy Australia (2014) refers to HSA systems as convective systems. Various minimum temperatures are given by different authors: 75 °C (cleanenergyaus, 2014), 130 °C (Huddlestone-Holmes and Russel, 2012), 140 °C (Barnet, 2009). Also, different depths are proposed: 1 to 3 km (cleanenergyaus, 2014); 2.5 to 3 km (newworldenergy, 2014). A maximum depth of 4.5 km was given by Huddlestone-Holmes and Russel (2012), reflecting that the likelihood that the permeability would be too low at greater depths. The Australian Energy Resource Assessment states that the depth should be "shallow enough for natural porosity and permeability to be preserved so that fluid circulation can occur without artificial enhancement". Although, stimulation techniques are not required, they might be applied to increase the near-wellbore permeability (Huddlestone-Holmes and Hayward, 2011). However, this statement does not clearly indicate which type of stimulation would be required below 4.5 km, although it is most likely to be hydraulic fracturing.

Specific values could neither be found for porosity, permeability nor for flow rates. The permeability of HSA systems can either be matrix permeability in sandstone or fracture permeability in tight limestones or fault zones (Huddlestone-Holmes and Hayward, 2011). Huddlestone-Holmes and Russel (2012) state that the rock density should be lower than the crystalline basement rocks, which are targeted for HDR or EGS resources, and should be around 2400 kg m$^{-3}$.

Another requirement is that the reservoir must be covered by a thick cap rock made of clay and/or coal rich sequences, which acts as a thermal insulator (Mortimer et al., 2010). This is also the case for volcanic HSs and true for all geothermal systems, as the cap rock significantly reduces heat loss.

For HSA systems in Australia, newworldenergy (2014) states that at least one of the following geological settings should be fulfilled:

- Radioactive decay in basement rocks acts as a heat source for overlying aquifers

- Remnant heat from old volcanic centres ensures an elevated geothermal gradient

- Hot water welling up from deep basins along thermal density and/or pressure gradients

- Rapid tectonic uplift brought a deep hot water formation closer to the surface and compressed the geothermal gradient.

## 3  Stimulation techniques

Stimulation techniques such as hydraulic stimulation, chemical stimulation, and thermally induced fracturing are commonly used to enhance the permeability of geothermal reservoirs, thereby increasing their productivity, to create new fractures and hence an artificial underground heat exchanger, or to clean the wells of drill cuttings. The selection of the most appropriate stimulation technique depends on, among other parameters, the desired depth of invasion, i.e. the radius of influence.

The most common stimulation technique is hydraulic stimulation, as it provides the largest depth of invasion and can be applied to re-open and/or create fractures up to several hundreds of metres away from the borehole (ENGINE, 2008b). Fractures generated by hydraulic stimulation can be tensile (perpendicular to minimum principal stress axis), shear (perpendicular to maximum principal stress axis), or a combination of both, and their orientation and distribution depends on the overall stress field (Zimmermann et al., 2010a). In some cases, it is recommended to isolate intervals in the wells and perform consecutive stimulations of these intervals rather than carrying out a massive hydraulic stimulation. This is an expedient to reduce the risk of creating

shortcuts and larger seismic events (ENGINE, 2008b). Hydraulic stimulation is a requirement for the creation of an artificial petrothermal heat exchanger.

An example of quantitative values for evaluating the impact of hydraulic fracturing in matrix-dominated formations and correlating input/output parameters is given by Groß Schönebeck. Three hydraulic stimulation treatments were carried out separately in a well over 6 days: the cyclic waterfrac treatment in the low permeable volcanic rocks and gel-proppant treatments in the lower and upper Dethlingen sandstones. For the waterfrac treatment, 13 170 m$^3$ of fluids and 24.4 tons of quartz sand (the latter as proppant) were injected. The maximum wellhead pressure of 58.6 MPa was reached at the maximum flow rate of 9 m$^3$ min$^{-1}$, with the total duration of the treatment being 6389 min. (Zimmermann et al., 2010a). After the isolation of this section with a bridge plug at 4300 m for the first and at 4123 m for the second treatment, two gel-proppant treatments in highly permeable sandstones were performed over 4 days. In total, 95 tons of proppants and 280 m$^3$ of cross-linked gel were injected into the lower Dethlingen formation with a flow rate of 4 m$^3$ min$^{-1}$ and 113 tons of proppants for the first treatment; 310 m$^3$ of cross-linked gel were injected into the upper Dethlingen formation at flow rates ranging from 3–3.5 m$^3$ min$^{-1}$ for the second treatment (Zimmermann et al., 2010b). The production test, which lasted 11.8 h and produced about 356 m$^3$ of fluids, showed an overall productivity increase after the stimulations by more than a factor of 4. 30 % of the total flow came from the volcanic rocks and 70 % from the sandstones (Zimmermann et al., 2010a). However, it can be argued that hydraulic fracturing in matrix-dominated formations is not the most common situation in deep geothermal projects.

Another example of hydraulic performance improvement of a PS through hydraulic fracturing is given by the Fenton Hill project, which has been referred to as a PS by Kruger (1990). In the second phase of this PS, a total fractured volume of 1 km$^3$ was created, flow rates were increased up to 18.5 L s$^{-1}$, and the permeability was improved to a value of 3 to 5 m$^2$ (MIT, 2006f).

Of course this is only one example; a case-by-case investigation of the geomechanics involved must be carried out to estimate the benefit of hydraulic stimulation. In some cases, the productivity index can be much higher than reported above. Schindler et al. (2010), for example, quote productivity improvement by a factor of 20 after massive hydraulic stimulations in crystalline rocks.

Jung (2013) presented an overview of different hydraulic stimulation techniques used for EGSs, such as multi-zone hydraulic fracturing in crystalline basements (based on the original HDR concept), multi-zone massive injection in naturally fractured crystalline rock formations (in order to generate multiple wing cracks), and open-hole massive injection in naturally fractured crystalline rock formations.

The second most common stimulation technique is chemical stimulation, which is applied to enhance the permeability in the near-wellbore region, i.e. up to a distance of few tens of metres (ENGINE, 2008b). This technique is also called acidizing as acids such as hydrochloric acid (HCL) and hydrofluoric acid (HF) are commonly used to react with carbonates and silicates, respectively. The only additives that might be used for geothermal systems are as follows: corrosion inhibitor, inhibitor intensifier, and high-temperature iron-control agent (ENGINE, 2008b). According to Schumacher and Schulz (2013), acidizing with HCL can significantly improve the performance of a geothermal well drilled into carbonate rock. In addition, it is an effective means to remove fine materials from the walls of the wells, i.e. to clean the well from drill cuttings and from scaled minerals that decrease permeability (Schumacher and Schulz, 2013; ENGINE, 2008b). The aim of acidizing in sandstones is to dissolve naturally occurring clay or material that originated from drilling and completion works and other plugging minerals in the near-wellbore region, thereby increasing the permeability (ENGINE, 2008b). In this case, the acidizing is performed in three stages: pre-flush (HCL), main flush (HCL-HF mixture) and overflush (HCL, or KCL, $NH_4CL$ or fresh water) (ENGINE, 2008b). Chemical stimulation can be applied to any of the following deep geothermal systems: HS, HSA, PS, EGS.

Schumacher and Schulz (2013) analysed improvements after several acidizing steps in a number of wells in the carbonate rocks of the south German Molasse Basin; their findings are relevant for analogue geothermal projects worldwide. The normalized flow rate for these wells was taken as $10\,L\,s^{-1}$, with an observed improvement of over 10 % per $m^3$ of 15 % HCL used. The analyses indicate that the first acid treatment significantly increased the productivity, whereas subsequent treatments did not have such a great impact any more, and in some cases resulting in deterioration of well performance.

Thermal fracturing is used in volcanic rock environments, such as found in Iceland, to increase the permeability of existing flow paths, to create new ones, and is achieved with a combination of induced temperature and pressure changes (ENGINE, 2008b). It is used when the temperature difference between injected fluid and rock formation is significant (Flores et al., 2005). Tulinius et al. (2000) provide some quantitative values for this type of stimulation for a 2500 m deep well in geothermal area of Bouillante, France, which was characterized by low steam output before stimulation. A 253 °C reservoir was stimulated in periods up to 72 h using seawater mixed with an inhibitor to prevent anhydrite scaling at a flow rate up to $25\,L\,s^{-1}$ and initial wellhead pressure of 2.5 MPa, which decreased gradually and was close to zero for maximum injection at the end of the programme. The thermally induced fracturing resulted in a 50 % increase of productivity.

## 4 Systematic overview of past and present deep geothermal systems

The following review consisting of PSs and HSA systems is not meant to be exhaustive, as it is based solely on information that is available in the public domain. This review excludes conventional HSs, because the focus of this paper is on PSs and HSA systems in relation to the widespread term EGS. The overview is divided into PS (see Tables 1–3) and HSA (see Tables 4–6). The PS database consists of 26 projects worldwide, whereas the HSA database consists of 10 projects. Conventional HSs, such as volcanic systems or vapour-dominated systems, are not presented in the tables.

Wherever the literature did not state whether a given project is a PS or an HSA system, and when the present authors did not agree with the classification offered by the literature, an independent view was taken.

Whether the heat source of a project was conduction-dominated or convection-dominated was difficult (and in most cases impossible) to find in the literature in order to differentiate HSs from HSA systems.

The databases for PSs and HSA systems are each divided into three parts: general information, petrophysical properties, and operational characteristics.

Table 1 (PS), resp. Table 4 (HSA), comprises general information about PSs, resp. HSA, such as location, operator, description, start date, end date, status, well depth, and distance between producer and injector at depth. The description contains the main goal of each project, whereas the status informs whether a project is still under development, ongoing, concluded, or abandoned.

Table 2 (PS), resp. Table 5 (HSA), presents petrophysical properties of the reservoir such as rock type, porosity, permeability or transmissivity, and temperature. However, only a few porosity and permeability values could be found in the literature. Permeabilities are given in $m^2$. In the case of permeabilities given in Darcy, the values were converted into $m^2$ under the assumption of the presence of fresh water and temperatures of only 10 °C, which the authors admit is an oversimplification. In some cases, only transmissivities were available, which were converted into permeabilities in the cases where reservoir thicknesses were available in the public domain.

Table 3 (PS), resp. Table 6 (HSA) shows operational characteristics such as flow rate, stimulation technique, seismic event, type of power plant, installed electrical capacity, thermal capacity, and flow assurance problems. In most cases, the production flow rate was given. However, in some cases, only injection flow rates could be found in the literature. Stimulation techniques state whether stimulation was applied or not and, in the cases where stimulation was performed, the method that was applied is given. Whenever the information was available in the public domain, it was differentiated in the tables which type of hydraulic stimulation was applied. In the case of missing differentiation in the refer-

**Table 1.** General Information about Petrothermal Systems.

| Project | Location[b] | Operator | Description | Start date | End date | Status | Well depth [m] | Distance between producer and injector [m] |
|---|---|---|---|---|---|---|---|---|
| Le Mayet | FR | Unknown | | 1978, Cornet (2012) | 1987, Cornet (2012) | Concluded experimental, Cornet (2012) | 200–800, Cornet (2012) | 100, Cornet (1987) |
| Soultz[a] | FR | European cooperation project, MIT (2006c) | Research, Cornet (2012); R&D, Genter (2012) | 1987, MIT (2006b) | Not ended* | Concluded experimental, Cornet (2012), power plant of stage 1.0 dismantled, planning stage Soultz 2.0 with different power plant, J. Scheiber, personal communication, 2015 | 3600, 5080, 5100 and 5270, Genter et al. (2010) | 450, 600 and 650, Genter et al. (2010) |
| GeneSys Hannover | DE | BGR[b], EGEC (2013) | Demonstrate single-well concepts, Tischner et al. (2010) | 2009, Tischner et al. (2010) | Not ended* | Under development, Tischner et al. (2013) | 3901, Tischner et al. (2013) | Single well, Tischner et al. (2010) |
| GeneSys Horstberg | DE | BGR[b], EGEC (2013) | Demonstrate single-well concepts, BGR (2014a) | 2003, Tischner et al. (2010) | 2007, BGR (2014b) | Concluded experimental, BGR (2014b) | 3800, Tischner et al. (2010) | Single well, Tischner et al. (2010) |
| Groß Schönebeck | DE | GFZ[c], Schmidt & Clemens GmbH, BINE (2012) | 1st in situ geothermal laboratory, EGS research, Zimmermann et al. (2009) | 2000, Zimmermann et al. (2009) | Not ended* | Under development, not generating electricity, GtV (2014a) | 4309–4400, Zimmermann et al. (2009); Henninges et al. (2012) | 470, Urpi et al. (2011) |
| Mauerstetten | DE | Exorka, GFZ[c], TUBAF[d], Schrage et al. (2012a) | Research, Schrage et al. (2012b) | 2011, Schrage et al. (2012a) | Not ended* | Under development, not generating electricity, GtV (2014a) | 4545, Exorka (2014) | Single-well drilled, Exorka (2014) |
| Falkenberg | DE | BGR[b] (coordinator), Kappelmeyer and Jung (1987) | Investigation of hydraulic fracturing at shallow depth, Tenzer (2001) | 1977, Tenzer (2001) | 1986, Tenzer (2001) | Concluded experimental, MIT (2006d) | 300–500, Kappelmeyer and Jung (1987) | eight wells within area of 100 m × 100 m, Kappelmeyer and Jung (1987) |
| Bad Urach | DE | Forschungs-Kollegium Physik des Erdkörper, MIT (2006g) | HDR demonstration pilot in Germany, Tenzer (2001) | 1977, Tenzer (2001); 2006, Wyborn (2011) | 1981, MIT (2006d); 2008, Wyborn (2011) | Single-well concept abandoned, consideration as deep geothermal probe, iTG (2010) | 4445, Tenzer (2001) and 2793, iTG (2010) | 100, iTG (2010) |
| Basel | CH | Geopower Basel, Giardini (2009) | Planning to develop EGS project, Ladner and Häring (2009) | 1996, Giardini (2009) | 2009, Giardini (2009) | Abandoned due to induced seismicity, Giardini (2009) | 5000, Ladner and Häring (2009) | Second well not drilled, Ladner and Häring (2009) |
| Fjällbacka | SE | University of Technology, Gothenburg, Sweden, Jupe et al. (1992) | Experimental project, Portier et al. (2007) | 1984, Jupe et al. (1992) | 1995, Wallroth et al. (1999) | Concluded experimental, Wallroth et al. (1999) | 70–500, Jupe et al. (1992) | 100, Wallroth et al. (1999) |
| Rosemanowes | GB | CSM[e], MIT (2006c) | Experimental project, MIT (2006c) | 1977, MIT (2006c) | 1992, MIT (2006c) | Concluded experimental, MIT (2006c) | 2000–2600, MIT (2006c) | 300 (vertically), MIT (2006c) |
| Eden | GB | EGS Energy Limited, Baria et al. (2013) | Commercial CHP, Baria et al. (2013) | 2010, Baria et al. (2013) | Not ended* | Early stage, under development, not generating electricity, Baria et al. (2013) | Target depth 4000, Baria et al. (2013) | Unknown at this stage |
| United Downs | GB | Geothermal Engineering ltd, EGEC (2013) | commercial CHP, three-well system, Bridgland (2011) | 2010, Atkins (2013) | Not ended* | Early stage, under development, not generating electricity, Atkins (2013) | Target depth 4500–5000, Atkins (2013) | Unknown at this stage |
| Litoměřice | CZ | Municipality of Litoměřice, Gryndler (2009) | Experimental, proof of concept, Stibitz et al. (2011) | 2007, Tym (2013) | Not ended* | Under development, not generating electricity, Stibitz et al. (2011) | Drilled up to 2111; target depth 5000, Tym (2011) | 600 planned, Tym (2011) |
| Ferenczfalús | HU | EU-FIRE kft. and Mannvit kft., Ministry of National Development, Sverrisson et al. (2013) | Commercial CHP, Sverrisson et al. (2013) | 2012, Sverrisson et al. (2013) | Not ended* | Early stage, under development, not generating electricity, Sverrisson et al. (2013) | Target depth 4000, Sverrisson et al. (2013) | Unknown at this stage |
| Newberry | USA | AltaRock Energy, Davenport Newberry, Cladouhos et al. (2012) | Demonstration for EGS Research, Cladouhos et al. (2012) | 2010, Cladouhos et al. (2012) | Not ended* | Under development, not generating electricity, Newberry EGS Demonstration (2014) | Over 3000, Sonnenthal et al. (2012) | Second well not yet drilled, Newberry EGS Demonstration (2014) |
| Northwest Geysers | USA | Calpine Corporation, Garcia et al. (2012) | EGS demonstration, Garcia et al. (2012) | 2009, Rutqvist et al. (2013) | Not ended* | Under development, not generating electricity, Garcia et al. (2012) | 3396, Garcia et al. (2012) | 525, Garcia et al. (2012) |
| Fenton Hill | USA | Los Alamos National Laboratory, MIT (2006) | First HDR in the world, MIT (2006) | 1974, MIT (2006) | 1993, MIT (2006f) | Concluded experimental, MIT (2006) | 2932–4390, MIT (2006) | 100, 380 (vertically), MIT (2006) |
| Paralana | AU | Petratherm Limited, Beach Energy, Reid and Messeiller (2013) | Commercial power development, Reid and Messeiller (2013) | 2005, Petratherm (2014) | Not ended* | Under development, not generating electricity, Petratherm (2014) | 4003, Bendall et al. (2014) | Second well not yet drilled, Petratherm (2014) |

**Table 1.**

| Project | Location[h] | Operator | Description | Start date | End date | Status | Well depth [m] | Distance between producer and injector [m] |
|---|---|---|---|---|---|---|---|---|
| Cooper Basin (Innamincka) | AU | Geodynamics Ltd., Origin Energy, Majer et al. (2007) | Largest demonstration project in the world, Stephens and Jiusto (2010) | 2003, Majer et al. (2007) | Not ended* | temporarily shut down, planning small scale project, Geodynamics (2014) | 4421, Majer et al. (2007) | Unknown |
| Olympic Dam | AU | Green Rock Energy Ltd., Lovelock (2011) | Commercial power development, Lovelock (2011) | 2005, Meyer et al. (2010) | Not ended* | Early stage, under development, not generating electricity, Lovelock (2011) | target depth 5500, Lovelock (2011) | Unknown at this stage |
| Parachilna | AU | Torrens Energy Ltd, Torrens Energy (2014) | Commercial power development, Torrens Energy (2014) | 2007, Canaris (2009) | Not ended* | Under development, not generating electricity, Torrens Energy (2014) | target depth 4500, Torrens Energy (2014) | Unknown at this stage |
| Frome | AU | Geothermal Resources Pty Limited, Geoscience Australia and ABARE (2010) | Commercial power development, Geothermal Resources (2014) | 2006, Geothermal Resources (2014) | Not ended* | Under development, not generating electricity, Geothermal Resources (2014) | target depth 3250, Goldstein et al. (2010) | Unknown at this stage |
| Pohang | KR | Nexgeo Inc., KIGAM, KICT, SNU, POSCO, Innogeo Tech. Inc., Lee et al. (2011) | Proof of concept power generation EGS, Lee et al. (2011) | 2010, Lee et al. (2011) | Not ended* | Under development, not generating electricity, Lee et al. (2011) | Target depth 5000, Lee et al. (2011) | Unknown at this stage |
| Hijiori | JP | Japan's New Energy, DiPippo (2012), NEDO[f], Sasaki (1998) | Developing HDR technologies, Sasaki (1998) | 1985, Sasaki (1998) | 2002, DiPippo (2012) | Abandoned due to failure to create a reservoir, Grant and Bixley (2011) | 1800–2200, DiPippo (2012) | 33, 38, 63 shallow reservoir, 90 and 130 deep reservoir, DiPippo (2012) |
| Ogachi | JP | CRIEPI[g], Kaieda et al. (2010) | Test run HDR project in shallow depth, Kaieda et al. (2005) | 1989, Kaieda et al. (2005) | 2002, Kaieda et al. (2005) | Concluded, experimental (Kaieda et al. 2010) | 1000–1300, Kaieda et al. (2005) | 50 shallow reservoir, 80 and 200 deep reservoir, Kaieda et al. (2005) |

* See status; [a] not clear whether Soultz project is hydro- or petrothermal; [b] BGR – Bundesanstalt für Geowissenschaften und Rohstoffe;

[c] GFZ – Helmholtz Centre Potsdam – GFZ German Research Centre;

[d] TUBAF – TU Bergakademie Freiberg; [e] CSM – Camborne School of Mines; [f] NEDO – New Energy and Industrial Technology Development Organization; [g] CRIEPI – Central Research Institute of the Electric Power Industry;

[h] country abbreviation after ISO 3166 Alpha-2; CHP – combined heat and power production; KIGAM – Korea Institute of Geoscience and Mineral Resources; KICT – Korea Institute of Construction Technology; SNU – Seoul National University; POSCO – Pohang Iron and Steel Company.

**Table 2.** Petrophysical properties of petrothermal systems.

| Project | Rock type | Porosity | Permeability (K) $[m^2]$/transmissivity (T) $[m^2 s^{-1}]$ | BHT/Reservoir temperature [°C] |
|---|---|---|---|---|
| Le Mayet | Granite, Cornet (2012) | Unknown | Unknown | 22, Wyborn (2011) |
| Soultz[a] | Granite, MIT (2006c) | Altered rock: 0.25, Ledésert et al. (2010); connected porosity: 0.0025–0.003, Portier and Vuataz (2009) | Fresh Soultz granite: $K = 4 \times 10^{-19}$, Ledésert et al. (2010) | 200, Genter et al. (2010) |
| GeneSys Hannover | Bunter sandstone, Tischner et al. (2013) | <0.1, ENGINE (2008a) | $K = 10^{-18}$, Tischner et al. (2013) | 169, Tischner et al. (2013) |
| GeneSys Horstberg | Bunter sandstone, Tischner et al. (2010) | 0.03–0.11, Orzol et al. (2005) | $K^i < 40 \times 10^{-15}$, GeneSys Hannover (2014a) | 150, Tischner et al. (2010) |
| Groß Schönebeck | Sandstone and andesitic volcanic rocks, Zimmermann et al. (2009) | 0.08 to 0.10, Zimmermann et al. (2010a) | $K^i = 10^{-14}$ to $10^{-13}$, Zimmermann et al. (2009); $K^i$ up to $16.5 \times 10^{-15}$, Zimmermann et al. (2010a) | 150, Henninges et al. (2012) |
| Mauerstetten | Limestone, Schrage et al. (2012a) | Unknown | Unknown | 130, Schrage et al. (2012a) |
| Falkenberg | Granite, MIT (2006d) | Unknown | Unknown | 13.5, Kappelmeyer and Jung (1987) |
| Bad Urach | Gneiss, Tenzer et al. (2000) | Unknown | T (rock matrix) $10^{-7}$ to $10^{-6}$, T (fractures) $10^{-4}$ to $10^{-3}$ at 3320–3488 m, Schanz et al. (2003) | 172 at 4445 m, Tenzer (2001); 112 at 3200 m, iTG (2010) |
| Basel | Granite, Ladner and Häring (2009) | Unknown | $K = 1 \times 10^{-17}$ estimated, Ladner and Häring (2009) | 174, Ladner and Häring (2009) |
| Fjällbacka | Granite, Portier et al. (2007) | Unknown | $K = 10^{-18}$ to $10^{-17}$, Jupe et al. (1992) T $= 10^{-8}$ to $10^{-7}$, Wallroth et al. (1999) | 16, Wallroth et al. (1999) |
| Rosemanowes | Granite, MIT (2006e) | Unknown | $K^i = 10^{-18}$ to $10^{-17}$, Parker (1999) | 79–100, MIT (2006e) |
| Eden | Granite, Baria et al. (2013) | 0.15 estimated, Atkins (2013) | $K = 9.9 \times 10^{-16}$ estimated, Atkins (2013) | 180 estimated, Baria et al. (2013) |
| United Downs | Granite, Atkins (2013) | 0.15 estimated, Atkins (2013) | $K = 9.9 \times 10^{-16}$ estimated, Atkins (2013) | 180–200 estimated, Atkins (2013) |
| Litoměřice | Sedimentary and granite, Stibitz et al. (2011) | Unknown | Unknown | 63.5, Stibitz et al. (2011); 178 to 207.5 estimated at 5 km, Stibitz et al. (2011) |
| Ferencszállás | Metamorphic schist and partly granitoid, Sverrisson et al. (2013) | Unknown | Unknown | 170 estimated, Sverrisson et al. (2013) |
| Newberry | Volcanic rocks, Fittermann (1988) | 0.01 to 0.20, Sonnenthal et al. (2012) | $K = 1.0 \times 10^{-18}$ to $1.5 \times 10^{-12}$, Sonnenthal et al. (2012) | 315, Cladouhos et al. 2012) |
| Northwest Geysers | Metasedimentary rocks (greywacke), Romero et al. (1995) | 0.01, Rutqvist et al. (2013) | $K = 2 \times 10^{-14}$, Rutqvist et al. (2013) | about 400, Garcia et al. (2012) |
| Fenton Hill | Crystalline rock, Brown (2009) | Unknown | Unknown | 180 to 327, MIT (2006f) |
| Paralana | Metasediments, granite, Petratherm (2014) | Unknown | Unknown | 190, Reid and Messeiller (2013) |
| Cooper Basin | Granite, Majer et al. (2007) | Unknown | Unknown | 243 to 264, Bendall et al. (2014) |
| Olympic Dam | Granite, Lovelock (2011) | Unknown | Unknown | 85.3 at 1934.2 m, Bendall et al. (2014) 190 estimated at target depth, Lovelock (2011) |
| Parachilna | Granite, Geoscience Australia and ABARE (2010) | Unknown | Unknown | 98.4 at 1807 m, 240 estimated at 4500 m, Torrens Energy (2014) |
| Frome | Granite, Geoscience Australia and ABARE (2010) | Unknown | Unknown | 93.5 at 1761 m, 200 estimated at 4080 m, Geoscience Australia and ABARE (2010) |
| Pohang | Paleozoic granodiorite, Lee et al. (2011) | Unknown | Unknown | 180 estimated, Lee et al. (2011) |
| Hijiori | Granodiorite, Sasaki (1998) | 0.01, Sasaki (1998) | K (Rock matrix) $10^{-19}$ to $10^{-21}$, Sasaki (1998) | 190, DiPippo (2012) |
| Ogachi | Granodiorite, Kaieda et al. (2010) | Unknown | $K = 0.8 \times 10^{-15}$ to $0.2 \times 10^{-13}$, Kaieda et al. (2005) | 228, Kaieda et al. (2005) |

[a] not clear whether Soultz project is petrothermal or HSA; [i] permeability calculated from Darcy into $m^2$ under assumption that water temperature is only 10 °C and fresh water; BHT – bottomhole temperature.

ences, the tables refer generically to hydraulic stimulation, which could mean either one of the hydraulic stimulation techniques, such as hydraulic fracturing, hydraulic shearing or a combination of both. Seismic events are given in Richter scale magnitudes. The type of power plant is commonly only available for those projects which are ongoing. All projects employ only binary power plants, such as organic rankine cycles (ORCs) or Kalina cycles. In the event of the information being available, it was possible to differentiate which type of binary power plant was used for each project. Installed elec-

trical and thermal capacities could only be provided for the ongoing projects.

In what follows, specific projects have been highlighted which presented ambiguity in their classification.

## 4.1 Petrothermal systems

The European HDR project Soultz-sous-Forêts in France was categorized as a PS, although there has been much debate among experts as to whether this system should be cate-

**Table 3.**

| Project | Flow rate [L s⁻¹] | Stimulation techniques | Seismic event (Richter scale) | Type of power plant | Installed electrical capacity [MWₑ] | Thermal capacity [MWth] | Flow assurance problem |
|---|---|---|---|---|---|---|---|
| Le Mayet | 5.2, Wyborn (2011) | Hydraulic fracturing with and without proppant, Cornet (2012); MIT (2006b) | Microseismic, not felt on surface, Cornet (2012) | None* | 0* | 0* | Unknown |
| Soultz[a] | 30, BMU (2011) | Hydraulic stimulation and acidizing, Genter et al. (2010) | Microseismic (M=−2 to 2.9), Genter (2012) | ORC, Genter et al. (2010) | 1.5, Genter et al. (2010) | Non-scheduled, Dumas (2010) | Corrosion due to high salt contents, BMU (2011) |
| GeneSys Hannover | 7 (planned), BGR (2014c) | Hydraulic fracturing, Tischner et al. (2013) | No seismic event due to geothermal activity, Tischner et al. (2013) | None* | 0* | 2 (planned), Tischner et al. (2013) | Salt precipitation removed with coiled tubing, GeneSys Hannover (2014b) |
| GeneSys Horstberg | 4, Tischner et al. (2010) | Hydraulic fracturing, Tischner et al. (2010) | No measured event, Kreuter (2011a) | None* | 0* | 1 to 1.4, Tischner et al. (2010) | Unknown |
| Groß Schönebeck | 4.4, Blöcher et al. (2012) | Hydraulic: gel proppant and fracturing, Zimmermann et al. (2009); Thermal, ENGINE (2008b); Chemical, Henninges et al. (2012) | Negligible (max. −1.8 to −1.0M), Blöcher et al. (2012) | ORC, BINE (2012) | 1, BINE (2012) | 0* | High salt content (265 g L⁻¹), BINE (2012) |
| Mauerstetten | – | Chemical, Schrage et al. (2012b); hydraulic stimulation, iTG (2013) | Unknown | Modular binary planned, Exorka (2014) | 0* | 0* | Unknown |
| Falkenberg | 3.5**, Kappelmeyer and Jung (1987) | Hydraulic fracturing, Tenzer (2001) | Microseismic, MIT (2006d) | None* | 0* | 0* | Unknown |
| Bad Urach | 50 for single-well; 15–25 estimated for doublet, iTG (2010) | Hydraulic fracturing, Schanz et al. (2003) | Microseismic, Schanz et al. (2003) | None* | 3 evaluated, Tenzer (2001) | 17 evaluated, Tenzer (2001) | High flow impedances in a single well, Schanz et al. (2003); Torn-off drill pipes in borehole, iTG (2010) |
| Basel | Unknown at this stage | Hydraulic fracturing, Ladner and Häring (2009) | Frequent earthquakes (max. 3.4M), Ladner and Häring (2009) | None* | 0* | 0* | Unknown at this stage |
| Fjällbacka | 0.9** to 1.8**, Wallroth et al. (1999) | Hydraulic fracturing and acidizing, Portier et al. (2007) | Microseismic, Wallroth et al. (1999) | None* | 0* | 0* | Too high fluid losses and flow impedance, Wallroth et al. (1999) |
| Rosemanowes | 4–15, MIT (2006d) | Hydraulic fracturing, MIT (2006d), viscous gel stimulation, Parker (1999) proppants, Parker (1999) | Max. magnitude 3.1, Bromley ane Mongillo (2008) | None* | 0* | 0* | Temperature drawdown, very low flow rate, pressure drop, MIT (2006e) |
| Eden | 55 estimated, Baria et al. (2013) | Hydraulic stimulation planned, Baria et al. (2013) | Unknown at this stage | None* | 4 estimated, Baria et al. (2013) | 3.45 EGEC (2013) | Unknown at this stage |
| United Downs | 150 estimated, Bridgland (2011) | Planned, Bridgland (2011) | Unknown at this stage | None* | 10 estimated, Atkins (2013) | 50 estimated, Bridgland (2011) | Unknown at this stage |
| Litoměřice | 120 estimated, Gryndler (2009) | No stimulation for a drilled exploration well, Stibitz et al. (2011) | Unknown at this stage | Kalina cycle, CHP[i] planned, Gryndler (2009) | 4.4 estimated, Gryndler (2009) | 50 estimated, Gryndler (2009) | Unknown at this stage |
| Ferencszállás | 160 estimated, Sverrisson et al. (2013) | Hydraulic fracturing and acidizing planned, Sverrisson et al. (2013) | Unknown at this stage | ORC, Sverrisson et al. (2013) | 5 estimated, Sverrisson et al. (2013) | 20 estimated, Sverrisson et al. 2013 | Unknown at this stage |
| Newberry | Unknown at this stage | Hydro-shearing, multi-zone isolation techniques, Cladouhos et al. (2012) | Microseismic, Cladouhos et al. (2012) | None* | 0* | 0* | Unknown |
| Northwest Geysers | 9.7, Garcia et al. (2012) | Thermal fracturing, Walters (2013) | Microseismic (0.9 to 2.87M), Garcia et al. (2012); Walters (2013) | None* | 0* | 0* | Corrosion in production well, Walters (2013) |
| Fenton Hill | 10.6 to 18.5 (test), MIT (2006f) | Hydraulic fracturing, MIT (2006f) | Microseismic, Brown (1995) | Binary, MIT (2006f) | 0.06, MIT (2006f) | 3–5, MIT (2006f) | Creating connection between wells, pressure drop in and near wellbore, MIT (2006f) |
| Paralana | 70 estimated, Reid and Messeiller (2013) | Hydraulic stimulation, Reid and Messeiller (2013) | Microseismic ≤ 2.6M, Petratherm (2014) | ORC planned, Reid and Messeiller (2013) | 3.5, Reid and Messeiller (2013) | 0* | Unknown |

**Table 3.** Operational characteristics of petrothermal systems.

| Project | Flow rate [L.s⁻¹] | Stimulation techniques | Seismic event (Richter scale) | Type of power plant | Installed electrical capacity [MWe] | Thermal capacity [MWth] | Flow assurance problem |
|---|---|---|---|---|---|---|---|
| Cooper Basin | 19 at 215°C, Geodynamics (2014) | Hydraulic stimulation, Majer et al. (2007); Holl (2012) | ≤3.7 M, Majer et al. (2007) | 1 MWe pilot plant, Geodynamics (2014) | 1 for proof of concept (Geodynamics (2014) | 0* | Unknown |
| Olympic Dam | Unknown | Hydraulic fracturing, Meyer et al. (2010) | Unknown | Probably binary, Lovelock (2011) | 400 planned, Green Rock Energy (2014) | 0* | Unknown |
| Paralhina | Unknown | Unknown | Unknown | ORC, Beardsmore and Matthews (2008) | 0* | 0* | Unknown |
| Frome | Unknown | Hydraulic fracturing, Goldstein et al. (2010) | Unknown | ORC, Giles (2009) | 0* | 0* | Unknown |
| Pohang | 40 estimated, Lee et al. (2011) | Hydraulic fracturing planned, Lee et al. (2011) | Unknown at this stage | Binary planned, Lee et al. (2011) | 1.5 estimated, Lee et al (2011) | 0* | Unknown at this stage |
| Hijiori | 16*, MIT (2006c) | Hydraulic fracturing, Sasaki (1998) | Microseismic, Sasaki (1998) | Binary, DiPippo (2012) | 0.13, DiPippo (2012) | 8, DiPippo (2012) | High water losses, precipitation of anhydrite, DiPippo (2012); rapid temperature drawdown in one production well, Grant and Bixley (2011) |
| Ogachi | 6.7** to 20**(test), Kaieda et al. (2005) | Multiple wells with multiple fracture zones; hydraulic fracturing, Kaieda et al. (2005) | Few microseismic, Kaieda et al. (2010) | None* | 0* | 0* | Low water recovery rate in circulation tests, Kaieda et al. (2005) |

* See status;
** Injection flow rate;
a not clear whether Soultz project is petrothermal or HSA. ORC – Organic Rankine Cycle; CHP – Combined heat and power production.

**Table 4.** General information about hot sedimentary aquifers.

| Project | Location | Operator | Description | Start date | End date | Status | Well depth [m] | Distance between producer and injector [m] |
|---|---|---|---|---|---|---|---|---|
| St. Gallen | CH | ITAG Tiefbohr GmbH, Geothermie Stadt St. Gallen (2014) | Hydrothermal heat production project, Hirschberg et al. (2015c) | 2009, Geothermie Stadt St. Gallen (2014) | 2014, Hirschberg et al. (2015c) | Abandoned, Hirschberg et al. (2015c) | 4450, Hirschberg et al. (2015d) | Single well, Hirschberg et al. (2015e) |
| Bruchsal | DE | EnBW[j], EWB[k], Rettenmaier (2012) | Commercial, Herzberger et al. (2010) | 1985, Herzberger et al. (2010) | Not ended* | On-going, generating electricity, GtV (2014a) | 1874 to 2542, BMU (2011) | 1500, Herzberger et al. (2010) |
| Landau | DE | BESTEC, Geox, Baumgärtner (2012) | First EGS in town in DE, Baumgärtner (2012) | 2004, Baumgärtner (2012) | 2014, GtV (2014b) | Abandoned due to groundwater contamination resulting from well damage, Geothermie-Nachrichten (2014) | 3170–3300, Baumgärtner (2012) | 1500, Bracke (2012) |
| Insheim | DE | Pfalzwerke geofuture GmbH, Pfalzwerke geofuture (2014) | New concept: side-leg injection well, Baumgärtner (2012) | 2007, Baumgärtner (2012) | Not ended* | On-going, generating electricity, Ganz et al. (2013) | 3600-3800, Pfalzwerke geofuture (2014); Baumgärtner (2012) | Unknown |
| Neustadt-Glewe | DE | WEMAG, Stadt Neustadt-Glewe, Geothermie Neubrandenburg, BMU (2011) | Commercial, Pilot plant for low-enthalpy, Broßmann and Koch (2005) | 1984, Broßmann and Koch (2005) | Not ended* | On-going, generating electricity, GtV (2014a) | 2320, Bracke (2012) | 1500, BMU (2011) |
| Unterhaching | DE | Geothermie Unterhaching, Rödl & Partner, Richter (2010) | First CHP[i] Kalina power plant in Germany, Richter (2010) | 2001, Richter (2010) | Not ended* | Ongoing, generating electricity, GtV (2014a) | 3350 to 3590, Richter (2010) | 4500, Bracke (2012) |
| Southampton*** | GB | SGHC[l], Southampton (2014a) | CHP station, district heating and chilling system, Smith (2000) | 1981, Smith (2000) | Not ended* | On-going, Southampton (2014a) | 1800, Smith (2000) | Single well, Smith (2000) |
| Altheim | AT | Municipality of Altheim, Pernecker, Pernecker (1999) | Commercial, Pernecker (1999) | 1989, Pernecker (1999) | Not ended* | On-going, generating electricity, Bloomquist (2014) | 2165-2306, City of Altheim (2014) | 1700, City of Altheim (2014) |
| Birdsville*** | AU | Ergon Energy, Ergon (2014) | The only operating HSA power station in Australia, Ergon (2014) | 1992, Ergon (2014) | Not ended* | On-going, generating electricity, Ergon (2014) | 1280, Ergon (2014) | Single well, Ergon (2014) |
| Penola*** | AU | Raya Group Limited, Panax Geothermal (2014) | Proof of concept for commercial power generation, Graaf et al. (2010) | 2010, Panax Geothermal (2014) | Not ended* | Under development, not generating electricity, Panax Geothermal (2014) | 4025, Graaf et al. (2010) | Single well; 10 wells planned, Proactive Investors (2009) |

* See status;
*** Project indicated as HSA in the literature;
j EnBW – Energie Baden-Württemberg AG;
k EWB – Energie und Wasserversorgung Bruchsal GmbH;
l SGHC – Southampton Geothermal Heating. Company

gorized as HDR, HFR or HSA (IGA R&R, 2013). This discussion has probably arisen because two different reservoirs are connected to the project: the upper reservoir being in a fractured granite formation with higher permeabilities ($3 \times 10^{-14}$ m$^2$) and the lower reservoir in a fresh granite formation with much poorer permeabilities ($1 \times 10^{-17}$ m$^2$) (Kohl et al., 2000). Although Soultz was initially planned as an HDR project and therefore created in crystalline basement rocks, it was found that the reservoir contains permeable structures with substantial volumes of natural brine. Hence, it differs from the classic definition of HDR and the geothermal anomaly is mainly controlled by natural fluid flow (Genter et al., 2010). However, the low hydraulic connection of the fracture system required a permeability enhancement using hydraulic stimulation. Following the definition in Sect. 2, this would indicate that Soultz is a PS as hydraulic stimulation was required to enhance the productivity index.

Some explanation is necessary also for the Northwest Geysers project. According to Walters (2013), this is an EGS demonstration project, launched in 2009 with the main goal of enhancing the permeability of hot, low-permeable rocks by means of thermal fracturing and creating an EGS doublet capable of producing 5 MW. Garcia et al. (2012) refer to the high temperature reservoir (HTR) of this EGS demonstration area as non-hydrothermal HDR due to conductive temperature gradients and the project not being part of the pre-existing HS. However, the same source mentions presence of steam entries in the HTR in previously abandoned wells after re-opening and deepening.

## 4.2  Hot sedimentary aquifers

The HSA database consists of 10 projects, whereof only 3 projects (Southampton, Birdsville, and Penola) were actually indicated as HSA in the literature. Since the term HSA was only invented recently and there is no international standard for the categorization of such a geothermal system, it has to be assumed that not all projects which are HSA are also indicated as such in the literature. Therefore, based on the geological setting, additional hydrothermal projects were added to the tables, where it can be assumed that they are HSA projects.

One could argue about the classification of the St. Gallen project in Switzerland. The project's aim was to use the naturally fractured Malm formation in a depth of 4 to 4.5 km for an HS. However, during the preparations for the production test, an unexpected high gas inlet in the well required interventions to secure the well, which in turn resulted in induced seismic activity. Therefore, the project was put on hold in order to evaluate the gathered data from the production test and to readjust further project steps (Geothermie Stadt St. Gallen, 2014). The encountered dissolved natural gas in the well indicates that St. Gallen might actually be a geo-pressured system. However, it is likely that the gas was coming from deep-seated, highly faulted permo-

carboniferous formations, which were penetrated by deep drilling. Hirschberg et al. (2015a), who do not differentiate between HS and HSA, classify St. Gallen as an HS. The analysed data with low flow rates of only 6 to 12 L s$^{-1}$, the existing gas inlet in the well, the increased risk of induced seismicity, and limited financial funds, eventually resulted in the abandonment of the project in May 2014 (Hirschberg et al., 2015c).

## 5  Results and discussion

### 5.1  Petrothermal Systems

For almost all PSs, hydraulic fracturing was applied (with the exception of Northwest Geysers, where thermal fracturing was conducted instead). For four projects, stimulation was either not yet performed or no information about it could be found in the public domain. For the projects Eden and United Downs, it was only stated that stimulation will be applied in the future. In the cases of Mauerstetten, Soultz-sous-Forêts, and Fjällbacka, not only was hydraulic fracturing carried out, but chemical stimulation of the near-wellbore region was also performed. Groß Schönebeck was the only project where all three stimulation techniques (hydraulic, chemical and thermal) were implemented.

The well depths of PSs vary widely within a range of 70 to 5000 m. However, most projects are deeper than 1800 m, with exception of the three shallow HDR systems Le Mayet, Falkenberg, and Fjällbacka, which were never operational, but were only implemented for research and demonstration purposes.

The temperature range of most of the PSs is 130 to 400 °C, excluding the three abovementioned shallow systems and Rosemanowes, which have a lower temperature range of 79 to 100 °C.

Rock types are usually crystalline and volcanic, with rocks such as granite and granodiorite with exception of GeneSys Hannover (Bunter sandstone), GeneSys Horstberg (sedimentary), Mauerstetten (limestone), and Northwest Geysers (metasedimentary rocks).

For those nine projects where porosity values were available in the public domain, the porosity shows a very wide range from 0.0025 to 0.25, depending on the type of porosity. For example, the former value represents the connected porosity, such as the fresh Soultz granite, and the highest value is related to the altered rock in Soultz. However, most projects have porosities in the range of 0.01 to 0.20.

Permeability values were available for 13 petrothermal projects: the lowest value was found for Hijiori in Japan with $10^{-21}$ m$^2$ and the highest one for Newberry with $1.5 \times 10^{-12}$ m$^2$. Hence, the permeability range is 9 orders of magnitude. In addition, the permeability changed significantly for one project: in the case of Newberry, permeability values from $10^{-18}$ to $1.5 \times 10^{-12}$ were found in the literature. The latter value is high enough for HSs, but considering

**Table 5.** Petrophysical properties of hot sedimentary aquifers.

| Project | Rock type | Porosity | Permeability (K) $[m^2]$/Transmissivity (T) $[m^2\,s^{-1}]$ | BHT/reservoir temperature [°C] |
|---|---|---|---|---|
| St. Gallen | Malm, shell limestone, Hirschberg et al. (2015e) | Unknown | Unknown | >145, Brunner and Huwiler (2014) |
| Bruchsal | Bunter sandstone, Herzberger et al. (2010) | Unknown | $T = 8.1 \times 10^{-5}$–$4.0 \times 10^{-3}$, Herzberger et al. (2010) | 120, Herzberger et al. (2010) |
| Landau | Sedimentary and igneous rocks, Atkins (2013) | Unknown | Unknown | 159, Baumgärtner (2012) |
| Insheim | Keuper, perm, bunter sandstone, granite, Baumgärtner (2012) | Unknown | Unknown | 160, Baumgärtner (2012) |
| Neustadt-Glewe | Sandstone, BMU (2011) | Well logging ~0.25, lab ~0.22, BMU (2011) | Well logging $K^i$ ~$1.4 \times 10^{-12}$, laboratory measurements $K^i$ ~$0.5 \times 10^{-12}$, BMU (2011) | 99, Bracke (2012) |
| Unterhaching | Limestone, Dumas (2010) | Unknown | Unknown | 122 and 133, Richter (2010) |
| Southampton*** | Triassic Sherwood Sandstone, Smith (2000) | Unknown | $K^m = 2.63$ to $5.26 \times 10^{-13}$, Atkins (2013), Southampton (2014c) | 76, Smith (2000) |
| Altheim | Limestone, City of Altheim (2014) | 0.08–0.28, ENGINE (2008b) | $T = 1 \times 10^{-4}$–$1 \times 10^{-2}$, ENGINE (2008b) | 106, Bloomquist (2014) |
| Birdsville*** | Unknown | Unknown | | 98, Ergon (2014) |
| Penola*** | Sandstone, Panax Geothermal (2014) | 0.14, Hot Rock Limited (2010) | $K^m = 5.96 \times 10^{-15}$–$1.2 \times 10^{-14}$, Graaf et al. (2010) | 171.4, Graaf et al. (2010) |

[i] permeability calculated from Darcy into $m^2$ under assumption that water temperature is only 10 °C and fresh water;
[m] Permeability calculated from transmissivity in case of known reservoir thickness;
*** Project indicated as HSA in the literature: Southampton, Atkins (2013); Penola, Graaf et al. (2010); Birdsville, RBS Morgans (2009).

the whole permeability range together with other factors such as water not being naturally available, the Newberry project should still be categorized as a PS.

The production flow rate ranges from 4 to 50 L s$^{-1}$. However, flow rates as high as 120 L s$^{-1}$ are expected in Litoměrice in the Czech Republic and 150 L s$^{-1}$ in the case of United Downs in Great Britain.

The most common flow assurance problems were high salt content, high fluid losses, pressure drop, and corrosion.

## 5.2 Hot sedimentary aquifers

For only 5 of the 10 HSA projects, information could be found that stimulation was applied to increase the permeability. For three projects hydraulic fracturing was applied; for two projects both hydraulic and chemical stimulation was conducted. Unterhaching in Germany was the only project where chemical stimulation alone was applied. No information as to whether stimulation techniques were conducted or not could be found for the remaining three projects, which are indicated as HSA in the literature.

The well depth ranges from 1280 to 4450 m for the HSA projects. The encountered rock types are mostly sandstone and limestone and other sedimentary rocks; this is, for instance, the case for Bruchsal. For Birdsville, no information about a rock type could be found. In the cases of Landau and Insheim, igneous rocks were found in addition to sandstone.

Porosity values were found for 3 of the 10 projects only, ranging from 0.08 to 0.28. For three projects, permeability values were given in the literature with a range of $5.96 \times 10^{-15}$ to $1.4 \times 10^{-12}$ m$^2$. The lowest value was found to be for the Penola project in Australia, which was indicated as HSA in the literature.

Production flow rates of 6 to 150 L s$^{-1}$ were found, whereof the highest flow rate was encountered in Unterhaching and the lowest one in St. Gallen (6 L s$^{-1}$). As mentioned before, one of the reasons for abandonment of the latter project was the overly low flow rates. Excluding St. Gallen, the lowest flow rate is 27 L s$^{-1}$.

The most common flow assurance problems were high salt content followed by overly low flow rate and high gas concentration.

## 6 Numerical criteria for classification of deep geothermal potential

Table 7 shows the most typical ranges for different parameters such as permeability, temperature, well depth, rock type, flow rate, stimulation technique, and porosity for both PSs and HSA systems. These values are based on the authors' database and are not meant to be exclusive. The values are quite similar to each other and sometimes the parameter ranges are even overlapping, suggesting that these quantita-

**Table 6.** Operational Characteristics of Hot Sedimentary Aquifers.

| Project | Flow rate [L s⁻¹] | Stimulation techniques | Seismic event (Richter scale) | Type of power plant | Installed electrical capacity [MWe] | Thermal capacity [MWth] | Flow assurance problem |
|---|---|---|---|---|---|---|---|
| St. Gallen | 6 to 12, Brunner and Huwiler (2014) | Chemical stimulation, Hirschberg et al. (2015e) | 3.5 M, Hirschberg et al. (2015d) | None* | 0* | 0* | Overly low flow rate; gas flow during production tests, Brunner and Huwiler (2014) |
| Bruchsal | 28.5, Herzberger et al. (2010) | Unknown | Microseismic, Rettenmaier (2012) | Kalina, Herzberger et al. (2010) | 0.55, Herzberger et al. (2010) | 5.5, GtV (2014a) | High salt contents (100 g/l); high $CO_2$ concentration, Herzberger et al. (2010) |
| Landau | 70 to 80, Baumgärtner (2012) | No stimulation for producer; hydraulic for injector, Baumgärtner (2012) | Microseismic (≤2.7 Mj), Baumgärtner et al. (2012) | ORC, Ganz et al. (2013) | Up to 3.6, Baumgärtner (2012) | 2 to 5, Baumgärtner (2012) | Well leakage resulting in groundwater contamination, Geothermie-Nachrichten (2014) |
| Insheim | 60–85, Baumgärtner (2012) | Yes, Baumgärtner (2012) | M: 2.0 to 2.4 and Micro-seismic, Groos et al. (2012) | ORC, Ganz et al. (2013) | 5, Ganz et al. (2013) | Planned; ca. 6 available, Baumgärtner et al. (2013); Ganz et al. (2013); Baumgärtner (2012) | Unknown |
| Neustadt-Glewe | 35, Bracke (2012) | Unknown | Unknown | ORC, Bracke (2012) | 0, Ganz et al. (2013) | 7, GtV (2014c) | High salt content, high gas concentration, Bracke (2012) |
| Unterhaching | 150, Richter (2010) | Acidizing, BMU (2011) | Unknown | Kalina, Ganz et al. (2013) | 3.36, Richter (2010) | 38, Richter (2010) | Unknown |
| Southampton*** | 35, Atkins (2013) | Unknown | Unknown | CHP plant, Southampton (2014b) | 2.7, Southampton (2014d) | Heat 23; chilled water 10.5, Southampton (2014d) | Unknown |
| Altheim | 81.7, City of Altheim (2014) | Chemical, Pernecker (1999), GINE (2008b) hydraulic, EN- | Unknown | ORC, City of Altheim (2014) | 1.0, Bloomquist (2014) | 12.4, City of Altheim (2014) | Clogging by a mixture consisting of stone material and bentonite, Pernecker (1999) |
| Birdsville*** | 27, Ergon (2014) | Unknown | Unknown | ORC, Ergon (2014) | 0.08, Ergon (2014) | Non scheduled, Ergon (2014) | Unknown |
| Penola*** | Unknown at this stage | Unknown | Unknown | None* | 59 MW (planned), Proactive Investors (2009) | 0* | Initial mud damage during drilling; solved with acidizing, Graaf et al. (2010) |

* See status;
*** Project indicated as HSA in the literature;
CHP – Combined Heat Power, ORC – Organic Rankine Cycle.

**Table 7.** Typical parameter ranges for petrothermal systems and hot sedimentary aquifers.

| Parameter | Petrothermal | Hot sedimentary aquifer |
|---|---|---|
| Permeability | $10^{-19}$–$10^{-14}$ m$^2$ | $10^{-15}$–$10^{-12}$ m$^2$ |
| Temperature | 130–400 °C | 76–171.4 °C |
| Well depth | 1800–5000 m | 1280–4450 m |
| Rock type | Igneous | Sedimentary |
| Stimulation | Hydraulic | Hydraulic and/or chemical |
| Porosity | 0.01–0.25 | 0.08–0.28 |
| Flow Rate | 4–50 L s$^{-1}$ | 27–150 L s$^{-1}$ |

tive parameters may not be used to differentiate PSs from HSA systems.

Additional important parameters such as productivity index and the productivity enhancement factor resulting from stimulation were unavailable in the public domain for most of the projects.

## 7 Conclusions

Over the past 40 years, more and more geothermal system classifications such as hot dry rock, enhanced or engineered geothermal systems, hot wet rock, hot fractured rock, and HSA systems have been defined in order to better characterise geothermal projects. However, some of these definitions are deceptive, such as that for deep heat mining, which suggests that the geothermal heat is mined and therefore not available anymore after the geothermal production. Other definitions (such as those for EGS) are not specific, as they provide only the information that the geothermal system was somehow enhanced by some technical measure such as water supply, stimulation of the reservoir etc.

This study recommends re-introducing three known definitions such as petrothermal, hydrothermal, as well as HSA, and abandoning the ambiguous terminology such as EGS. This threefold classification provides more information compared to the defined EGS, which is unfortunately quite common nowadays.

The definition of petrothermal already includes the information that not enough water is contained in the subsurface and thus water has to be supplied and re-injected after geothermal production. Hence, more than one well is required for the project. However, this is not a distinctive criterion, as most of HSs and HSA systems consist of two wells, with exception of Birdsville and Southampton. In addition, the permeability is too low for the production well and therefore hydraulic fracturing has to be applied as stimulation in order to create an artificial reservoir. PSs, which sit in the first-proposed category, indicate a conduction-dominated heat source. Based on the authors' own database, typical permeability ranges are in the order of $10^{-19}$ to $10^{-14}$ m$^2$ , the most common formation type is igneous such as granite, well depth is typically more than 1800 m, and hydraulic stimulation has to be applied in order to create an artificial reservoir. The temperature of investigated petrothermal projects varies significantly with typical ranges between 130 and 400 °C.

On the other hand, the definition of hydrothermal informs us that a geothermal reservoir, with high enough permeability and sufficient water supply, is already available and that (usually) no stimulation needs to be applied, but the project might be improved if formation damage is reduced via more careful drilling or the near-wellbore region is stimulated. HSs, which occupy the second proposed category, can be managed with only one well if the water is additionally used for other purposes such as balneology. However, for sustainability and to maintain high pressure in the reservoir, it might be required to re-inject the produced water, which would mean that a second well would be necessary. Re-injection might also be necessary in case of the water being saline to avoid environmental risks. HSs indicate a convection-dominated or an advection-dominated heat source.

The third proposed category is HSA systems. These systems are similar to common and conventional HSs with the difference being that the heat supply is conduction-dominated and the heat source is similar to PSs, such as high heat producing granites seen in the Australian HSA systems. The analysis of HSA projects resulted in the following typical parameter ranges: permeability from $10^{-15}$ to $10^{-12}$ m$^2$, temperature from 76 to 171.4 °C, well depth between 1280 and 4450 m,; reservoir rock types are typically sedimentary, such as sandstone and limestone.

**Acknowledgements.** The authors would like to acknowledge the financial support of the Open Access Publishing Fund at Clausthal University of Technology to publish this paper.

## References

AGRCC: Australian Geothermal Reporting Code Committee: Geothermal lexicon for resources and reserves definition and reporting, 2nd edn. Australian Geothermal Reporting Code Committee, Adelaide, p. 67, 2010.

Atkins Ltd: Deep Geothermal Review Study, Version 5.0, 21 October 2013.

Baria, R., Bennett, T., Macpherson-Grant, G., Baumgaertner, J., and Jupe, A.: Cornish Rocks – Hotting Up? European Geothermal Congress (EGC) 2013, Pisa, Italy, 3–7 June 2013.

Barnet, P.: Large scale hot sedimentary aquifer (HSA) geothermal projects, Presentation to Victoria Energy Conference, Melbourne, 27 August 2009.

Baumgärtner, J.: Insheim and Landau – recent experiences with EGS technology in the Upper Rhine Graben, oral presentation presented at ICEGS 2012, Freiburg, 25 May 2012.

Beardsmore, G. and Matthews, C.: Parachilna Geothermal Play, Statement of Inferred Geothermal Resources, Torrens Energy Limited, 15 August 2008.

Bendall, B., Hogarth, R., Holl, H., McMahon, A., Larking A., and Reid, P.: Australian Experiences in EGS Permeability Enhancement – A Review of 3 Case Studies. PROCEEDINGS, Thirty-Ninth Workshop on Geothermal Reservoir Engineering, SGP-TR-202, Stanford University, Stanford, California, 24–26 February 2014.

Bertani, R.: Geothermal power generation in the world 2005-2010 update report, Geothermics, 41, 1–29, 2012.

BGR (2014a): GeneSys Horstberg, http://www.genesys-hannover.de/Genesys/EN/Horstberg/horstberg_node_en.html, last access: 2 June 2014.

BGR (2014b): Milestones of the Genesys project, http://www.bgr.bund.de/Genesys/EN/Meilensteine/meilensteine_inhalt_en.html, last access: 2 June 2014.

BGR (2014c): GeneSys, http://www.genesys-hannover.de/Genesys/DE/Home/genesys_node.html, last access: 10 June 2014

BINE (2012): Korrosion in geothermischen Anlagen. http://www.bine.info/service/bestellen/download-print/publikation/korrosion-in-geothermischen-anlagen/korrosion-und-materialqualifizierung/, last access: 2 June 2014.

Blöcher, G., Zimmermann, G., Moeck, I., and Huenges, E.: Groß-Schönebeck (D) – the development of the learning curve: experience from the projects of recent years. Oral presentation presented at ICEGS 2012, Freiburg, 25 May 2012.

Bloomquist, R.: Integrating small power plants into agricultural projects, pangea.stanford.edu/ERE/pdf/IGAstandard/EGC/szeged/I-8-01.pdf, last access: 3 June 2014.

BMU: Tiefe Geothermie – Nutzungsmöglichkeiten in Deutschland, Beltz Bad Langensalza GmbH, BT Weimar, Bundesministerium für Umwelt, Naturschutz und Reaktorsicherheit, Referat Öffentlichkeitsarbeit, Berlin, 2011.

Bracke, R.: Geothermal energy – low enthalpy technologies. Oral presentation presented at Congreso Nacional de Energia 2012, CICR, San Jose/Costa Rica, 15–16 February 2012.

Breede, K., Dzebisashvili, K., Liu, X., and Falcone, G.: A systematic review of enhanced (or engineered) geothermal systems: past, present and future, Geotherm Energ., 1, doi:10.1186/2195-9706-1-4, 2013.

Bridgland, D.: United Downs Deep Geothermal Project, Progress Summary of Geothermal Engineering Ltd., 17 November 2011.

Bromley, C. J. and Mongillo, M. A.: Geothermal energy from fractured reservoirs – dealing with induced seismicity. In: IEA OPEN Energy Technology Bulletin, Issue No. 48, IEA, http://www.iea.org/impagr/cip/pdf/Issue48Geothermal.pdf (last access: 02 June 2014), 2008.

Broßmann, E. and Koch, M.: First Experiences with the Geothermal Power Plant in Neustadt-Glewe (Germany) – Proceedings World Geothermal Congress 2005, Antalya, Turkey, 24–29 April 2005.

Brown, D.: The US Hot Dry Rock program – 20 years of experience in reservoir testing. Paper presented at world geothermal congress 1995, Firenze, 18–31 May 1995.

Brown, D. (2009): Hot Dry Rock Geothermal Energy: important lessons from Fenton Hill. Paper presented at thirty-fourth workshop on geothermal reservoir engineering, Stanford University, Stanford, 9–11 Feb 2009.

Brunner, F. and Huwiler, M.: Geothermie-Projekt St.Gallen, Ergebnis Produktionstests, weitere Schritte,Medienmitteilung der Stadt St. Gallen, retrieved from: http://www.geothermie.stadt.sg.ch/fileadmin/downloads/medienmitteilungen/140213_MK_Geothermie.pdf (ast access: 17 March 2015), 13 February 2014.

Canaris, J. (2009): Company Overview. Torrens Energy Limited. November 2009

City of Altheim (2014): City of Altheim – Geothermal energy supply. http://engine.brgm.fr/web-offlines/conference-Electricity_generation_from_Enhanced_Geothermal_Systems_-_Strasbourg,_France,_Workshop5/other_contributions/27-slides-0-2_Altheim_pdf.pdf last access: 11 December 2014.

Cladouhos, T. T., Osborn, W. L., Petty, S., Bour, D., Iovenitti, J., Callahan, O., Nordin, Y., Perry, D., and Stern, P. L.: Newberry volcano EGS demonstration – phase I results. In: Proceedings of thirty-seventh workshop on geothermal reservoir engineering, Stanford University, Stanford, 30 January–1 February 2012.

cleanenergyaus: Geothermal Energy Sources, http://cleanenergyaus.com.au/technology.htm, last access: 18 June 2014.

Cornet, F. H.: Results from Le Mayet De Montagne Project, Geothermics, 16, 355–374, 1987.

Cornet, F. H.: The learning curve: the Le Mayet de Montagne experiment (1978–1987), Oral presentation presented at ICEGS 2012, Freiburg, 25 May 2012.

Cuenot, N.: Overview of Microseismic activity observed at the Soultz-sous-Forêts EGS power plant under stimulation and circulation conditions, 75th EAGE conference and exhibition, Workshop 16, Microseismicity – What now? What next?, London, UK, 13–16 June 2013.

Cummings, R. G. and Morris, G. E.: Economic modelling of electricity production from Hot Dry Rock geothermal reservoirs: methodology and analysis. EA-630, Research Project 1017 LASL (LA-7888-HDR). OSTI Information Bridge., http://www.osti.gov/bridge/servlets/purl/5716131-wg4gUV/native/5716131.pdf (last access: 31 May 2013), 1979.

DiPippo, R.: Geothermal power plants, 3rd edn, Elsevier, New York, 451–456, 2012.

Duchane, D.: The history of HDR research and development, in: Draft proceedings of the 4th international HDR forum, Strasbourg, 28–30 September 1998.

Dumas, P.: NER300: what for geothermal? Oral presentation held at second EGEC TP geoelec meeting, Brussels, 24 March 2010.

EGEC: EGEC Market report 2013/2014, third edition, December 2013.

ENGINE (): ENGINE – geothermal lighthouse projects in Europe, last update: April 2008, http://engine.brgm.fr/mediapages/lighthouseProjects/LH-Quest_EGS_4_GeneSys.pdf (last access: 27 May 2014), 2008a.

ENGINE: ENGINE coordination action. Best practice handbook for the development of unconventional geothermal resources with a focus on enhanced geothermal system, BRGM, Orleans, ISBN 978-2-7159-2482-6, 40–41, 2008b.

Ergon: Birdsville Organic Rankine Cycle Geothermal Power Station. https://www.ergon.com.au/__data/assets/pdf_file/0008/

4967/EGE0425-birdsville-geothermal-brochure-r3.pdf, last access: 13 May 2014.

Exorka: F&E-Projekt Geothermie Allgäu 2.0., http://cif-ev.de/pdf/block2/P1.pdf, last access: 28 May 2014.

Fittermann, D. V.: Overview of the structure and geothermal potential of Newberry Volcano, Oregon, J. Geophys. Res., 93, 10059–10066, 1988.

Flores, M., Davies, D., Couples, G., and Palsson, B.: Stimulation of Geothermal Wells, Can We Afford It?, Proceedings World Geothermal Congress 2005, Antalya, Turkey, 24–29 April 2005.

Ganz, B., Schellschmidt, R., Schulz, R., and Sanner, B.: Geothermal Energy Use in Germany, European Geothermal Congress 2013 Pisa, Italy, 3–7 June, 2013.

Garcia, J., Walters, M., Beall, J., Hartline, C., Pingol, A., Pistone, S., and Wright, M.: Overview of the Northwest Geysers EGS Demonstration Project. In: Proceedings of the thirty-seventh workshop on geothermal reservoir engineering (ed) Proceedings of the thirty-seventh workshop on geothermal reservoir engineering. Stanford University, Stanford, 30 January–1 February 2012.

GeneSys Hannover: In-situ Experiments. http://www.genesys-hannover.de/Genesys/EN/Forschung/In-Situ/in-situ_inhalt_en.html (last access: 28 April 2014), 2014a.

GeneSys Hannover: Aktuelles. http://www.genesys-hannover.de/Genesys/DE/Aktuelles/aktuelles_node.html (last access: 28 May 2014), 2014b.

Genter, A.: Lessons learned from projects in the early stage of EGS development: Soultz-sous-Fôrets (F), oral presentation held at ICEGS 2012, Freiburg, 25 May 2012.

Genter, A., Guillou-Frottier, L., Feybesse, J.-L., Nicol, N., Dezayes, C., and Schwartz, S.: Typology of potential hot fractured rock resources in Europe, Geothermics, 32, 701–710, 2003.

Genter, A., Evans, K., Cuenot, N., Fritsch, D., and Sanjuan, B.: Contribution of the exploration of deep crystalline fractured reservoir of Soultz to the knowledge of enhanced geothermal systems (EGS), CR GEOSCI, 342, 502–516, 2010.

Geodynamics: Innamincka (EGS) Project, urlhttp://www.geodynamics.com.au/Our-Projects/Innamincka-Deeps.aspx, last access: 30 May 2014.

Geoscience Australia and ABARE: Australian Energy Resource Assessment, Canberra, Commonwealth of Australia, 215–216, 2010.

Geothermal Resources: Frome Project (North Eastern SA), http://www.geothermal-resources.com.au/project_frome.html, last access: 26 May 2014.

Geothermie-Nachrichten: Landau: Grundwasser am Geothermie-Kraftwerk verunreinigt, http://www.geothermie-nachrichten.de/landau-grundwasser-am-geothermie-kraftwerk-verunreinigt, last access: 13 June 2014.

Geothermie Stadt St. Gallen: Das Geothermie-Projekt der Stadt St. Gallen, http://www.geothermie.stadt.sg.ch/projekt.html, last access: 28 May 2014.

Giardini, D.: Geothermal quake risks must be faced, Nature, 462, 848–849, 2009.

Giles, C.: Frome Project – Statement of Estimated Geothermal Resources, Media Release on 13th of July 2009 from Geothermal Resources Limited, retrieved from: http://www.geothermal-resources.com.au/pdf/Resource_Statement_130709.pdf (last access: 17 March 2015), 2009.

Goldstein, B. A., Bendall, B., and Long, A.: Australian 2009 Annual Country Report for the International Energy Agency Geothermal Implementing Agreement, Appendix A: Company Status Updates for 2009, Australian Geothermal Energy group; retrieved from: http://www.geothermal.dmitre.sa.gov.au/__data/assets/pdf_file/0018/156114/2009_Australian_Country_Rpt_Final_.pdf (last access: 16 March 2015) 13 August 2010.

Graaf, B., Reid, I., Palmer, R., Jenson, D., and Parker, K.: Salamander-1 – a geothermal well based on petroleum exploration results, edited by: Gurgenci, H. and Weber, R. D., Proceedings of the 2010 Australian Geothermal Energy Conference, Geoscience Australia, Record 2010/35, 2010.

Grant, M. A. and Bixley, P. F.: Geothermal Reservoir Engineering. 2nd edition, Elsevier Inc, 478–479, 2011.

Grassiani, M., Krieger, Z., and Legmann, H.: Advanced power plants for use with HDR/enhanced geothermal technology, Bulletin D'Hydrogéologie No. 17, 165–172, 1999.

Green Rock Energy: Olympic Dam (100 % owned), http://www.greenrock.com.au/assetsSAOlympicDam.php, last access: 25 May 2014.

Groos, J. C., Grund, M., and Ritter, J. R. R.: Automated detection of microseismic events in the Upper Rhine valley near the city of Landau/South Palatinate. Geophys. Res. Abstr., 14, EGU2012-10482, EGU General Assembly 2012.

Gryndler, P.: City of Litoměřice Geothermal Project http://www.eeportal.mk/gallery/files/Day_4_-_Geothermal_and_solar_energy.pdf, 2009.

GtV (Bundesverband Geothermie): Liste der tiefen Geothermieprojekte in Deutschland, http://www.geothermie.de/wissenswelt/geothermie/in-deutschland.html (last access: 3 June 2014), 2014a.

GtV (Bundesverband Geothermie) (): Zügige Ursachenforschung in Landau, 11 April 2014 – Tiefe Geothermie, http://www.geothermie.de/news-anzeigen/2014/04/11/zugige-ursachenforschung-in-landau.html (last access 29 May 2014), 2014b.

GtV (Bundesverband Geothermie) (2014c): Petrothermale Geothermie, http://www.geothermie.de/wissenswelt/geothermie/technologien/petrothermale-systeme.html (last access: 13 June 2014), 2014c.

GtV (Bundesverband Geothermie) (2014d): Hydrothermale Geothermie, http://www.geothermie.de/wissenswelt/geothermie/technologien/hydrothermale-systeme.html (last access: 13 June 2014), 2014d.

Häring, M. O.: Geothermische Stromproduktion aus Enhanced Geothermal Systems (EGS) Stand der Technik, www.geothermal.ch/fileadmin/docs/downloads/egs061207.pdf (last access: 30 January 2013), 2007.

Henninges, J., Brandt, W., Erbas, K., Moeck, I., Saadat, A., Reinsch, T., and Zimmermann, G.: Downhole monitoring during hydraulic experiments at the in-situ geothermal lab Groß Schönebeck, in: Proceedings of the thirty-seventh workshop on geothermal reservoir engineering, Stanford University, SProceedings, Thirty-Seventh Workshop on Geothermal Reservoir Engineering, SGP-TR-194, Stanford University, Stanford, California, 30 January–1 February 2012.

Herzberger, P., Münch, W., Kölbel, T., Bruchmann, U., Schlagermann, P., Hötzl, H., Wolf, L., Rettenmaier, D., Steger, H., Zorn,

R., Seibt, P., Möllmann, G.-U., Sauter, M., Ghergut, J., and Ptak, T.: The Geothermal Power Plant Bruchsal. Proceedings World Geothermal Congress 2010, Bali, Indonesia, 25–29 April 2010.

Hirschberg, S., Wiemer, S., and Burgherr, P. (eds.): Energy from the Earth – Deep Geothermal as a Resource for the Future? Vdf Hochschulverlag AG an der ETH Zürich, ISBN 978-3-7281-3655-8/, p. 32, doi:10.3218/3655-8, 2015a.

Hirschberg, S., Wiemer, S., and Burgherr, P. (eds.): Energy from the Earth – Deep Geothermal as a Resource for the Future? Vdf Hochschulverlag AG an der ETH Zürich, ISBN 978-3-7281-3655-8/, p. 274, doi:10.3218/3655-8, 2015b.

Hirschberg, S., Wiemer, S., and Burgherr, P. (eds.): Energy from the Earth – Deep Geothermal as a Resource for the Future? Vdf Hochschulverlag AG an der ETH Zürich, ISBN 978-3-7281-3655-8/, p. 19, doi:10.3218/3655-8, 2015c.

Hirschberg, S., Wiemer, S., and Burgherr, P. (eds.): Energy from the Earth – Deep Geothermal as a Resource for the Future? Vdf Hochschulverlag AG an der ETH Zürich, ISBN 978-3-7281-3655-8/, p. 130, doi:10.3218/3655-8, 2015d.

Hirschberg, S., Wiemer, S., and Burgherr, P. (eds.): Energy from the Earth – Deep Geothermal as a Resource for the Future? Vdf Hochschulverlag AG an der ETH Zürich, ISBN 978-3-7281-3655-8/, p. 278–282 doi:10.3218/3655-8, 2015e.

Holl, H.: Geodynamics Update: Innamincka Deeps EGS Project, oral presentation held at ICEGS 2012, Freiburg, 25 May 2012.

Hot Rock Limited (2010): Annual Report 2010.

Huddlestone-Holmes, C. R. and Hayward, J.: The potential of geothermal energy, http://www.garnautreview.org.au/update-2011/commissioned-work/potential-of-geothermal-energy.pdf (last access: 17 June 2014), 2011.

Huddlestone-Holmes, C. R. and Russel, C.: AEMO 100 % Renewable Energy Study: Geothermal Energy, CSIRO, Newcastle, Australia, p. 10, 2012.

Huenges, E.: Geothermal energy systems – exploration, development, and utilization. Wiley, Weinheim, p. 22, 2010a.

Huenges, E.: Geothermal energy systems – exploration, development, and utilization. Wiley, Weinheim, 27–30, 2010b.

Huenges, E.: Geothermal energy systems – exploration, development, and utilization. Wiley, Weinheim, 184–185, 2010c.

IGA R&R: International Geothermal Association, Resources and Reserves adhoc-committee, internal committee discussion, November 2013.

iTG: HDR-Projekt in Bad Urach vor dem Aus. 15 December 2010. http://www.tiefegeothermie.de/news/hdr-projekt-in-bad-urach-vor-dem-aus (last access: 30 May 2014), 2010.

iTG: Mauerstetten soll als Forschungsprojekt neu erschlossen werden., http://www.tiefegeothermie.de/news/mauerstetten-soll-als-forschungsprojekt-neu-erschlossen-werden (last access: 2 June 2014), 2013.

Jung, R.: EGS – Goodbye or Back to the Future, Effective and Sustainable Hydraulic Fracturing, Dr. Rob Jeffrey (Ed.), ISBN: 978-953-51-1137-5, InTech, doi:10.5772/56458, available from: http://www.intechopen.com/books/effective-and-sustainable-hydraulic-fracturing/egs-goodbye-or-back-to-the-future-95, 2013.

Jupe, A. J., Green, A. S. P., and Wallroth, T.: Induced microseismicity and reservoir growth at the Fjällbacka hot dry rocks project,

Sweden. Int J Rock Mechanics Mining Sci Geomechanics Abstracts 29, 343–354, 1992.

Kaieda, H., Ito, H., Kiho, K., Suzuki, K., Suenaga, H., and Shin, K.: Review of the Ogachi HDR project in Japan, in: Proceedings of the world geothermal congress 2005, Antalya, 24–29 April 2005.

Kaieda, H., Sasaki, S., and Wyborn, D.: Comparison of characteristics of micro-earthquakes observed during hydraulicstimulation operations in Ogachi, Hijiori and Cooper Basin HDR projects, in: Proceedings of the World Geothermal Congress 2010, Bali, 25–29 April 2010.

Kappelmeyer, O. and Jung, R.: HDR experiments at Falkenberg/Bavaria, Geothermics, 16, 375–392, 1987.

Kohl, T., Bächler, D., and Rybach, L.: Steps towards a comprehensive thermo-hydraulic analysis of the HDR test site Soultz-sous-Forêts, Proceedings World Geothermal Congress 2000, Kyushu-Tohoku, Japan, 28 May–10 June 2000.

Kreuter, H.: Deep geothermal projects in Germany – status and future development, Paris, 5 April 2011, oral presentation, 5 April 2011; retrieved from: http://www.renewablesb2b.com/data/shared/GEO_PPT_Kreuter.pdf (last access: 17 March 2015), 2011a.

Kreuter, H.: Definition on EGS. http://egec.info/wp-content/uploads/2011/08/3-TP-GEOELEC-PISA-Kreuter-.pdf (last access: 17 June 2014), 2011b.

Kruger, P.: Heat extraction from Microseismic estimated geothermal reservoir volume, GRC Transactions, 14, Part II, 1225–1232, 1990.

Ladner, F. and Häring, M. O.: Hydraulic characteristics of the basel 1 enhanced geothermal system, GRC Trans. 33, 199–203, 2009.

Ledésert, B., Hebert, R., Genter, A., Bartier, D., Clauer, N., and Grall, C.: Fractures, hydrothermal alterations and permeability in the Soultz Enhanced Geothermal System, Compt. Rend. Geosci., 342, 607–615, 2010.

Lee, T. J., Song, Y., Yoon, W. S., Kim, K., Jeon, J., Min, K., and Cho, Y.: The first Enhanced Geothermal System Project in Korea. Proceedings of the 9th Asian Geothermal Symposium, 7–9 November 2011.

Lovelock, B.: Statement of the Updated Geothermal Reserves and Resources Estimates as at 31 October 2011, Green Rock Energy Ltd report, 2011.

Majer, E., Baria, R., Stark, M., Oates, S., Bommer, J., Smith, B., and Asanuma, H.: Induced seismicity associated with enhanced geothermal systems, Geothermics, 36, 185–222, 2007.

Meyer, G., Larking, A., Jeffrey, R., and Bunger, A.: Olympic Dam EGS Project. Proceedings World Geothermal Congress 2010, Bali, Indonesia, 25–29 April 2010.

MIT (Massachusetts Institute of Technology), Tester, J. W., Anderson, B. J., Batchelor, A. S., Blackwell, D. D., DiPippo, R., Drake, E. M., Garnish, J., Livesay, B., Moore, M. C., Nichols, K., Petty, S., Toksöz, M. N., and Veatch Jr., R. W.: The future of geothermal energy – impact of enhanced geothermal systems on the United States in the 21st Century, US Department of Energy, Washington, D.C., complete report, 2006a.

MIT (Massachusetts Institute of Technology), Tester, J. W., Anderson, B. J., Batchelor, A. S., Blackwell, D. D., DiPippo, R., Drake, E. M., Garnish, J., Livesay, B., Moore, M. C., Nichols, K., Petty, S., Toksöz, M. N., and Veatch Jr., R. W.: The future of geothermal energy – impact of enhanced geothermal systems on the United

States in the 21st Century, US Department of Energy, Washington, D.C., 4–40, 2006b.

MIT (Massachusetts Institute of Technology), Tester, J. W., Anderson, B. J., Batchelor, A. S., Blackwell, D. D., DiPippo, R., Drake, E. M., Garnish, J., Livesay, B., Moore, M. C., Nichols, K., Petty, S., Toksöz, M. N., and Veatch Jr., R. W.: The future of geothermal energy – impact of enhanced geothermal systems on the United States in the 21st Century, US Department of Energy, Washington, D.C., 4-22–4-26, 2006c.

MIT (Massachusetts Institute of Technology), Tester, J. W., Anderson, B. J., Batchelor, A. S., Blackwell, D. D., DiPippo, R., Drake, E. M., Garnish, J., Livesay, B., Moore, M. C., Nichols, K., Petty, S., Toksöz, M. N., and Veatch Jr., R. W.: The future of geothermal energy – impact of enhanced geothermal systems on the United States in the 21st Century, US Department of Energy, Washington, D.C., 4–37, 2006d.

MIT (Massachusetts Institute of Technology), Tester, J. W., Anderson, B. J., Batchelor, A. S., Blackwell, D. D., DiPippo, R., Drake, E. M., Garnish, J., Livesay, B., Moor,e M. C., Nichols, K., Petty, S., Toksöz, M. N., and Veatch Jr., R. W.: The future of geothermal energy – impact of enhanced geothermal systems on the United States in the 21st Century, US Department of Energy, Washington, D.C., 4-14–4-18, 2006e.

MIT (Massachusetts Institute of Technology), Tester, J. W., Anderson, B. J., Batchelor, A. S., Blackwell, D. D., DiPippo, R., Drake, E. M., Garnish, J., Livesay, B., Moore M. C., Nichols, K., Petty, S., Toksöz, M. N., and Veatch Jr., R. W.: The future of geothermal energy – impact of enhanced geothermal systems on the United States in the 21st Century, US Department of Energy, Washington, D.C., chapter 4.3, 4-7–4-13, 2006f.

MIT (Massachusetts Institute of Technology), Tester, J. W., Anderson, B. J., Batchelor, A. S., Blackwell, D. D., DiPippo, R., Drake, E. M., Garnish, J., Livesay, B., Moore, M. C., Nichols, K., Petty, S., Toksöz, M. N., and Veatch Jr., R. W.: The future of geothermal energy – impact of enhanced geothermal systems on the United States in the 21st Century. US Department of Energy, Washington, D.C., 4–38, 2006g.

MIT (Massachusetts Institute of Technology), Tester, J. W., Anderson, B. J., Batchelor, A. S., Blackwell, D. D., DiPippo, R., Drake, E. M., Garnish, J., Livesay, B., Moore, M. C., Nichols, K., Petty, S., Toksöz, M. N., and Veatch Jr., R. W.: The future of geothermal energy – impact of enhanced geothermal systems on the United States in the 21st Century. US Department of Energy, Washington, D.C., p. 4, 2006h.

Mortimer, L., Cooper, G., and Beardsmore, G.: As assessment of the Geothermal Energy Potential of Hot Sedimentary Aquifers (HSA). http://www.hotdryrocks.com/component/option,com_docman/task,doc_download/gid,79/Itemid,71/ (last access: 17 June 2014), 2010.

Nag, P. K.: Power plant engineering, 3rd edition. Tata McGraw-Hill Publishing Company Limited, New Delhi, 2008.

Newberry EGS Demonstration: Stimulation and Microseismicity, 12 November 2014, http://blog.newberrygeothermal.com/, last access: 10 December 2014.

newworldenergy: Styles of geothermal energy. http://newworldenergy.com.au/index.php/about-geothermal/styles-of-geothermal-energy/, last access: 17 June 2014.

Orzol, J., Jung, R., Jatho, R., Tischner, T., and Kehrer, P.: The GeneSys-Project: Extraction of Geothermal Heat from Tight Sediments. Proceedings World Geothermal Congress 2005, Antalya, 2005.

Panax Geothermal: Projects. Australia. Penola. http://www.panaxgeothermal.com.au/projects-domestic-otway-penola.htm, last access: 27 May 2014.

Parker, R.: The Rosemanowes HDR project 1983–1991, Geothermics, 28, 603-615, 1999.

Pernecker, G.: Altheim geothermal plant for electricity production by ORC-turbogenerator, edited by: Peter. L., in: Bulletin d'hydrogéologie No 17. Centre d'Hydrogéologie. Université de Neuchâtel, Altheim, Austria, 1999.

Petratherm: Paralana, http://www.petratherm.com.au/projects/paralana last access: 30 May 2014.

Pfalzwerke-geofuture: Projekt Insheim., http://www.geothermie-insheim.de/index.php/das-kraftwerk, last access: 29 May 2014.

PK Tiefe Geothermie: Nutzungen der geothermischen Energie aus dem tiefen Untergrund (Tiefe Geothermie) – Arbeitshilfe für Geologische Dienste, http://www.infogeo.de/dokumente/download_pool/tiefe_geothermie_arbeitshilfe_08022007.pdf (last access: 17 June 2014), 2007.

Portier, S. and Vuataz, F.-D. (eds.): Studies and support for the EGS reservoirs at Soultz-sous-Forêts, final report April 2004–May 2009, Project financed by State Secretariat for Education and Research (SER/SBF) and Swiss Federal Office of Energy (OFEN/BFE), 2009.

Portier, S., André, L., and Vuataz, F.-D.: Review on chemical stimulation techniques in oil industry and applications to geothermal systems. Technical report in enhanced geothermal innovative network for Europe, CREGE – Centre for Geothermal Research, Neuchâtel, 2007.

Potter, R., Robinson, E., and Smith, M.: Method of extracting heat from dry geothermal reservoirs, US Patent No. 3,786–858, USA, Los Alamos, New Mexico, 1974.

Proactive Investors: Panax Geothermal study confirms commerciality of Penola Geothermal project to supply power. 20 August 2009, http://www.proactiveinvestors.com.au/companies/news/2322/ (last access: 27 May 2014), 2009.

RBS Morgans Ltd; retrieved from: http://newworldenergy.com.au/wp-content/uploads/2010/02/RBS-Morgans-Research-Report.pdf (last accessed: 17 March 2015), 2009.

Reid, P. W. and Messeiller, M.: Paralana Engineered Geothermal Systems Project 3.5 MW Development Plan, Proceedings Australian Geothermal Energy Conferences 2013, Brisbane, Australia, 14–15 November 2013.

Rettenmaier, D. (): Lessons Learned - Reservoirmanagement Bruchsal., http://www.ta-survey.nl/pdf/GU2012-Detlev_Rettenmaier.pdf (last access: 3 June 2014), 2012.

Richter, B.: Geothermal Energy Plant Unterhaching, Germany, Proceedings World Geothermal Congress 2010, Bali, Indonesia, 25–29 April 2010.

Roberts, V. and Kruger, P.: Utility Industry Estimates of Geothermal Electricity – Geothermal power production to continue rapid growth through the year 2000, Geothermal Resources Council Bulletin, 7–10, 1982.

Romero Jr., A., McEvilly, T. V., Majer, E., and Vasco, D.: Characterization of the geothermal system beneath the Northwest Geysers

Steam Field, California, from seismicity and velocity patterns, Geothermics, 24, 471–487, 1995.

Rutqvist, J., Dobson, P. F., Garcia, J., Hartline, C., Jeanne, P., Oldenburg, C. M., Vasco, D. V., and Walters, M.: The Northwest Geysers EGS Demonstration Project, California: Pre-stimulation Modeling and Interpretation of the Stimulation, Math. Geosci., 47, 3–29, 2013.

Sasaki, S.: Characteristics of microseismic events induced during hydraulic fracturing experiments at the Hijiori hot dry rock geothermal energy site, Yamagata, Japan, Tectonophysics, 289, 171–188, 1998.

Sass, I. and Goetz, A. E.: The thermofacies concept. Proceedings, Thirty-Sixth Workshop on Geothermal Reservoir Engineering, SGP-TR-191, Stanford University, Stanford, California, 31 January–2 February 2011.

Schanz, U., Stang, H., Tenzer, H., Homeier, G., Hase, M., Baisch, S., Weidler, R., Macek, A., and Uhlig, S. Hot dry rock project Urach – a general overview, in: Proceedings of the European geothermal conference, Szeged, 25–30 May 2003.

Schill, E., Genter, A. and Soultz team: EGS geothermal challenges within the Upper Rhine Valley based on the Soultz experience. Third European Geothermal Review (TEGR), Mainz, Germany, 24–26 June 2013.

Schindler, M., Baumgärtner, J., Gandy, T., Hauffe, P., Hettkamp, T., Menzel, H., Penzkofer, P., Teza, D., Tischner, T., and Wahl, G.: Successful Hydraulic Stimulation Techniques for Electric Power Production in the Upper Rhine Graben, Central Europe, Proceedings World Geothermal Congress 2010, Bali, Indonesia, 25–29 April 2010.

Schrage, C., Bems, C., Kreuter, H., Hild, S., and Volland, S.: Overview of the enhanced geothermal energy project in Mauerstetten, Germany, http://ta-survey.nl/pdf/120213_Geothermie_Update_EGS_Mauerstetten.pdf (last access: 2 June 2014), 2012a.

Schrage, C., Bems, C., Kreuter, H., Hild, S., and Volland, S.: Geothermie Allgäu 2.0 – overview of the enhanced geothermal energy project in Mauerstetten, oral presentation held at Amsterdam, Germany, 18 April 2012, 2012b.

Schulz, R.: Nutzung petrothermaler Technik – Vorschlag für eine Definition für die Anwendung des EEG, http://www.liag-hannover.de/fileadmin/produkte/20081126095553.pdf (last access: 17 June 2014), 2008.

Schulz, R.: Nutzung petrothermaler Technik – Entwurf 2.0. https://www.clearingstelle-eeg.de/files/private/active/0/GtV_Bonus_EEG28-3-vers2.pdf (last access: 17 Ju8ne 2014), 2009.

Schumacher, S. and Schulz, R.: Effectiveness of acidizing geothermal wells in the South German Molasse Basin, Geoth. Energ. Sci., 1, 1–11, 2013.

Smith, M.: Southampton Energy Scheme – Proceedings World Geothermal Congress 2000, Kyushu-Tohoku, Japan, 28 May–10 June 2000.

Sonnenthal, E., Spycher, N., Callahan, O., Cladouhos, T., and Petty, S.: A Thermal-Hydrological-Chemical Model for the Enhanced Geothermal System Demonstration Project at Newberry Volcano, Oregon. Proceedings, Thirty-Seventh Workshop on Geothermal Reservoir Engineering, SGP-TR-194, Stanford University, Stanford, California, 30 January–1 February 2012.

Southampton: Geothermal and CHP scheme, http://www.southampton.gov.uk/s-environment/energy/Geothermal/ (last access: 27 May 2014), 2014a.

Southampton: Southampton District Energy Scheme Features, http://www.southampton.gov.uk/Images/SouthamptonDistrictEnergySchemeFeatures_tcm46-314480.pdf (last access: 27 May 2014), 2014b

Southampton: Southampton Geothermal Well in Use, http://www.southampton.gov.uk/Images/Southampton-Geothermal-Diagram_tcm46-305517.jpg (last access: 27 May 2014), 2014c.

Southampton: SGHC network facts 2011, http://www.southampton.gov.uk/s-environment/energy/Geothermal/ (last access: 27 May 2014), 2014d.

Stephens, J. C. and Jiusto, S.: Assessing innovation in emerging energy technologies: socio-technical dynamics of carbon capture and storage (CCS) and enhanced geothermal systems (EGS) in the USA, Energ. Pol., 38, 2020–2031, 2010.

Stibitz, M., Jiráková, H., and Frydrych, V.: Deep geothermal exploration drilling in the Bohemian massif (Litoměřice, Czech Republic). 1st Sustainable Earth Sciences Conference & Exhibition – Technologies for Sustainable Use of the Deep Sub-surface, Valencia, Spain, 8–11 November 2011.

Sverrisson, H., Gudlaugsson, S. Th., Holm, S. L., Ingason, K., Tolnai, Z., Adam, L., Albertsson, O., and Tryggvadottir, L.: Case Study of an EGS Power Plant in Southern Hungary. European Geothermal Congress 2013, Pisa, Italy, 3–7 June 2013.

Tenzer, H.: Development of hot dry rock technology, GHC Bulletin, December, 2001.

Tenzer, H., Schanz, U., and Homeier, G.: HDR research programme and results of drill hole Urach 3 to depth of 4440 m – the key for realisation of a HDR programme in Southern Germany and Northern Switzerland. In: Proceedings of the world geothermal congress, Kyushu, Tohoku, 25–30 April 2000.

Tester, J. W., Brown, D. W., and Potter, R. M.: Hot dry rock geothermal energy – a new energy agenda for the 21st century, Los Alamos National Laboratory report, LA-11514-MS, US Department of Energy, Washington D.C., 1989.

Tischner, T., Evers, H., Hauswirth, H., Jatho, R., Kosinowski, M., and Sulzbacher, H.: New concepts for extracting geothermal energy from one well: the GeneSys-Project. In: Proceedings of the world geothermal congress, Bali, 25–30 April 2010.

Tischner, T., Kurg, S., Pechan, E., Hesshaus, A., Jatho, R., Bischoff, M., Wonik, T.: Massive Hydraulic Fracturing in Low Permeable Sedimentary Rock in the Genesys Project. In: Proceedings, Thirty-Eights Workshop on Geothermal Reservoir Engineering, SGP-TR-198, Stanford University, Stanford, California, 11–13 February 2013.

Torrens Energy: Projects, Parachilna. http://www2.torrensenergy.com/projects/parachilna.html, last access: 26 May 2014.

Tulinius, H., Correia, H. and Sigurdsson, O.: Stimulating a high enthalpy well by thermal cracking. - Proceedings, World Geothermal Congress, Japan, 1883–1887, 2000.

Tym, A.: City of Litoměřice Geothermal Project. EGEC Meeting, Brussels, 2 September 2011.

Tym, A.: Do have cities really a choice in planning sustainable energy future? 7th European Conference on Sustainable Cities & Towns, Break-Out Session: Sustainable energy infrastructures: what role for cities?, Geneva, oral presentation; retrieved from: http://www.sustainablegeneva2013.org/wp-content/uploads/2013/04/A10_Antonin-Tym.pdf (last access: 17 March 2015), 2013.

Urpi, L., Zimmermann, G., Blöcher, G., and Kwiatek, G.: Microseismicity at Groß Schönebeck – A Case Review. - PROCEEDINGS, Thirty-Sixth Workshop on Geothermal Reservoir Engineering, SGP-TR-191, Stanford University, Stanford, California, 31 January–2 February 2011.

VDI-Richtlinie 4640: Thermal use of the underground. Verein Deutscher Ingenieure-Gesellschaft Energie und Umwelt (GEU), Fachbereich Energiewandlung und -anwendung, Beuth Verlag GmbH, Berlin, 2010.

Wallroth, T., Eliasson, T., and Sundquist, U.: Hot dry rock research experiments at Fjällbacka, Sweden, Geothermics, 28, 617–625, 1999.

Walters, M.: Demonstration of an Enhanced Geothermal System at the Northwest Geysers Geothermal Field, CA. Geothermal Technologies Office 2013 Peer Review, Presentation hold at the April 2013 peer review meeting in Denver, Colorado, 2013.

Williams, C. F., Reed, J. J., and Anderson, A. F.: Updating the classification of geothermal resources, in: Proceedings of the thirty-sixth workshop on geothermal reservoir engineering, Stanford University, Stanford, 31 January–2 February 2011.

Wyborn, D.: Hydraulic stimulation of the Habanero enhanced geothermal system (EGS), South Australia. Presentation held at the 5th BC unconventional gas technical forum, April 2011.

Zimmermann, G., Tischner, T., Legarth, B., and Huenges, E.: Pressure-dependent production efficiency of an enhanced geothermal system (EGS): stimulation results and implications for hydraulic fracture treatments, Pure Appl. Geophys., 166, 1089–1106, 2009.

Zimmermann, G., Moeck, I., and Blöcher, G.: Cyclic waterfrac stimulation to develop an enhanced geothermal system (EGS): Conceptual design and experimental results, Geothermics, 39, 59–69, 2010a.

Zimmermann, G., Reinicke, A., Blöcher, G., Moeck, I., Kwiatek, G., Brandt, W., Regenspurg, S., Schulte, T., Saadat, A., and Huenges, E.: Multiple Fracture Stimulation Treatments to Develop an Enhanced Geothermal System (EGS), Conceptual Design and Experimental Results, Proceedings World Geothermal Congress 2010, Bali, Indonesia, 25–29 April 2010, 2010b.

# 9

# Thermodynamic and thermoeconomic analysis of combined geothermal space heating and thermal storage using phase change materials

**V. Chauhan**[1,3] **and Á. Ragnarsson**[2]

[1]Reykjavik University, Reykjavik, Iceland
[2]IcelandGeoSurvey (ISOR), Reykjavik, Iceland
[3]UNU Geothermal Training Programme, Reykjavík, Iceland

*Correspondence to:* V. Chauhan (vijay30008@gmail.com)

**Abstract.** The present work discusses the utilization of phase change materials for energy storage in geothermal space heating systems. Thermodynamics and thermoeconomics of the combined heating and thermal storing system were studied to show the scope of energy storage and cost savings. A computational model of the combined space heating and thermal storage system was developed and used to perform thermodynamic studies of the heat storage process and heating system efficiency at different times and ambient temperatures. The basis for these studies is daily variations in heating demand that is higher during the night than during the day. The results show the scope of the utilization of phase change material for low ambient temperature conditions. Under proper conditions a sufficient amount of exergy is stored during the charging period at a low ambient temperature to fulfill the daytime heat load requirement. Under these conditions the cost flow rate of exergy storage is found to be lower than the radiator heating cost flow rate. Thus, the use of exergy storage at low ambient temperatures for heating at higher ambient temperatures makes a significant contribution to cost savings.

## 1 Introduction

Space heating has been one of the most well-known applications of geothermal energy utilization for decades. The use of geothermal energy for space heating provides an economical and a non-polluting method for achieving human comfort. In order to study a thermodynamic system's performance that can either involve heating or power generation, the second law of thermodynamics plays an important role. The second law helps to better understand energy flow processes alongside the first law of thermodynamics. Exergy is the maximum theoretical useful work obtainable as the systems interact to equilibrium, the heat transfer occurring with the environment only. Several studies have been conducted on the exergy analysis of buildings. The concept of low exergy systems for heating and cooling was proposed in Annex 37 (2000). An exergetic life cycle assessment for resource evaluation in the built environment was conducted by Meester et al. (2009). Shukuya and Komuro (1996) applied concepts of entropy and exergy for investigating the relationships between buildings, passive solar heating, and the environment. Various results about patterns of human exergy consumption in relation to various heating and cooling systems were given by Saito and Sukaya (2001). Conclusions were made about the inadequacy of the energy conservation concept for understanding important aspects of energy utilization processes by Yildiz and Gungör (2009). The second-law analysis is important for an efficient utilization of the available resources.

### 1.1 Phase change material

One of the best solutions for the problem when supply and demand are out of phase is the use of energy storage materials. The key to the effective utilization of renewable energy sources is efficient and economical energy storage systems. One of the most efficient ways of storing heat is the application of phase change materials (PCM). Such systems absorb and release heat energy as latent heat of the storing material with a change of phase. Various advantages of the application of PCM over other storage systems have helped the method to gain importance over the years. The study done by Adebiyi and Russell (1987) concluded two advantages derived from using a phase change material in a thermal energy storage system design. The first advantage was the increase in second-law efficiency of the system as compared to systems that use sensible heat storage. The second major advantage was concluded to be the reduced amount of storage material required. Since PCM is based on the principle of storing heat as latent energy of phase change, the energy stored is far higher than that stored by sensible heating systems that store energy equal to their specific heat capacity.

Various analyses of the latent heat storage based on the first law of thermodynamics can be found in various literature. The first-law analysis helps us to obtain a workable design but not an optimum one. In order to consider the effect of time and the temperature at which heat is supplied, second-law analysis is required. According to Bejan et al. (1996) an optimal system which a designer can develop with the least irreversibility is based on the minimization of entropy generation. The application of second-law analysis for studying latent heat storage was studied by Bjurstrom and Carlsson (1985) and Adebiyi and Russell (1987), and was later added to by Bejan (1996). A study was done by El-Dessouky and Al-Juwayhel (1997) which investigated the effect of different variables on the entropy generation number defined by Bejan et al. (1996). The analysis considered the case of the storage material exchanging heat at a constant melting point. For analysis, two commonly available storing materials, paraffin wax and calcium chloride, were considered, with air or water as the heating fluid.

A number of materials exist which can be used as PCM over a wide range of temperatures. A list of such materials can be found in the literature (Abhat, 1983). The selection of a PCM for energy storage is based on the fulfillment of criteria such as high latent heat capacity, non-corrosiveness and high thermal conductivity; and it should be non-toxic without deposition or supercooling. The transition temperature of the phase change material is decided by the room temperature required.

## 1.2　Importance of PCM for geothermal energy storage

Studies on the analysis of PCM, both numerical and experimental, have been reported in the literature (Farid et al., 2004). The cited work is mainly focused on solar energy storage. It is important to be aware of the necessity of storing geothermal energy where the mass flow itself can be controlled. The answer to the question depends upon the conditions for storage application. For an existing heating system using radiators and heat exchangers, an important fact discussed in the literature (Karlsson and Ragnarsson, 1995) is the variation of geothermal exit temperature with outside ambient temperature for a fixed design network. It is found that

as the ambient temperature decreases, the exit temperature of the geothermal heating network increases. The increase in exit temperature reflects an increase in useful heat loss. In terms of second-law efficiency, the decrease in ambient temperature causes an increase in exergy loss to the atmosphere. Surely the application of energy storage can provide a means of saving energy which can be used off-peak when the demand is low. Stored heat from the geothermal exit fluid at night can be used for fulfilling the daytime requirement. The other advantage of thermal storage for geothermal application is found in areas with limited flow and high heat demand. Heat energy in such cases can be stored during off-peak or daytime periods and can be utilized for peak load demands during the night. Managing the heating from limited flow will also help save drilling costs by reducing the need for new wells.

## 1.3　Thermoeconomics

Thermoeconomic or exergoeconomic analysis considers both thermodynamic and economic principles for improving systems' performance. Exergoeconomic analysis helps in quantifying the exergy losses in terms of monetary losses. According to Bejan et al. (1996) exergoeconomics is defined as the branch of engineering that combines exergy analysis and economic principles to provide the system designer or operator with information that is not available through conventional energy analysis and economic evaluations, yet is crucial to the design and operation of a cost-effective system. Application of exergoeconomic analysis can be found in the literature for processes involving heat transfer and power generation. Exergetic cost analysis for space heating using a ground source heat pump system was done by Jingyana et al. (2010). The work shows the sensitivity of various subsystems to unit exergy cost. The application of exergoeconomic analysis to a geothermal district heating system was presented by Oktay and Dincer (2009) and reviewed by Hepbasli (2010) using energy, exergy, and economic aspects. Performance evaluation of the geothermal heating system and case studies were conducted by Kecebas (2011). The effect of reference temperature on the thermoeconomic evaluation of geothermal heating systems was studied by Kecebas (2013).

In this context the current work proposes the use of phase change material for geothermal energy storage combined with a radiator heating system. Thermodynamic and thermoeconomic aspects of the proposed system are studied, taking second-law efficiency of the system into consideration.

## 2　Thermodynamic modelling

Figure 1 shows the schematic diagram of the combined space heating and thermal storage system. The geothermal water is first passed through a heat exchanger, passing heat to the secondary fluid. The secondary fluid passes through the radiator heating system, transferring heat to the room, and

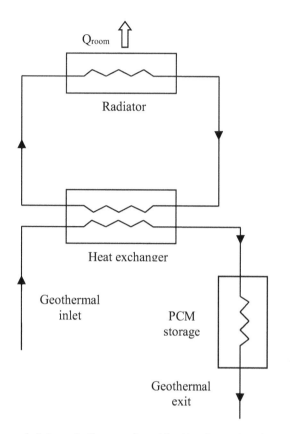

**Figure 1.** Schematic diagram of combined heating and storing system.

then returns back to the heat exchanger, forming a closed loop. The geothermal fluid, after exiting the heat exchanger, passes through the phase change thermal storage system. The geothermal fluid passes heat to the phase change material and then exits out of the system. Thermodynamic modelling of the combined system is described below.

## 2.1  Heating system

The current analysis assumes the use of an indirect method of space heating with radiators and plate-type heat exchangers as mentioned in Karlsson and Ragnarsson (1995). The geothermal fluid transfers heat to the secondary fluid which is passed through the radiators for room heating. The radiators manufactured are designed for fixed design conditions at the assumed room temperature. For temperatures other than the design temperature, the heat load is calculated as

$$\dot{Q}_T = \dot{Q}_{des} \left( \frac{T_{amb} - T_{room}}{T_{des} - T_{room}} \right). \tag{1}$$

The value of logarithmic mean temperature difference (LMTD) between the radiator and the room under design conditions is given as

$$\mathrm{LMTD_R} = \frac{T_{R,in} - T_{R,out}}{\ln\left( \frac{T_{R,in} - T_{room}}{T_{R,out} - T_{room}} \right)}. \tag{2}$$

Knowing the logarithmic mean temperature difference for the design conditions, the logarithmic temperature difference at different conditions is calculated as

$$\frac{\dot{Q}_T}{\dot{Q}_{des}} = \left( \frac{\mathrm{LMTD_R}}{\mathrm{LMTD_{R,des}}} \right)^n. \tag{3}$$

The value of the exponent $n$ is 1.3 (Anon, 1977).

On obtaining the logarithmic mean temperature difference, the value of the radiator outlet temperature can be calculated using Eq. (2).

Obtaining the temperature at the radiator outlet, the mass flow rate of fluid required through the radiator is found using the equation given below:

$$\dot{m} = \frac{\dot{Q}_T}{C\left(T_{in} - T_{out}\right)}. \tag{4}$$

Calculation of the above parameters for the radiators allow us to be able to fully describe the radiator or secondary fluid side of the system. The output parameters from the radiators as well as the geothermal fluid inlet temperature are input parameters for the heat exchanger. In general, plate-type heat exchangers are used for residential heating due to their advantages of compactness, high heat transfer rate, and ease of construction and maintenance over shell- and tube-type heat exchangers. For calculation of heat transfer coefficients through plate heat exchangers, the current analysis uses the relation suggested by Incropera et al. (2007) for flow through circular pipes with a diameter equal to the hydraulic diameter of a non-circular channel of the heat exchanger, through which the geothermal and the secondary fluids pass. The relation is given as

$$\mathrm{Nu} = 0.0296 \mathrm{Re}^{0.8} \mathrm{Pr}^{0.333}. \tag{5}$$

Calculating the convective heat transfer coefficient for both radiator side and geothermal side fluid, the overall heat transfer coefficient can be determined.

The amount of heat transferred through the heat exchanger is then given as

$$\dot{Q}_T = A_{HX} \cdot U_{HX} \cdot \mathrm{LMTD_{HX}}. \tag{6}$$

Knowing the value of heat exchange required and the area of the heat exchanger selected for the specific design, the logarithmic mean temperature difference of the heat exchanger can be calculated from the above equation. The geothermal exit temperature from the heat exchanger can then be calculated from the following equation:

$$\mathrm{LMTD_{HX}} = \frac{\left(T_{geo,in} - T_{R,in}\right) - \left(T_{geo,out} - T_{R,out}\right)}{\ln\left( \frac{T_{geo,in} - T_{R,in}}{T_{geo,out} - T_{R,out}} \right)}. \tag{7}$$

Knowing the return water temperature from the above equation, the mass flow rate of geothermal fluid can be found using Eq. (4).

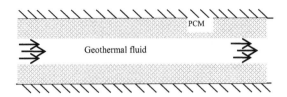

**Figure 2.** Configuration of storage system.

## 2.2 Phase change material

The current work focuses on the analysis of PCM combined with a geothermal space heating system, taking fixed values of design parameters such as storage component length, diameters, and mass of phase change material. The aim of the study is to calculate the amount of energy and exergy saved using PCM combined with geothermal space heating. The mass of phase change material was assumed to be 400 kg. The mass of the storage material and other physical properties was kept constant during the analysis. The heat storage system has two concentric cylinders with inner and outer diameters of 0.02 and 0.045 m respectively. The heat source fluid flows through the inner cylinder and the storage material is filled in the annulus between the cylinders, as shown in Fig. 2.

The small outer diameter of the cylinders allows such an arrangement to be laid along the floor or walls of a room in a loop, similar to a radiator floor heating arrangement. The parameters such as diameters, length, and phase change material weight can have optimized values as per the room dimensions and heat load. It is to be made clear that the analysis does not claim the current cylindrical arrangement of storage system as the optimum arrangement. The PCM material used is calcium chloride. The properties of the PCM material and the heating fluid (water) are taken from El-Dessouky and Al-Juwayhel (1997). Analysis for the PCM using air and water as heating fluid was discussed in the corresponding literature that considered water and air as the ideal working fluid. For the calculation of exergy terms, this assumption is not fully validated for water. Exergy analysis for PCM was also done by Bjurstrom and Carlsson (1985). Relations were derived for the amount of exergy stored in the PCM and final temperature of the storage material. The current work uses the equations derived by Bjurstrom and Carlsson (1985) for analysis of the storage system. The main assumptions used in the analysis are as follows.

- There is no heat exchange between the storage system and the surrounding area during the charging process.

- The heat exchanger wall is assumed to have no resistance to heat transfer.

- Thermophysical properties of the PCM and flowing fluid are assumed to be constant.

- Since the analysis is done using phase change material with a low transition temperature, the temperature difference between the transition temperature and room temperature is small; hence all stored exergy is assumed to be used for room heating.

- The heat transfer coefficient of the fluid (geothermal water) side ($h_f$) is calculated using Eq. (5), assuming flow through the cylinder to be turbulent.

The heat transfer coefficient between the wall and storage material is calculated according to the relationship developed by Yanadori and Masuda (1989) based on experimental data.

$$h_m = \frac{2\lambda_s}{0.4\left(D_{inner}\ln\left(\frac{D_{outer}}{D_{inner}}\right)\right)} \tag{8}$$

Calculating the convective heat transfer coefficient for the geothermal fluid and the phase change material side, the overall heat transfer coefficient for the PCM storage is calculated as

$$\frac{1}{U_s} = \frac{1}{h_f} + \frac{1}{h_m}. \tag{9}$$

In the above equation, resistance due to the wall is neglected as it is small in comparison to the convective heat transfer coefficients.

The number of transfer units (NTU) on the fluid side is given by the following equation:

$$\text{NTU} = \frac{-U_s\pi D_{inner}L_s}{\dot{m}_{geo}c_{geo,f}}. \tag{10}$$

For the analysis the initial temperature of the storage material and inlet temperature of the heat source is assumed. The procedure used for calculation of the final storage temperature and the amount of stored exergy is found from equations given by Bjurstrom and Carlsson (1985). The procedure used is detailed as follows.

A dimensionless time $\Omega$ is defined as

$$\Omega = \frac{\dot{m}_{geo}c_{geo,f}}{M_{pcm}C_{pcm}}t. \tag{11}$$

For the analysis, temperatures are made dimensionless using the following equation:

$$\theta = \frac{T - T_o}{T_o}, \tag{12}$$

where $T_o$ is the reference ambient temperature (K).

The temperature efficiency ($r$) of the heat exchanger is given as

$$r = 1 - \exp(-\text{NTU}). \tag{13}$$

For a stratified source with NTU on the fluid side, the relation can be approximated as

$$r = \frac{\text{NTU}}{1+\text{NTU}}. \tag{14}$$

Since the phase change materials exhibit three stages, absorbing sensible heat in solid phase from initial temperature to the transition stage, latent heat during the transition phase, and then sensible heat in the liquid phase, an equivalent heat capacity is defined by the relation:

$$M_s C_{\text{pcm}}(T_l - T_b)$$
$$= M_s C_s (T_m - T_b) + M_s H + M_s C_l (T_l - T_m), \tag{15}$$

or

$$\theta_l - \theta_b = \alpha(\theta_m - \theta_b) + \omega(\theta_l - \theta_b) + \sigma(\theta_l - \theta_m). \tag{16}$$

The constants $\alpha$, $\omega$ and $\sigma$ are given as

$$\alpha = \frac{C_s}{C_{\text{pcm}}}, \tag{17}$$

$$\omega = \frac{\Delta H}{C_{\text{pcm}}(T_l - T_b)}, \tag{18}$$

$$\sigma = \frac{C_l}{C_{\text{pcm}}}. \tag{19}$$

For heating at temperatures below the phase transition temperature, the instantaneous storage temperature as a function of time is given as

$$\theta_s = \theta_b + (\theta_i - \theta_b)\left(1 - \exp^{-r\Omega'}\right), \tag{20}$$

where $\Omega'$ is given as

$$\Omega' = \frac{\Omega}{\alpha}. \tag{21}$$

For heating the PCM above the transition temperature, the equation for instantaneous storage temperature and geothermal fluid outlet temperature is given as

$$\theta_s = \theta_m + (\theta_i - \theta_m)\left(1 - \exp^{-r\Omega''}\right) \tag{22}$$

$$\theta_e = \theta_i(1-r) + \theta_m r, \tag{23}$$

where $\Omega''$ is given as

$$\Omega'' = \frac{1}{\sigma}\left(\theta - \left(\frac{\alpha}{r}\ln\left(\frac{\theta_i - \theta_b}{\theta_i - \theta_m}\right)\right) - \frac{\omega}{r}\left(\frac{\theta_l - \theta_b}{\theta_i - \theta_m}\right)\right), \tag{24}$$

where second and third terms in the brackets represent the dimensionless terms required for bringing PCM to the phase transition temperature and its completion.

For PCM where the storage temperature exceeds the transition temperature, the amount of energy stored in the PCM is given by

$$Q_{\text{stored}}$$
$$= M_{\text{pcm}} C_s (\theta_m - \theta_b) + M_{\text{pcm}} H + M_{\text{pcm}} C_l (\theta_s - \theta_m). \tag{25}$$

## 3 System analysis

### 3.1 Exergy analysis

#### 3.1.1 Heating system

On determining the heat load, the amount of exergy required for the heating is determined. In order to determine the exergy required, the quality factor of the room air is to be estimated. Since heat energy is a form of low-grade energy, the amount of exergy present in the heat energy is determined by the quality factor which is to be estimated by means of Carnot efficiency, given as

$$Y_{q,\text{room}} = 1 - \frac{T_{\text{amb}}}{T_{\text{room}}}. \tag{26}$$

The amount of exergy required for satisfying the heat load demand is given as

$$\dot{\varepsilon}_{\text{room}} = Y_{q,\text{room}} \cdot \dot{Q}_{\text{T}}. \tag{27}$$

The amount of exergy given by the geothermal fluid is calculated as

$$\dot{\varepsilon}_{\text{geo}} = \dot{m}_{\text{geo}}\left[(h_{\text{geo,in}} - h_o) - T_{\text{amb}}(s_{\text{geo,in}} - s_o)\right]. \tag{28}$$

The value of enthalpy and entropy of the secondary radiator fluid as a function of temperature and pressure are calculated using the relations given by Cooper and Dooley (2007).

#### 3.1.2 Storage system

The amount of exergy stored in the PCM is given by Bjurstrom and Carlsson (1985):

$$\varepsilon_{\text{stored}} = M_{\text{pcm}} C_{\text{pcm}} T_{\text{amb}} \alpha \left[(\theta_m - \theta_b) - \ln\left(\frac{1+\theta_m}{1+\theta_b}\right)\right]$$
$$+ \omega(\theta_l - \theta_b)\frac{\theta_m}{1+\theta_m} + \sigma\left[(\theta_s - \theta_m) - \ln\left(\frac{1+\theta_s}{1+\theta_m}\right)\right]. \tag{29}$$

### 3.2 Exergoeconomic analysis

The total exergy given to a system as fuel exergy is transformed into product exergy and remaining energy is lost in the form of exergy destruction and exergy loss. The increase in efficiency causes an increase in exergy output for a given input but it also causes an increase in cost that is required to improve the system. For exergoeconomic analysis, cost is assigned for every exergy flow in a system. The cost is proportional to the amount of exergy the flow contains. The product cost is defined according to the fuel cost, capital expenditure, and other operation and maintenance costs required for production or services. The general equation is given as

$$\dot{C}_{\text{P}} = \dot{C}_{\text{F}} + \dot{Z}^{\text{CI}} + \dot{Z}^{\text{OM}}. \tag{30}$$

The equation signifies that the total cost associated with the product is the sum of the fuel cost, capital investment, and the other costs related to the operation and maintenance of the system that produces the product.

A general cost balance equation in terms of cost per unit exergy for an $i$th component with heat and work interactions with the surroundings can be represented in terms of cost per unit exergy as

$$\sum c_{\text{out},i} \dot{\varepsilon}_{\text{out},i} + c_{w,i} \dot{W}_i$$
$$= \sum c_{\text{in},i} \dot{\varepsilon}_{\text{in},i} + c_{Q,i} \dot{\varepsilon}_{Q,i} + \dot{Z}_i^{\text{CI}} + \dot{Z}_i^{\text{OM}}. \qquad (31)$$

The inlet cost in the above equation is obtained from the exit cost of the previous component. For the first component, the inlet cost is the cost at which the fuel is supplied. Hence the above equation can be solved for the unknowns.

### 3.2.1　Purchase equipment cost

For real-life applications, equipment cost can be obtained from a vendor's catalogue. Generally, it is not possible to obtain detailed cost of every component for every design condition. For such cases the literature provides some useful sources. Mathematical charts and relationships established from past experiences are available, giving cost value in terms of different parameters such as design and geometry. The simplest way of estimating product cost is using exponential law that defines the product cost as an exponential function of the size of the component. The relation is given as follows:

$$I = I_r \left( \frac{S}{S_r} \right)^e. \qquad (32)$$

Such a relation is assumed to be valid for a given range of equipment size. A general sixth-tenth rule is used for any equipment by taking the value of the exponent $e$ as 0.6. Different values of reference costs and their size along with the type of component were given by Boehm (1987). Because of various economic factors, the cost always changes with time. The above relation is for finding the cost from the reference cost of the indexed year. The obtained cost is brought to the current year cost by using a conversion relation given as follows:

reference year cost = original cost
$$\times \frac{\text{reference year cost index}}{\text{cost index for year for which calculation was made}}. \qquad (33)$$

The cost index used in the above equation takes into consideration the inflation in the cost of material, equipment, and labour. Various cost indices are available in the literature. The current analysis uses the Marshall and Swift cost index (Marshall et al., 2009) for indexing the equipment cost. The exergoeconomic analysis requires levelized costs of the equipment. For converting investment cost into levelized costs, the capital recovery factor (CRF) calculated for the interest rate

of $i\,\%$ with $N$ years of life and $t$ hours of annual operation is used, given by the following equation:

$$\text{CRF} = \left( \frac{i(1+i)^N}{(1+i)^N - 1} \right) \left( \frac{1}{t \cdot 3600} \right) s^{-1}. \qquad (34)$$

The other factor to be taken into consideration while calculating component cost is the operation and maintenance cost. The analysis assumes 2 % of the investment cost of each component to be the operation and maintenance cost ($\beta$). The cost flow rate for a given $i$th component is then calculated as

$$\dot{Z}_i = \frac{I_i(\text{CRF} + \beta)}{t}. \qquad (35)$$

### 3.2.2　Exergy costing

Guidelines for obtaining equations for different streams can be found in the literature (Bejan et al., 1996). The present case mainly involves the use of heat exchanging components. The balancing equations for the components used are described below.

#### Heat exchanger

In this system a heat exchanger is used to deliver heat coming from the geothermal inlet stream to the radiator exit stream, which is the product stream. Hence the following equations are obtained:

$$\dot{C}_{\text{geo,in}} - \dot{C}_{\text{HX,out}} + \dot{C}_{\text{R,out}} - \dot{C}_{\text{R,in}} = -\dot{Z}_{\text{HX}}. \qquad (36)$$

Also, since the fuel side is that of the geothermal stream, its cost per unit exergy remains constant. The equation is given as

$$\frac{\dot{C}_{\text{geo,in}}}{\dot{\varepsilon}_{\text{geo,in}}} = \frac{\dot{C}_{\text{HX,out}}}{\dot{\varepsilon}_{\text{HX,out}}}. \qquad (37)$$

#### Radiator

The cost associated with the radiator is charged to the product which is the exergy flowing out from the radiator for room heating. The equation obtained is

$$\dot{C}_{\text{R,in}} - \dot{C}_{\text{R,out}} - \dot{C}_{\text{heating}} = -\dot{Z}_{\text{R}}. \qquad (38)$$

Also, no exergy is added to the inlet stream; hence the cost per unit exergy remains the same, and the equation obtained is

$$\frac{\dot{C}_{\text{R,in}}}{\dot{\varepsilon}_{\text{R,in}}} = \frac{\dot{C}_{\text{R,out}}}{\dot{\varepsilon}_{\text{R,out}}}. \qquad (39)$$

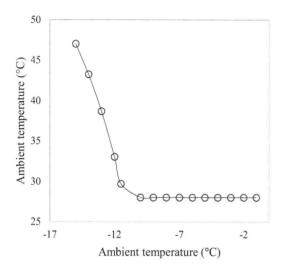

**Figure 3.** Variation of storage temperature with ambient temperature after an hour of operation.

### Storage system

The cost associated with the storage is charged to the product which is the exergy stored in the system. The equation obtained is

$$\dot{C}_{\text{HX,in}} - \dot{C}_{\text{geo,out}} - \dot{C}_{\text{storage}} = -\dot{Z}_{\text{storage}}. \tag{40}$$

Also, no exergy is added to the inlet stream; hence the cost per unit exergy remains same, and the equation obtained is

$$\frac{\dot{C}_{\text{HX,out}}}{\dot{\varepsilon}_{\text{HX,out}}} = \frac{\dot{C}_{\text{geo,out}}}{\dot{\varepsilon}_{\text{geo,out}}}. \tag{41}$$

The above linear equations can be solved simultaneously to obtain the unknown variables.

### 4  Results and discussions

In order to achieve an optimized heating system, a low temperature of geothermal fluid at the exit of the heat exchanger is required, signifying high heat exchange in the radiator. With the aim of improving thermodynamic efficiency, the heating system design was simulated by parallel addition of heat storage using phase change material for different ambient temperatures and for different time durations of the heat storing process in the phase change material. The design load for the radiator heating arrangement assumed was 35 kW at a temperature of $-15\,°\text{C}$. The room temperature is assumed to be $20\,°\text{C}$. The inlet temperature of the geothermal water was assumed to be $80\,°\text{C}$.

Figure 3 shows the variation of storage temperature with ambient temperature after an hour of the heat transfer process. It is found that at a lower ambient temperature, heat gained by phase change material is high enough to reach the temperature above transition quickly. This occurs due to high

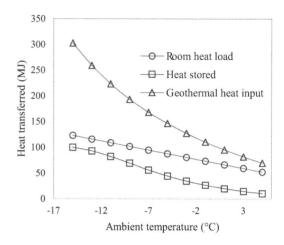

**Figure 4.** Variation of heat transferred with ambient temperature after an hour of operation.

temperature and mass flow rate of geothermal fluid at the storage inlet. With an increase in ambient temperature, storage inlet temperature and mass flow rate of the geothermal fluid decreases, causing less heat transfer rate between the geothermal fluid and the phase change material. This causes phase change material to still be in the transition stage after the same interval of time as that for lower temperature.

Figure 4 shows the variation of heat changes for different ambient temperatures after an hour of the heat storing process that keeps all other parameters constant. With the increase in ambient temperature, heat input from the geothermal fluid decreases. This occurs due to a low geothermal fluid exit temperature at the heat exchanger exit at the high ambient temperature. The mass flow rate of the geothermal fluid required also decreases with the increase in ambient temperature which also causes a decrease in heat input from the geothermal fluid. Room heat load shows continuous decrease as the ambient temperature increases. The amount of heat gained by phase change storage is higher at lower temperatures and decreases with an increase in ambient temperature. At a lower ambient temperature, the phase change material is above the transition stage. The amount of heat stored is higher as a high mass flow rate of geothermal fluid at a lower temperature also adds to high heat storage. The amount of heat stored at a lower ambient temperature is the summation of latent heat of material and the sensible heat stored up to the storage temperature of material at that time period. The exergy changes also show similar trends as shown in Fig. 5. Figures 4 and 5 show the significant contribution made by phase change storage in energy saved at low ambient temperatures. The graphs show that the energy and exergy stored are small in comparison to the total input, but are significant in comparison to the room heat load.

Figure 6 shows the variation of the phase change material storage temperature with time during the charging process, assuming an outside ambient temperature of $-10\,°\text{C}$.

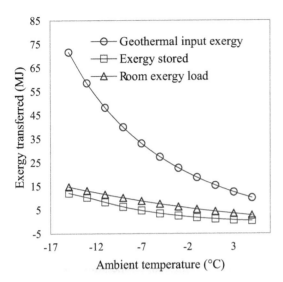

**Figure 5.** Variation of exergy changes with ambient temperature.

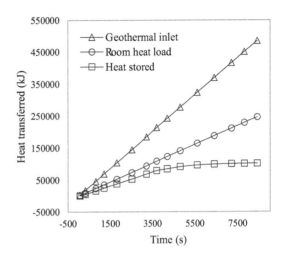

**Figure 7.** Variation of total heat transferred with time at −10 °C ambient temperature.

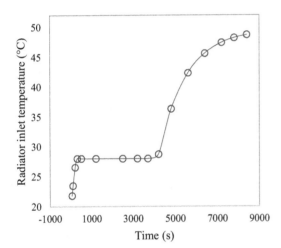

**Figure 6.** Variation of storage temperature with time at −10 °C ambient temperature.

The initial temperature of the phase change material is assumed to be equal to the room temperature. With an increase in storage time, the storage temperature increases until the transition temperature occurs. Since the temperature difference between the heat source fluid and the storage temperature is high, heat transfer is high, causing a rapid increase in temperature. On reaching the transition temperature, heat transfer occurs at constant temperature until the latent heat of the material is absorbed. After the transition phase is over, storage temperature starts increasing again. The initial phase of the storage temperature shows a high rate of increasing temperature and then starts decreasing as the storage temperature becomes closer to the heat transfer fluid inlet temperature.

Figure 7 shows the variation of total heat given by the geothermal heat source fluid, room heat load, and stored heat with time. With an increase in time duration the room heat increases constantly as the ambient temperature is fixed; hence we have a fixed heat load. Total geothermal heat input from the geothermal heat source also increases linearly since for a constant room heat load mass, the flow rate of geothermal fluid also remains constant and the inlet hot fluid temperature is also fixed. The constant increase is reflected by the constant slopes of the geothermal heat input and room heat load in Fig. 7. On the other hand the total heat stored increases constantly until the transition stage is complete and after that it decreases until the total heat stored becomes constant. The total heat stored shows a constant slope during the latent heat absorption. After that, sensible heat storage starts taking place and the rate of heat storage decreases as the storage temperature starts approaching the heat source fluid temperature. It can be seen from the graph that the room heat load requirement is much smaller than the total heat input from the geothermal fluid. The addition of the phase change storage system makes a significant heat saving comparable to the room heat load requirement. The significant amount of heat stored can be enough for satisfying the heat load requirement during the daytime when the heat load is less.

Figure 8 shows the variation of total exergy stored in phase change material with time. The initial phase of the storage process shows constant increase in exergy accumulation in the phase change storage process. This exergy accumulation process increases constantly until latent heat storage takes place. After the phase transition is over, sensible heating starts. The exergy accumulation in the phase change storage then starts decreasing as the phase change storage temperature approaches geothermal fluid inlet temperature.

Figure 9 shows the variation of fractional exergy stored as a function of fractional heat stored. The fractional exergy stored represents the ratio of rate of exergy storage to the total inlet exergy flow rate. The fractional heat stored is de-

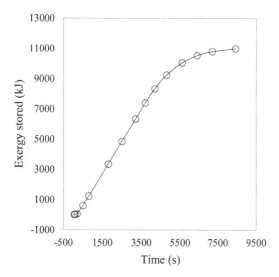

**Figure 8.** Variation of total exergy stored with time at $-10\,°C$ ambient temperature.

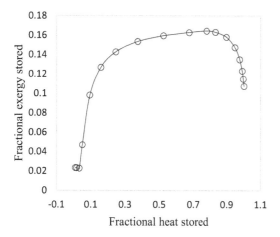

**Figure 9.** Variation of fractional exergy stored with fractional heat stored.

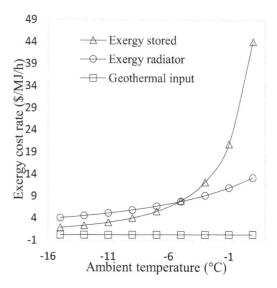

**Figure 10.** Variation of exergy cost flow rate per unit exergy with ambient temperature.

fined as the ratio of cumulative heat stored to the total heat stored. The initial and end part of the curve represents heat storage due to sensible heating and the middle constant range represents latent heat storage. It is seen from the graph that major exergy accumulation takes place only in the latent heat change process and sensible heating of phase change material does not contribute much to exergy storage. Hence it can be concluded from the graph that latent heat storage makes a greater contribution to heat storage than sensible heat does.

Figure 10 shows the variation of cost flow per unit exergy at different ambient temperatures. The inlet cost flow rate of the geothermal fluid is associated with that of well and pumping cost. Since the well cost varies from place to place, the current analysis assumes a well cost of USD 0.5 million and 0.1 million kWh$^{-1}$ for the pumping cost with a total required head of 200 m from the pump and 7000 h of operation annually. The discharge from the well is assumed to be 12 kg s$^{-1}$.

The exergy flow rate from a geothermal fluid depends upon the ambient temperature. With an increase in ambient temperature, exergy per unit mass flow rate of geothermal fluid decreases as the reference temperature increases. Also, the increase in ambient temperature causes a decrease in mass flow rate through the well as heat load decreases. These two factors cause cost flow rate of the geothermal fluid to increase with ambient temperature. Cost flow rate of exergy supplied by the radiator to the room increases with ambient temperature. As the ambient temperature increases, heat load required decreases, causing the required exergy flow rate to decrease. The decrease in exergy flow rate does not affect the purchase cost associated with the heat exchanger and radiator since the size of the equipment depends upon the design conditions for the minimum ambient temperature. A similar variation is found for the cost flow rate of exergy stored in the phase change storage system. The increase is found to be greater at a higher ambient temperature. This occurs because as the ambient temperature increases, the heat exchanger exit temperature and the mass flow rate decreases, causing a lower exergy flow rate at the heat exchanger exit. The decrease in exergy flow rate causes an increase in cost flow per unit exergy as the purchasing cost remains the same as that for minimum ambient design conditions. An important observation as noted from the graph is the difference in cost flow rates of the radiator heating and the stored exergy. It can be seen from the graph that the cost flow rate of stored exergy is lower than the radiator heating at a lower ambient temperature. Since with the increase in ambient temperature, the cost flow rate of radiator heating increases, the amount of exergy stored at lower temperature with a low cost flow rate can be used at higher temperatures, resulting in cost savings.

**Table 1.** Exergy and cost flow rates at $-5\,^{\circ}\mathrm{C}$ ambient temperature.

| State point | Exergy flow rate (kW) | Cost per unit exergy (USD MJ$^{-1}$ h$^{-1}$) |
|---|---|---|
| Geothermal inlet | 7.61 | 0.32 |
| Storage inlet | 0.32 | 0.32 |
| Storage exit | 1.6 | 0.32 |
| Radiator inlet | 6.04 | 0.69 |
| Radiator exit | 1.77 | 0.69 |
| Room heating | 2.08 | 7.83 |
| Storage | 0.69 | 7.98 |

Table 1 shows the exergy and cost flow rate per unit exergy at different state points in the combined heating and storing system at $-5\,^{\circ}\mathrm{C}$ ambient temperature. The values of cost flow rates per unit exergy show that each system, including geothermal wells and pumping, the heating system, and the heat storage system, makes a significant contribution to the overall product cost flow rate, that is, the cost of room heating and storing exergy. Also, the low value of exergy storage cost in comparison to room heating using a radiator system at lower ambient temperatures has the advantage of storing exergy at low ambient temperatures.

## 5  Conclusion

Thermodynamic and exergoeconomic analysis of space heating was performed using geothermal energy. The cost involved with drilling and pumping also contributes to the total cost for space heating using geothermal energy. The cost becomes significant when the flow rate requirement is high due to a high heat load requirement. The variation of the heating system efficiency with ambient temperature was studied. The heating system efficiency was found to be low at a lower ambient temperature; hence thermal losses become significant at lower temperatures.

A parallel combination of phase change storage system with heating system was studied. Heat supplied to the storage system was provided from the exit of the heat exchanger. Thermodynamic studies of the heat storage process at different time and ambient temperatures were conducted. Higher heat storage was observed at low ambient temperatures. The rate of exergy storage during the latent heat transition period was found to be larger than sensible heat storage. A sufficient amount of exergy was stored during the charging period at low temperatures which could be used to fulfill the daytime heat load requirement.

An exergoeconomic analysis of the combined heating and storage system was carried out. At a low ambient temperature, the cost flow rate of the exergy storage was found to be lower than the radiator heating cost flow rate. For the assumed value of investment cost flow rates, the heat storage using phase change material was found to be more economical than radiator heating to temperatures of $-5\,^{\circ}\mathrm{C}$. The cost flow rate of radiator heating exergy increases with increases in ambient temperature. The use of exergy storage at low ambient temperatures for heating at higher ambient temperatures makes a significant contribution to cost savings. Future work shall focus on the optimization of design parameters which influence the thermodynamic and thermoeconomic performance of the system.

## Appendix A

**Table A1.** Nomenclature.

| | |
|---|---|
| $A$ | Area ($m^2$) |
| $c$ | Cost flow per unit exergy ($USD\,kJ^{-1}\,h^{-1}$) |
| $C$ | Specific heat capacity ($kJ\,kg^{-1}\,K^{-1}$) |
| $\dot{C}$ | Cost flow rate ($USD\,s^{-1}$) |
| $D$ | Diameter (m) |
| $h$ | Specific enthalpy ($kJ\,kg^{-1}$), convective heat transfer coefficient ($W\,(m^2\,K)^{-1}$) |
| $H$ | Latent heat capacity of phase change material ($kJ\,kg^{-1}$) |
| $I$ | Component cost (USD) |
| $L$ | Length (m) |
| LMTD | Logarithmic mean temperature difference |
| $\dot{m}$ | Mass flow rate ($kg\,s^{-1}$) |
| $M$ | Mass (kg) |
| Nu | Nusselt number |
| NTU | Number of transfer units |
| $P$ | Power (kW) |
| Pr | Prandtl number |
| $\dot{Q}$ | Heat flow rate (kW) |
| Re | Reynolds number |
| $s$ | Specific entropy ($kJ\,K^{-1}$) |
| $S$ | Equipment size |
| $t$ | Time (s) |
| $T$ | Temperature (K) |
| $U$ | Overall heat transfer coefficient ($W\,(m^2\,K)^{-1}$) |
| $\dot{W}$ | Rate of work (kW) |
| $x$ | Thickness (m) |
| $\dot{Z}$ | Levelized cost |

| Greek letters | |
|---|---|
| $\dot{\varepsilon}$ | Rate of exergy (kW) |
| $\lambda$ | Thermal conductivity ($W\,m\,K^{-1}$) |
| $\theta$ | Dimensionless temperature |
| $\Omega$ | Dimensionless time |

| Subscripts and superscripts | |
|---|---|
| amb | Ambient |
| $b$ | Initial state of phase change material |
| Cl | Capital |
| D | Destruction |
| des | Design |
| e | Exit |
| f | Fluid |
| F | Fuel |
| geo | Geothermal |
| HX | Heat exchanger |
| $i$, $i$th | Component |
| in | Inlet |
| inner | Inner |
| $l$ | Liquid phase |
| L | Loss |
| $m$ | Transition phase |
| $o$ | Reference state |
| out | Outlet |
| outer | Outer |
| OM | Operation and maintenance |
| P | Product |
| $q$ | Heat |
| pcm | Phase change material |
| r | Reference |
| room | Room |
| R | Radiator |
| $s$ | Solid phase |
| storage | Phase change material storage system |
| $T$ | Temperature |

**Acknowledgements.** The authors would like to thank the anonymous reviewers for their valuable comments and suggestions that greatly improved the paper. A special acknowledgment goes to the GtES chief-executive editor, Horst Rüter, and the United Nations University Geothermal Training Programme in Iceland for their support.

# References

Abhat, A.: Low temperature latent heat thermal energy storage: heat storage materials, Solar Energy, 30, 313–332, 1983.

Adebiyi, G. A. and Russell, L. D.: Second law analysis of phase-change thermal energy storage systems, Proceedings ASME, WA-HTD-80, Boston, MA, 9–20, 1987.

Annex 37: Low Exergy Systems for Heating and Cooling of Buildings, IEA Energy Conservation in Buildings and Community Systems, Technical Presentations, Zurich, Switzerland, 12 July 2000, available at: http://virtual.vtt.fi/virtual/proj6/annex37/, 2000.

Bejan, A.: Entropy generation minimization: The new thermodynamics of finite size devices and finite time processes, J. Applied Physics, 79, 1191–1218, 1996.

Bejan, A., Tsatsaronis, G., and Moran, M.: Thermal Design and optimization, John Wiley & Sons, New York, 542 pp., 1996.

Bjurstrom, H. and Carlsson, B.: An exergy analysis of sensible and latent heat storage, J. Heat Recov. Sys., 5, 233–250, 1985.

Boehm, R. F.: Design and Analysis of thermal systems, John Wiley and Sons, New York, 259 pp., 1987.

Cooper, J. R. and Dooley, R. B.: Revised Release on the IAPWS Industrial Formulation 1997 for the Thermodynamic Properties of Water and Steam, Lucerne, Switzerland, 2007.

DIN 4703-1:1999-12: Raumheizkörper – Teil 3: Umrechnung der Norm-Wärmeleistung, Beuth Verlag, Berlin, Germany, 1977 (in German).

El-Dessouky, H. and Al-Juwayhel, F.: Effectiveness of a thermal energy storage system using phase change materials, Energ. Convers. Manage., 38, 601–617, 1997.

Farid, M. M., Khudhair, A. M., Razack, S. A. K., and Al-Hallaj, S.: A review on phase change energy storage: materials and applications, Energ. Convers. Manage., 45, 1597–1615, 2004.

Hepbasli, A.: A review on energetic, exergetic and exergoeconomic aspects of geothermal district heating systems (GDHSs), Energ. Convers. Manage., 51, 2041–2061, 2010.

Incropera, F. P., Dewitt, D. P., Bergman, T. L., Lavine, A. S., and Middleman, S.: Fundamentals of Heat and Mass Transfer: An Introduction to Mass and Heat Transfer, 6th Edn., John Wiley and Sons Inc., 1720 pp., 2007.

Jingyana, X., Juna, Z., and Na, Q.: Exergetic cost analysis of a space heating system, Energ. Buildings, 42, 1987–1994, 2010.

Karlsson, T. and Ragnarsson, A.: Use of very low temperature geothermal water in radiator heating system, Proceedings of the World Geothermal Congress, Florence, Italy, 2193–2198, 18–31 May 1995.

Keçebaş, A.: Performance and thermo-economic assessments of geothermal district heating system: a case study in Afyon, Turkey, Renew. Energ., 36, 77–83, 2011.

Keçebaş, A.: Effect of reference state on the exergoeconomic evaluation of geothermal district heating systems, Renew. Sustain. Energ. Rev., 25, 462–469, 2013.

Marshall, R. J., Lozowski, D., Ondrey, G., Torzewski, K.. and Shelley, S. A. (Eds.): Marshall and Swift cost index, Chem. Eng. Mag., 64 pp., 2009.

Meester, B. D., Dewulf, J., Verbeke, S., Janssens, A. and Langenhove, H. V.: Exergetic life cycle assessment (ELCA) for resource consumption evaluation in the built environment, Build Environment, 44, 11–17, 2009.

Oktay, Z. and Dincer, I.: Exergoeconomic analysis of the Gonen geothermal district heating system for buildings, Energ. Buildings, 41, 154–163, 2009.

Saito, M. and Shukuya, M.: The human body consumes exergy for thermal comfort, LowEx News, 2, 6–7, 2001.

Shukuya, M. and Komuro, D.: Exergy–entropy process of passive solar heating and global environmental systems, Sol. Energ., 58, 25–32, 1996.

Yanadori, M. and Masuda, T.: Heat Transfer Study on a Heat Storage Container with Phase Change Materials: Part 2, Sol. Energ., 42, 27–34, 1989.

Yildiz, A. and Gungör, A.: Energy and exergy analyses of space heating in buildings, Appl. Energ., 86, 1939–1948, 2009.

# 10

# Assessing the prospective resource base for enhanced geothermal systems in Europe

**J. Limberger[1], P. Calcagno[2], A. Manzella[3], E. Trumpy[3], T. Boxem[4], M. P. D. Pluymaekers[4], and J.-D. van Wees[1,4]**

[1]Department of Earth Sciences, Utrecht University, Utrecht, the Netherlands
[2]BRGM, Orléans, France
[3]Institute of Geosciences and Earth Resources, CNR, Pisa, Italy
[4]TNO – Geological Survey of the Netherlands, Utrecht, the Netherlands

*Correspondence to:* J. Limberger (j.limberger@uu.nl)

**Abstract.** In this study the resource base for EGS (enhanced geothermal systems) in Europe was quantified and economically constrained, applying a discounted cash-flow model to different techno-economic scenarios for future EGS in 2020, 2030, and 2050. Temperature is a critical parameter that controls the amount of thermal energy available in the subsurface. Therefore, the first step in assessing the European resource base for EGS is the construction of a subsurface temperature model of onshore Europe. Subsurface temperatures were computed to a depth of 10 km below ground level for a regular 3-D hexahedral grid with a horizontal resolution of 10 km and a vertical resolution of 250 m. Vertical conductive heat transport was considered as the main heat transfer mechanism. Surface temperature and basal heat flow were used as boundary conditions for the top and bottom of the model, respectively. If publicly available, the most recent and comprehensive regional temperature models, based on data from wells, were incorporated.

With the modeled subsurface temperatures and future technical and economic scenarios, the technical potential and minimum levelized cost of energy (LCOE) were calculated for each grid cell of the temperature model. Calculations for a typical EGS scenario yield costs of EUR $215\,\mathrm{MWh}^{-1}$ in 2020, EUR $127\,\mathrm{MWh}^{-1}$ in 2030, and EUR $70\,\mathrm{MWh}^{-1}$ in 2050. Cutoff values of EUR $200\,\mathrm{MWh}^{-1}$ in 2020, EUR $150\,\mathrm{MWh}^{-1}$ in 2030, and EUR $100\,\mathrm{MWh}^{-1}$ in 2050 are imposed to the calculated LCOE values in each grid cell to limit the technical potential, resulting in an economic potential for Europe of $19\,\mathrm{GW_e}$ in 2020, $22\,\mathrm{GW_e}$ in 2030, and $522\,\mathrm{GW_e}$ in 2050. The results of our approach do not only provide an indication of prospective areas for future EGS in Europe, but also show a more realistic cost determined and depth-dependent distribution of the technical potential by applying different well cost models for 2020, 2030, and 2050.

## 1 Introduction

Enhanced or engineered geothermal systems (EGS) have increased the number of locations that could be suitable for geothermal power production. In the past, geothermal power production was limited to shallow high-enthalpy reservoirs ($>180\,°\mathrm{C}$) in volcanic areas, whereas current EGS technologies facilitate exploitation of medium-enthalpy reservoirs ($80$–$180\,°\mathrm{C}$) situated at greater depth in sedimentary basins or in the crystalline basement.

Breakthroughs in binary power plant technology (e.g., organic Rankine cycle and Kalina plants) have enabled the use of medium enthalpy heat sources by using a binary working fluid to power the turbines (Astolfi et al., 2014a, b; Coskun et al., 2014). Innovations from the oil and gas industry such as directional drilling and techniques to enhance the reservoir properties, including hydraulic stimulation, provide a way to exploit these deeper reservoirs and, in theory, decrease the dependency on their natural permeability (Huenges, 2010).

Consequently, these developments should allow for more flexibility and a significant increase in the number of suitable locations for geothermal power production. In practice, development of EGS is not straightforward and so far in Europe most of the EGS power plants currently operational are limited to areas around the failed rift system of the Rhine Graben and the Molasse Basin of the northern Alpine foreland (e.g., Gérard et al., 2006; Baumgärtner, 2012; Breede et al., 2013).

For EGS and other geothermal systems, flow rate $Q$ and the temperature of the reservoir fluid $T$ are the key parameters that control the power output $P$ of a geothermal power plant. For large-scale resource assessments the temperature in the subsurface is a relatively convenient parameter to work with. In most nonvolcanic areas in Europe, conduction is the dominant heat transfer mechanism in the lithosphere. Temperatures can therefore be estimated with a steady-state conductive model based on assumptions and inferences on the thermal conductivity structure of the lithosphere, the heat flow at the base of the lithosphere, and on the content of heat-producing elements in the lithosphere (Cloetingh et al., 2010).

$Q$ depends strongly on the (enhanced) reservoir permeability, determined by lithological properties such as porosity, and the presence, distribution and permeability of natural fractures. These properties are dynamic and will change when the area of the reservoir is subjected to changes in temperature, pressure and the state of stress. Therefore, the reservoir permeability can easily vary by several orders of magnitude. Without knowledge of the geological history and thorough reservoir characterization, extreme caution should be taken when predicting $Q$ for a prospect.

This European resource assessment for EGS was conducted as part of the GeoElec European project to favor the development of geothermal electricity production in Europe (Dumas et al., 2013). The study covers the continental Europe plus the UK, Ireland, and Iceland but does not take into account the European overseas territories.

The first large-scale resource assessment for EGS was conducted for the United States (Tester, 2006; Blackwell et al., 2007). More recently, an updated resource assessment for the United States from (Williams et al., 2008) has been combined with a development cost model to create resource supply curves (Augustine et al., 2010; Augustine, 2011).

The most important input for the resource assessment in this study is a 3-D subsurface temperature model of Europe. The basic methodology of this temperature model is given in Sect. 2. The most recent and comprehensive regional temperature models available are incorporated, and combined with lithosphere-scale models to construct the model geometry and distribute thermal properties.

For the resource assessment of Europe we propose an approach similar to (Augustine et al., 2010) that extends the protocol from Beardsmore et al. (2010). As a starting point, the electrical power that could be technically produced from the theoretical capacity of thermal energy stored in the subsurface was estimated from the subsurface temperature model, with a set of assumptions such as flow rate, plant lifetime, conversion efficiency, and a recovery factor. This approach is extended, evaluating the levelized cost of energy (LCOE) with a discounted cash-flow model. The LCOE are subsequently used to assess the effect on the economic potential by restricting the technical potential to an economically recoverable subvolume. Technical scenarios for 2020, 2030, and 2050 time lines were used to estimate the different techno-economic scenarios for future EGS in 2020, 2030, and 2050. The resource assessment approach, the cash-flow model with the underlying assumptions for the different future scenarios, and the results for the economic potential are presented in Sect. 3.

The results of this approach do not only delineate prospective areas for future EGS in Europe, but also show an economically constrained depth-dependent distribution of the technical potential. Finally, implications of the results and potential improvements are discussed.

## 2 Temperature model

### 2.1 Methodology, model geometry and property distribution

This model mainly relies on temperature and heat flow values measured at Earth's surface and on a simple distribution of thermal properties in the upper crust. The modeling routine is designed in a way that it can easily be extended with additional information such as local temperature models.

The model assumptions of the temperature model in this study are similar to the protocol proposed by Beardsmore et al. (2010), but a 3-D finite difference method is used to solve the boundary value problem and to generate a steady-state solution for the temperature. The methodology of the protocol was based on earlier work of Tester (2006) and has been used to assess the geothermal potential of the USA (Blackwell et al., 2007). When data are scarcely available it is a fast way to generate an adequate temperature model for a large area such as Europe or the USA. It makes optimal use of data that are relatively easy to acquire and the variability of the model parameters can be easily adjusted whenever more data have become available. For this method, considering the European scale of the application, local convection is neglected and it is assumed that heat is transported via thermal conduction.

The model works on a voxel (a regular 3-D grid representation), which for the European assessment was chosen at a resolution of 10 by 10 km in northing and easting and by 0.25 km in depth. Depending on the location, each vertical column of stacked grid cells can represent two layers: one layer that represents sedimentary cover and the other layer that represents the crustal basement (Fig. 1, Table 1). Both

**Table 1.** Model geometry and boundary conditions.

| Model | | Reference |
|---|---|---|
| Topography | ETOPO1 | Amante and Eakins (2009) |
| Basement depth and crustal thickness | EuCRUST-07; CRUST2.0 | Tesauro et al. (2008); Bassin et al. (2000) |
| Surface temperature | WorldClim | Hijmans et al. (2005) |
| Surface heat flow | – | Cloetingh et al. (2010) |

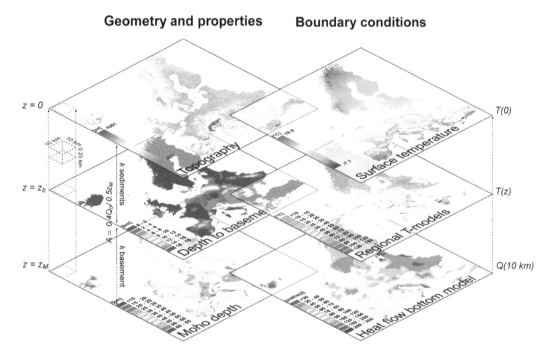

**Figure 1.** In this model a two-layer setup is used to assign values for $k$. For each $xy$ column, values of radiogenic heat production $A$ were calculated and assigned assuming that 40 % of the surface heat flow $Q_0$ has been generated by radiogenic heat production in the crust (Eq. 1). Following the same assumption, the heat flow at the base of the model at 10 km depth was calculated (Eq. 2). As boundary conditions for the top and bottom of the model, annual surface temperatures and heat flow at 10 km depth were used, respectively. Along the vertical edges of the model zero heat flow was assumed. Temperatures from regional temperature models are set as fixed values in the corresponding grid cells (Table 2).

layers have two thermal properties: thermal conductivity ($k$) and radiogenic heat production ($A$).

Values for $k$ are assigned according to the vertical position relative to the boundary between the sedimentary cover and the crustal basement. This boundary represents the depth of the sediment–basement interface ($S$) that divides the two layers.

The sediment thickness or the depth of $S$ is created by using the sediment thickness map from the high-resolution (0.25° by 0.25°) EuCRUST-07 model from Tesauro et al. (2008). This model is a compilation of existing sediment thickness maps that, where possible, have been improved by using seismic profiles. Because the EuCRUST-07 model does not fully cover the area of interest (eastern Turkey and eastern Ukraine are missing) the CRUST 2.0 model from Bassin et al. (2000) (with the sediment maps from Laske and Mas-

ters, 1997) is used. This model is largely based on the sediment thickness from the *Tectonic Map of the World*, created by Exxon Production Research Company (1995).

For the sediments, an average value for $k$ of $2.0 \, \text{W m}^{-1} \, \text{K}^{-1}$ was used, based on basin modeling predictions for lithologies which have not been subject to metamorphism (e.g., Hantschel and Kauerauf, 2009; Van Wees et al., 2009). For the European crystalline basement, dominated by plutonic and metamorphic rocks, a value of $2.6 \, \text{W m}^{-1} \, \text{K}^{-1}$ was adopted (e.g., Hantschel and Kauerauf, 2009; Van Wees et al., 2009).

To obtain values for $A$ in each grid cell the partition model of Pollack and Chapman (1977) was applied. Using the surface heat flow $Q_0$ and depth $Z_M$ of the Moho, $A$ was calculated for every grid cell by

$$A_z = \frac{0.4 Q_0}{0.5 \, z_M}, \tag{1}$$

which forces $A_z$ ($\mathrm{W\,m^{-3}}$) to be constant with depth, but to vary laterally according to $Q_0$ ($\mathrm{W\,m^{-2}}$) and $z_M$ (m). It was assumed that the upper crust forms half of the thickness of the total crust, which is approximately half of the depth of the Moho. The Moho depth in Europe varies from 15 to 63 km and was also derived from the EuCRUST-07 model from Tesauro et al. (2008) and is complemented by the CRUST 2.0 model from Bassin et al. (2000) to cover eastern Turkey and parts of Ukraine.

In nature, radiogenic heat production can show variations of up to several orders of magnitude even in samples that have been taken within a 1 km distance from each other (Vilà et al., 2010). A constant heat production with depth may not be realistic, but the advantage of the adopted model is that it reduces $A$ to a simple function of $Q_0$, which is capable of capturing the most important cause for regional heat flow variations, as reflected by correlation of regional variations of the surface heat flow $Q_0$ and the average radiogenic heat production observed in upper parts of the crust (Hasterok and Chapman, 2011).

The model works generally well in stable cratonic areas but, in more tectonically active regions, heat flow measurements can be severely affected by transient effects (Artemieva, 2011).

### 2.1.1  Boundary conditions

For the top of the model, constant values for the surface temperature ($T_0$), of the WorldClim Global Climate Database from Hijmans et al. (2005), are imposed as a Dirichlet boundary condition. This data set contains mean temperatures from 24 542 locations that represent the 1950–2000 time period. As reference level for the top, the ETOPO1 1 arc-minute Global Relief Model of Amante and Eakins (2009) is used.

As a Neumann boundary condition for the base of the model at 10 km below ground level, constant heat flow values are imposed. The heat flow at 10 km ($Q_{10\,\mathrm{km}}$) is obtained by subtracting the sum of the total radiogenic heat production of a column of stacked grid cells from the surface heat flow (Eq. 2).

$$Q_{10\,\mathrm{km}} = Q_0 - \sum_{z=0}^{z=10\,\mathrm{km}} A_z \tag{2}$$

For the surface heat flow ($Q_0$), the heat flow model of Europe from Cloetingh et al. (2010) is used, except for Iceland where the geothermal atlas was used (Hurter and Haenel, 2002).

At the vertical edges of the model, values of zero heat flow are imposed, which can be considered as a special case of a

**Table 2.** Input depth slices of subsurface temperature models (b.g.l. – below ground level)

| Area | Depth (km b.g.l.) | Reference |
|---|---|---|
| France | 1, 2, 3, 4, 5 | Bonté et al. (2010) |
| Germany | 1, 2, 3, 4, 5 | Agemar et al. (2012) |
| Ireland | 1, 5 | Goodman et al. (2004) |
| The Netherlands | 1, 2, 3, 4, 5, 6 | Bonté et al. (2012) |
| United Kingdom | 1 | Busby et al. (2011) |
| Europe | 1, 2 | Hurter and Haenel (2002) |

Neumann boundary condition. Finally, this model calculates temperature values in the 3-D grid, given the 3-D thermal conductivity and radiogenic heat production structure.

### 2.1.2  Input temperature models

Subsurface temperature models were collected from several geologic surveys, including France, Germany, Ireland, the UK and the Netherlands (Table 2). Apart from the UK, which only provided a map of 1 km depth, the subsurface temperature models provide constraints of up to a depth of 5 km. All of these models are based on bottom-hole temperature (BHT) or drill-stem test (DST) data, but their methodologies to compute them differ.

The French model from Bonté et al. (2010) and the German model from Agemar et al. (2012) are based on 3-D kriging geostatistical estimation. The Irish model from Goodman et al. (2004) is based on 2-D natural neighbor interpolation and the deeper temperature intervals have been generated by simple extrapolation of the average geothermal gradients observed in the boreholes. The UK model from Busby et al. (2011) is based on a 2-D interpolation of BHT data using a minimum curvature algorithm.

The Dutch temperature model from Bonté et al. (2012) uses the most comprehensive approach based on a three-step Runge–Kutta finite difference approach with a finite volume approximation. This model approach incorporates the effects of petrophysical parameters, including thermal conductivity and radiogenic heat production, as well as transient effects that affect temperature. Examples of transient effects are the accumulation of sediments, erosion and crustal deformation.

To use as much reliable temperature data as possible, we merged the regional temperature models and incorporated them in the modeling routine. To have constraints for the areas where no temperature models were available, the digitized subsurface temperature maps of 1 and 2 km depths from the geothermal atlas of Hurter and Haenel (2002) were also included.

For areas where more than one temperature value was available, we preferred to use values from integrated models over values derived by interpolation. For regional models where a similar methodology was used, we looked at the amount of measurements that were incorporated near the

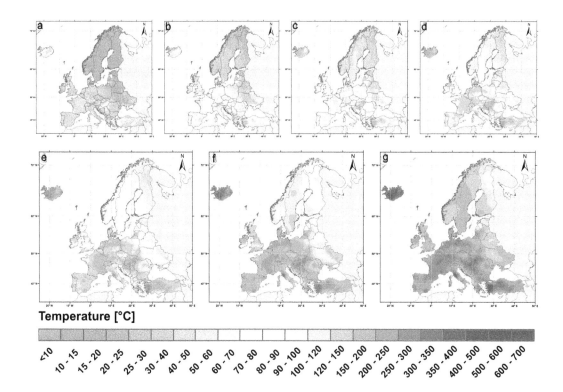

**Figure 2.** Depth slices of the modeled temperature voxet. Depths are below ground level. (**a**) 1 km, (**b**) 2 km, (**c**) 3 km, (**d**) 4 km, (**e**) 5 km, (**f**) 7 km and (**g**) 10 km.

shared boundary. We have chosen the Dutch model for the overlapping areas between the Dutch and German models and the German model for the overlapping areas between France and Germany.

Next, we replaced the calculated temperature values with the values from the merged temperature models without any smoothing. This approach could potentially cause discrepancies along shared borders between countries, as well as inconsistencies between the imported temperatures and the calculated heat flow. However, it enables the inclusion of more reliable data based on temperature measurements.

## 2.2 Modeling results

The outcome of the temperature modeling routine is a 3-D temperature voxet which contains values for every 10 by 10 by 0.25 km cell. Depth slices of the model taken at shallow to intermediate depth levels of 1–10 km are shown in Fig. 2.

The model shows high average geothermal gradients of up to 60 °C in volcanically active regions such as Iceland, parts of Italy, Greece and Turkey. Especially in Iceland and around volcanic regions in Italy, temperatures can reach more than 300 °C at a depth of 5 km and more than 500 °C at a depth of 10 km. What really stands out, apart from the regions with elevated temperatures, is the profound division between relatively high temperatures in the southwestern part of Europe and low temperatures in the northeastern part. These colder

zones are mostly constrained to the East European Craton and to the Fennoscandian or Baltic Shield.

This dichotomy fits with the Trans-European Suture Zone (TESZ), which marks a clear division between the stable Precambrian Europe and the dynamic Phanerozoic Europe (Pharaoh, 1999; Jones et al., 2010; Artemieva, 2011). The Precambrian zone has large lithosphere thicknesses and the Moho lies deeper, while in the Phanerozoic part of Europe the lithosphere is thinner and the Moho lies more shallow (Tesauro et al., 2008).

At 5 km depth (Fig. 2e) the model has a mean temperature of 111 °C and a total range varying between 40 and 310 °C and a standard deviation $\sigma$ of 44. At 10 km depth (Fig. 2g) the mean temperature is 201 °C, a total range between 80 and 590 °C and $\sigma = 74$.

The lowest temperatures at 10 km depth are around 80 °C, which is in line with geothermal gradients of 5–10 °C km$^{-1}$ that are observed in old cratonic crust (Artemieva, 2011). In the model, large anomalies between the observed and modeled temperature could be an indication for the presence of thermal convection (Bonté et al., 2012). These temperature anomalies can be used as a proxy for high permeability as was shown for the Netherlands by Van Oversteeg et al. (2014).

## 3 Techno-economic model

### 3.1 Methodology

To develop a geothermal system it is necessary to have favorable geological conditions, including a high temperature and appropriate reservoir properties. However, favorable geological conditions alone are not enough to initiate any commercial development. Because the development of a geothermal system involves high upfront costs and high financial risks (mostly related to drilling), it is vital to assess the financial feasibility for different scenarios. For the GeoElec project we applied a methodology that incorporates economic parameters in the estimation of geothermal resources in Europe. The main outputs from this method are the minimum LCOE and the economic power potential. Both are calculated on the basis of the temperature model described earlier. Because it is difficult to constrain the flow rate without information from a well, fixed flow rates have been used for the calculations, building from the generalized assumption that natural permeability can be enhanced – through stimulation – to sustain the assumed flow rates.

The techno-economic model uses the 3-D temperature voxel derived from the temperature modeling routine as input for its calculations. The complexity of this techno-economic model lies in the large quantity of variables inherent to economic problems, rather than in the mathematical solution. The model is based on a combination of the volumetric approach of Beardsmore et al. (2010) and a discounted cash-flow model from Lako et al. (2011) and Van Wees et al. (2012). The model is digitally available as an Excel spreadsheet. As depicted schematically in Fig. 3, the temperature model voxel is used to generate voxels for the heat in place $H$ (J), the theoretical potential $P_{\text{theory}}$ (MW$_e$), the technical potential $P_{\text{technical}}$ (MW$_e$) and finally the LCOE (EUR MWh$^{-1}$). The LCOE values are used to restrict $P_{\text{technical}}$ to obtain the economic potential $P_{\text{economic}}$ (MW$_e$). It is important to keep in mind that this methodology is based on a number of assumptions and that these potentials provide only an indication of the global European prospective resource base. In these following subsections the main concepts and assumptions used in this methodology are described.

### 3.1.1 Heat in place

Following the protocol of Beardsmore et al. (2010), the theoretically available thermal energy or heat in place $H$ (J) is calculated by combining Eq. (3a) with Eq. (3b):

$$H = V_{\text{rock}} \times \rho_{\text{rock}} \times C p_{\text{rock}} (T_z - T_r), \tag{3a}$$

$$T_r = T_0 + T_i. \tag{3b}$$

Where $V_{\text{rock}}$ is the volume of the rock (m$^3$), $\rho_{\text{rock}}$ is the density of the rock (kg m$^{-3}$) and $C p_{\text{rock}}$ is the heat capacity of the rock (J kg$^{-1}$ K$^{-1}$). The temperature difference that is

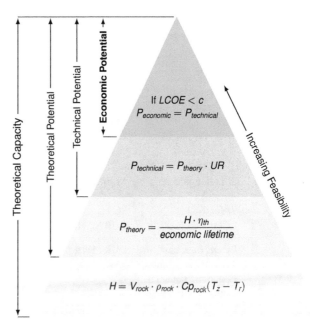

**Figure 3.** Assessment of the potential power output from a geothermal system. The theoretical capacity is the amount of thermal energy physically present in the reservoir rocks of a certain area or prospect. The theoretical potential describes the total amount of power that can be converted from the theoretical capacity within a certain period of time using a conversion efficiency. The technical potential is that part of the theoretical potential that can be exploited with current technology available calculated using a recovery factor. The economic potential describes the part of the technical potential that can be commercially exploited for a range of economic conditions. In this study we used different cutoff values for the LCOE ($c$) so the total costs of the system would fall in the same range as existing geothermal energy systems or other competing forms of energy production.

available for geothermal power is assumed to be the difference between $T_z$ (°C) (the temperature at depth $z$) and the base or reinjection temperature $T_r$ (°C). $T_r$ is the temperature to which the reservoir can theoretically be cooled using a surface temperature and $T_i$. For $T_i$ we assumed a default value of 80 °C (Williams et al., 2008).

### 3.1.2 Theoretical potential and efficiency

Next, $H$ is converted into the theoretical potential $P_{\text{theory}}$ (W$_e$) which is the power that could be theoretically produced during the expected lifetime of the system. $P_{\text{theory}}$ marks the upper limit of the theoretically realizable power output and is calculated from $H$ by combining Eq. (4a) with Eq. (4b):

$$P_{\text{theory}} = \frac{H \times \eta_{\text{th}}}{t}, \tag{4a}$$

$$\eta_{\text{th}} \approx \frac{T_z - T_0}{T_z + T_0} \times \eta_{\text{relative}}. \tag{4b}$$

**Table 3.** Important assumptions on economic parameters and the main results, including LCOE, theoretical potential, economic potential and the effective ultimate recovery.

| Parameter/result | Unit | 2020 | 2030 | 2050 |
|---|---|---|---|---|
| Maximum depth | m | 7000 | 7000 | 10 000 |
| Flow rate | $L\,s^{-1}$ | 75 | 100 | 100 |
| COP | $MW_{th}\,MW_e^{-1}$ | 30 | 50 | 1000 |
| Well cost model: | – | – | ThermoGIS (1.5) | EUR $1500\,m^{-1}$ |
|   <5200 m | - | ThermoGIS (1.5) | – | – |
|   >5200 m | – | WellCost Lite (1.0) | – | – |
| Stimulation costs | M EUR per well | 10 | 10 | 10 |
| $\eta_{relative}$ | – | 0.6 | 0.6 | 0.7 |
| $T_i$ for $T_r$ | °C | 80 | 80 | 50 |
| LCOE cutoff ($c$) | EUR $MWh^{-1}$ | 200 | 150 | 100 |
| LCOE Base case | EUR $MWh^{-1}$ | 215 | 127 | 70 |
| Theoretical potential | $TW_e$ | 14 | 14 | 22 |
| Economic potential | $GW_e$ | 19 | 22 | 522 |
| Effective UR | % | 0.1 | 0.2 | 2.4 |

$\eta_{th}$ describes the estimated thermal efficiency of the power plant and $t$ the expected lifetime of the geothermal system. The efficiency $\eta_{th}$ is also known as the cycle thermal efficiency of a power plant and is the efficiency at which the heat energy is converted to electrical energy, from Eq. (4b) it follows that high $\eta_{th}$ values are realized when there is a large difference between the inlet temperature, assumed to be equal to the temperature at depth $T_z$, and the outlet temperature, assumed to be equal to the surface temperature $T_0$. Typical values for $\eta_{th}$ range between 0.1 and 0.2. (DiPippo, 2007; Clauser, 2011).

To calculate $\eta_{th}$ we made use of $\eta_{relative}$ to convert from the ideal to the practical $\eta_{th}$. DiPippo (2007) documented values of $\eta_{relative}$ ranging from 0.44 to 0.85 with an average of 0.58. We have chosen similar values based on observed relative efficiencies for low to medium enthalpy binary systems (Table 3).

### 3.1.3 Technical potential

The technical potential $P_{technical}$ ($W_e$) is the fraction of $P_{theory}$ that can be theoretically produced, within the limits of current technology. In this case, the broadest definition for technical limits is used and includes geological limitations and technical limitations, such as drilling and power plant technology, but also environmental and political limitations. For example, some geothermal energy systems with promising geological potential cannot be developed because they are located in nature reserves, densely populated areas or areas where (sub)surface exploitation has been (temporarily) prohibited for political or legal reasons. Consequently, some areas can have a limited $P_{technical}$ permanently, while for other areas limits on $P_{technical}$ can be temporary.

Because it is difficult to precisely quantify all the different types of limitations, $P_{technical}$ is often derived by multiplying

$P_{theory}$ with an ultimate recoverability factor UR (Eq. 5a).

$$P_{technical} = P_{theory} \times UR \quad (5a)$$
$$UR = R_{av} \cdot R_f \times R_{TD} \quad (5b)$$

Equation (5b) describes how UR can be determined by combining the recovery factors corresponding to limitations in available land area ($R_{av}$), limitations in the recovery of heat from a fracture network ($R_f$) and limitations caused by the effect of temperature drawdown ($R_{TD}$). Beardsmore et al. (2010) recommended using a value between 1 and 20 % for UR, based on the work of Tester (2006) and Williams et al. (2008). This range of values for UR is based on the average recovery factor for all the layers with temperatures exceeding the base temperature within the total rock column beneath a selected area. For the volumetric estimation of the resource base, no distinction is made between (known) good reservoir rocks (e.g., coarse sandstones or karstified carbonates) and poor reservoir rocks (e.g., tight shales or low-permeability igneous rocks). For actual geothermal reservoirs, values for UR can be much higher and typically vary between 10 and 50 % (Dumas et al., 2013). We decided to omit $R_{av}$ at the scale of each individual grid cell, but for the European-scale assessment, the total land area available in each country was limited to 25 % ($R_{av} = 0.25$). For $R_f$, the lower bound value of 0.14 proposed by the protocol of Beardsmore et al. (2010) was used and for $R_{TD}$ fraction we assumed a value of 0.9. This results in an UR of ca. 12.5 % ($0.14 \times 0.9$) for each individual grid cell. Consequently, for calculating the potential of each country, the UR of a grid cell is limited to ca. 3 % ($0.25 \times 0.14 \times 0.9$).

## 3.2 Levelized cost of energy

For potential investors it is essential to quantify the economic potential of a geothermal energy system. Any economic potential study should base its calculations on the investment costs, also known as capital expenditures (CAPEX), and the operational costs, known as operational expenditures (OPEX). Both are usually expressed in EUR cents or USD cents per kilowatt or megawatt of installed capacity.

CAPEX is the sum of all the initial capital, $I_t$, that needs to be invested in a geothermal energy project in the year $t$. $I_t$ includes the investment costs such as the exploration costs, the drilling costs for the wells, the construction costs of the power plant including grid connection, the costs of reservoir stimulation and the costs of financing and commissioning the whole system. OPEX is the sum of all the operational and maintenance expenses, $M_t$, that are made in the year $t$. $M_t$ includes costs of personnel and equipment, costs of maintenance and, if required, costs of re-stimulation of the reservoir or the drilling of replacement wells. $Tax_t$ is the amount of tax that is paid in the year $t$, $E_t$ is the amount of electricity produced in the year $t$, and $r$ is the discount rate or inflation.

$$E_t = Q \times \rho_{water} \times Cp_{water}(T_z - T_r) \times \eta_{th} \times 10^{-3} \times t_{load} \quad (6)$$

Following Eq. (6), $E_t$ is calculated with flow rate $Q$ ($L\,s^{-1}$), water density $\rho_{water}$ ($kg\,m^{-3}$) and the heat capacity of water $Cp_{water}$ ($J\,kg^{-1}\,K^{-1}$). $\eta_{th}$ is used to convert from thermal power to electrical power and $t_{load}$ is the time (hours per year) at which the plant is fully operational. $E_t$ will only be calculated for grid cells for which the temperature is sufficiently high ($T_z > T_r$).

Once $I_t$, $M_t$ and $E_t$ are calculated and discounted using $r$, it is possible to calculate the expected costs per unit power or LCOE (Van Wees et al., 2012). The LCOE is calculated using Eq. (7):

$$\text{LCOE} \quad (7)$$
$$= \frac{\text{cumulative discounted yearly net revenue}}{\text{cumulative discounted yearly electricity production}}$$
$$= \frac{\sum_{t=1}^{T} \frac{I_t + M_t + Tax_t}{(1+r)^t}}{\sum_{t=1}^{T} \frac{E_t}{(1+r)^t}}.$$

Here the total discounted expenditures made during the project's lifetime in EUR are divided by the total discounted energy produced in megawatt hours for an expected lifetime $T$ in years. The total discounted life cycle costs are equal the sum of all the discounted CAPEX, OPEX, and taxes from year $t = 1$ to $t = T$. The total lifetime energy production is the sum of the discounted energy produced from year $t = 1$ to $t = T$. The discount is imposed by dividing the sum of the CAPEX and OPEX in year $t$ and the energy production in year $t$ by $(1+r)^t$. LCOE can only be calculated for grid cells where $E_t > 0$.

The outcome of this calculation are the levelized costs per unit of energy produced over time in EUR cents per kilo-watt hour or EUR per megawatt hour, which represent the costs that an energy provider would need to charge to break even. The LCOE for future EGS were calculated for techno-economic scenarios in 2020, 2030, and 2050 for which the full list of input parameters and default values can be found in Appendix A. Changes to the default values for the specific scenarios can be found in Table 3.

The LCOE is an economic parameter that is commonly used to describe the costs of energy for conventional and emerging power producing technologies, and provides an easy way to compare the costs between different energy systems. However caution must be taken when comparing the LCOE between sources of power that are dispatchable or nondispatchable. Enhanced geothermal systems that use pumps to produce geothermal fluids can be considered dispatchable since the power output can be adjusted by varying the pumping pressure. Power sources that are nondispatchable cannot simply adjust the power output on demand because they are dependent on energy sources that are strongly variable, such as the wind or the sun.

Besides dispatchability, an important factor for replacing conventional power plants with an alternative form is the capacity factor CF. CF is the ratio of the actual energy output and the maximum energy output that a power plant could produce when always operating at full capacity. The actual energy output is always lower than the maximum energy output since a power plant can be out of service or operating at a lower capacity due to equipment maintenance or failure or, in the case of solar or wind power, due to lack of resources. According to Goldstein et al. (2011), the average CF of all operational geothermal power plants is 74.5 %, while new plants often reach 90 % or more. This is higher than the CF of coal- and gas-fired power plants and much higher than the CF of other renewable energy technologies that are dependent on weather, such as solar and wind power (Goldstein et al., 2011; U.S. Energy Information Administration, 2012; REN21, 2014). Conventional geothermal energy systems have proven to be generally reliable and are able to provide base-load electricity. Because EGS systems work on the same principles as conventional geothermal systems, it is assumed that the CF will be in the same range as for conventional systems. For this study we assumed $t_{load} = 8000\,h$, corresponding to CF $\approx 91\%$.

### 3.2.1 Well cost models

To estimate the economic potential we combine the volumetric resource assessment with the techno-economic model described earlier in Sect. 3.2. A great portion of the CAPEX is determined by the costs that are related to the drilling of the wells. Three different well cost models were used to calculate the investment costs (EUR per well):

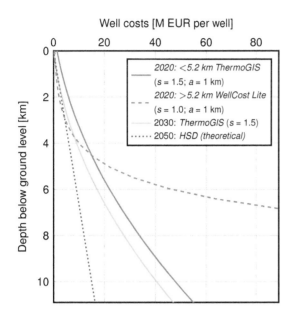

**Figure 4.** Well costs in million EUR for 2020, 2030, and 2050. (Eq. 8). For the 2020 scenario a combination of two well cost models is used. Above 5200 m the ThermoGIS model from Kramers et al. (2012) is applied, while below 5200 m the WellCost Lite model from Tester (2006) is used. For the 2020 scenario an additional 1000 m horizontal along hole length $a$ was added to replicate the divergent well layout normally used for an EGS doublet. For the 2030 scenario the same ThermoGIS model is used with $a = 0$. For 2050 it is assumed that HSD is possible and well costs increase linearly with depth.

$$\text{WellCost Lite} = s \times 10^{-0.67+0.000334(z+a)} \times n, \qquad (8)$$

$$\text{ThermoGIS} = s \times (0.2(z+a)^2 + 700(z+a) + 25\,000) \times n \times 10^{-6},$$

$$\text{HSD} = 1500 \times z \times n \times 10^{-6}.$$

Where $s$ is the well cost scaling factor, $z$ is the depth (m), $a$ the possible extra horizontal length of the well (m) and $n$ the number of wells (see also Table 3).

The WellCost Lite model has been proposed by Tester (2006) and has been derived from historical records between 1976 and 2004 of well costs in the United States. The well costs in this model increase exponentially with depth, reflecting the increase in time and cost required for bit replacement at greater depths. The ThermoGIS well cost model is developed by TNO for the ThermoGIS project and is based on historical well costs in the Netherlands (Kramers et al., 2012).

For 2020, the WellCost Lite model and the ThermoGIS model were combined with $a = 1000$ m. Up to a depth of 5200 m, the ThermoGIS model is used with $s = 1.5$, while below 5200 m the more exponential WellCost Lite model uses $s = 1$. For 2030, only the ThermoGIS model is used with $s = 1.5$ and $a = 0$. For 2050 a linear well cost model is applied, based on the prediction that new drilling techniques such as hydrothermal spallation drilling (HSD) or plasma drilling will emerge (Augustine, 2009). Compared to exponential well cost models, the assumed drilling costs for HSD of EUR 1500 m$^{-1}$ could especially lower the LCOE at greater depths (Fig. 4). Additionally, the higher temperatures that are expected at greater depths should increase the thermal efficiency $\eta_{\text{th}}$, which will also lower the LCOE.

Another advantage of geothermal energy is that the OPEX are relatively low and do not depend on fuel costs, contrary to the OPEX of conventional power plants, which can vary strongly due to the erratic development of coal and gas prices. The problems encountered with the development of geothermal energy systems are mostly related to the high upfront costs and the related finances. The high upfront costs are usually caused by the costs involved with the drilling of the wells. The problems with financing geothermal projects relate to the substantial uncertainties in the performance of the wells. EGS technology is still in a research and development stage since only a handful of projects have been realized (Breede et al., 2013). More experience with EGS needs to be gained and solutions for potential problems need to be developed before costs of EGS are expected to decline.

### 3.2.2 Sensitivity analysis of the LCOE for EGS and comparison with other LCOE estimates

Because most EGS are relatively new and commercial exploitation has just started it is difficult to assess the LCOE (Breede et al., 2013). According to (Goldstein et al., 2011) most existing conventional geothermal systems have LCOE that vary between USD 31 MWh$^{-1}$ and USD 170 MWh$^{-1}$. In the work of (Huenges, 2010) the LCOE is estimated at EUR 260 MWh$^{-1}$ and EUR 340 MWh$^{-1}$ for two hypothetical EGS in Europe. LCOE calculated by Tester (2006) for potential EGS projects in the US, range between ca. USD 100 MWh$^{-1}$ and USD 1000 MWh$^{-1}$. However, for the same cases 20 years into the future, assuming mature and cheaper technology, the calculated LCOE could be much lower, ranging between USD 36 MWh$^{-1}$ and USD 92 MWh$^{-1}$ (Tester, 2006). (Augustine, 2011) estimates the range of costs for present-day deep EGS between USD 140 MWh$^{-1}$ and USD 310 MWh$^{-1}$, with a mean of USD 210 MWh$^{-1}$. For 2020 the LCOE are estimated to be between 89 and 93 % of the present-day values.

These costs should enable EGS in the near future to become competitive with conventional power sources, such as coal and gas, currently priced at USD 65 MWh$^{-1}$–USD 95 MWh$^{-1}$ in the US and EUR 38 MWh$^{-1}$–EUR 100 MWh$^{-1}$ in Europe (U.S. Energy Information Administration, 2014; Kost et al., 2013).

To make a comparison we applied our techno-economic model on a hypothetical EGS project situated near the Rhine-Graben, with a reservoir depth at 5000 m and a default temperature of 200 °C. For this hypothetical case, combined with our assumptions for future scenarios (Table 3), the model calculates LCOE of EUR 215 MWh$^{-1}$ in 2020,

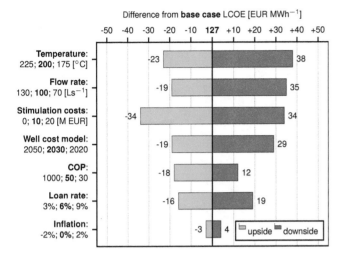

Difference from **base case** LCOE [EUR MWh$^{-1}$]

**Figure 5.** Tornado plot showing the sensitivity of the calculated LCOE to changes in a selection of parameters. The default settings of the 2030 scenario (bold) were applied to a reservoir at 5 km depth with a temperature of 200 °C resulting in a LCOE of EUR 127 MWh$^{-1}$. For each of the selected parameters, we assumed values for what the upside and downside scenarios could be and calculated the effect on the LCOE compared to the base case.

EUR 127 MWh$^{-1}$ in 2030 and EUR 70 MWh$^{-1}$ in 2050. The LCOE calculated for the 2020 scenario is in range with the estimates described earlier. The costs show a strong decline for the 2030 and 2050 scenarios and are comparable to the future scenarios from (Tester, 2006).

Figure 5 shows the sensitivity of the calculated LCOE to variations in a selection of parameters. For each of the selected parameters, we assumed values for what the upside and downside scenarios could be and calculated the difference compared to the LCOE for the 2030 scenario (EUR 127 MWh$^{-1}$). Temperature and flow rate have the largest uncertainty and variations in these parameters have a strong impact on the LCOE. Improving geothermal exploration is therefore essential to decrease the financial risks and to lower the LCOE. The effect on the LCOE of selecting different well cost models, together with variations in the stimulation costs and COP, reveal the importance of drilling technologies and stimulation techniques. The effect of drilling costs on the LCOE would have been even more profound for deeper reservoirs (Fig. 4). Lowering these costs is crucial for reaching higher temperatures at greater depths and increase the number of suitable locations for EGS, while enabling the installation of higher capacity power plants.

## 3.3 Economic potential

The economic potential describes the part of the technical potential that can be commercially exploited for a range of economic conditions. The total costs of the system should ideally fall within the same range as the costs for operational geothermal energy systems. The developable potential

is the part of the economic potential that can actually be developed taking into account all economic and noneconomic circumstances (Rybach, 2010). It is usually smaller than the economic potential, but it can be larger if governments have policies to promote renewable energy, including geothermal energy, such as feed-in tariffs, favorable taxes and favorable risk insurances.

Important for utilizing EGS for periods longer than the initial life time, is the sustainable potential (Sanyal, 2005; Rybach, 2010). It describes the fraction of the economic potential that can be used with sustainable production levels, while taking into account the resource degradation over time caused by pressure drawdown or by declining reservoir temperatures (Sanyal, 2005). We did not account for this in our study, but the effect of reduced temperatures and flow rates on the LCOE is shown in Fig 5. The effect of stimulation costs on the LCOE in Fig. 5 can also be used to assess the effect of measures countering resource degradation (e.g., additional stimulation or drilling of new/relieve wells).

For this resource assessment, we restricted the technical potential to grid cells where the LCOE was lower than a given threshold $c$ (Eq. 9).

$$\text{If LCOE} < c: P_{\text{economic}} = P_{\text{technical}} \tag{9}$$

For the 2020, 2030, and 2050 scenarios, values for $c$ of 200, 150 and 100 were chosen, respectively. These numbers were adopted to reflect the likely reduction of feed-in tariffs in the future beyond 2020 and renewable energy prices that will eventually become compatible with current fossil fuel-based energy prices (Tester, 2006; Goldstein et al., 2011).

To visualize the spatial distribution of the LCOE we compiled maps (Fig. 6) for each future scenario (Table 3), depicting the minimum value for the LCOE for each stacked $xy$ column of grid cells. Since fixed flow rates are assumed for the three different scenarios, the subsurface temperature automatically becomes the most important parameter. The LCOE distribution in the maps of 2020 and 2030 (Fig. 6a, b), therefore correspond largely to the areas where elevated temperatures are present at shallower depth. These mainly consist of volcanic areas such as Iceland, Italy and Turkey, but also sedimentary basins such as the Rhine-Graben, Pannonian Basin and the Southern Permian Basin. The LCOE map for the 2050 scenario clearly shows that the cost of drilling is a determining factor for the LCOE. Due to the use of a linear well cost model for the 2050 scenario, LCOE are lower than EUR 100 MWh$^{-1}$ for almost all of Europe southwest of the the TESZ.

To calculate the total economic potential for each stacked $xy$ column Eq. (10) was used:

$$P_{\text{economic}} = \sum_{z=0}^{z=7-10 \text{ km}} (P_{\text{economic}})_z. \tag{10}$$

For the country outlooks, the $P_{\text{economic}}$ for each country is summed and then multiplied by 0.25 to limit the economic

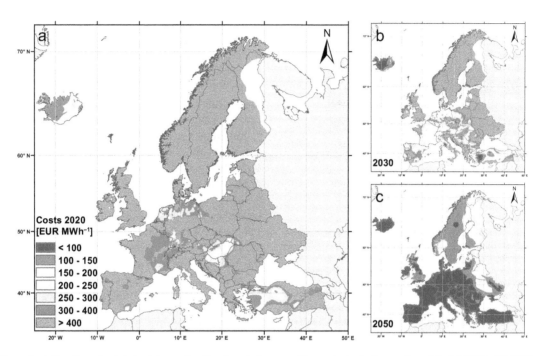

**Figure 6.** Maps depicting the calculated minimum levelized cost of energy (for each stacked $xy$ column) in **(a)** 2020, **(b)** 2030 and **(c)** 2050.

potential for land use restrictions. For 2020 and 2030, only the potentials up to a depth of 7 km are considered, while for 2050 the maximum depth is extended to 10 km. The economic potentials for 2020, 2030, and 2050 are 19, 22 , and 522 $GW_e$, respectively.

The effect of the different values of the LCOE threshold $c$ is illustrated in Fig. 7, where the economic potentials are plotted for $c$ values varying between EUR 300 MWh$^{-1}$ and EUR 50 MWh$^{-1}$. The economic potentials for the whole area considered in this study can be found in Table 3, along with the most important assumptions for each scenario.

By dividing $P_{economic}$ by $P_{technical}$, we calculated the effective UR. This results in an UR of 0.1 % for 2020, 0.2 % for 2030, and 2.4 % for 2050. The large difference of the $P_{economic}$ and UR of 2020 and 2030 compared to 2050, can mostly be ascribed to the use of a linear well cost model combined with the increase in the maximum drilling depth from 7 to 10 km, enabling exploitation of deeper reservoirs with higher temperatures.

We also assumed that all wells will be self-flowing in 2050, by adopting a COP of 1000 (Table 3). From theoretical considerations, it can be argued that well pressures in production and injection wells can be self-flowing, provided the reservoir temperature is in excess of 220 °C and the reservoir is located sufficiently deep (e.g., Sanyal et al., 2007). Due to the assumed lower drilling costs in 2050, it becomes financially feasible to drill for these deeper reservoirs with higher temperatures. Furthermore, by adopting a threshold value $c$ of EUR 100 MWh$^{-1}$ for 2050, these higher temperatures and associated larger depths are also implicitly required.

## 4  Discussion and conclusions

The economic resource assessment clearly demonstrates the strong sensitivity of the spatial and depth distribution of economic potential to both subsurface and cost parameters. Temperature and flow rate are the most important constraints for the development of an EGS project. These parameters are also the most uncertain since their exact values can only be determined by drilling a well and successfully creating a reservoir. For the LCOE, costs of drilling is the most important parameter, and the models clearly demonstrate the significant impact in the economic potential through a lowered cost curve for the 2020, 2030, and 2050 scenarios.

To reduce the uncertainty for the temperature, the temperature model should be improved. For this work, a simple two-layer conductive model is used where values for $k$ are distributed according to their location in respect to the sediment–basement interface for the basement. This could be improved by adopting a higher resolution for the thermal properties $k$ and $A$, based on lithological information and well data (e.g., Clauser, 2011; Bonté et al., 2012).

The underlying cause for variations in radiogenic heat production in the upper crust are lithological variations (e.g., Hasterok and Chapman, 2011). Inclusion of lithological interpretations of crustal composition for thermal properties ($k$ and $A$) could strengthen the geological interpretation and robustness of the models. This has been considered beyond the scope of the present study as little detailed information is readily available on the crustal lithology at a European scale (Tesauro et al., 2008) and – in the adopted workflow – would most likely not affect first-order temperature vari-

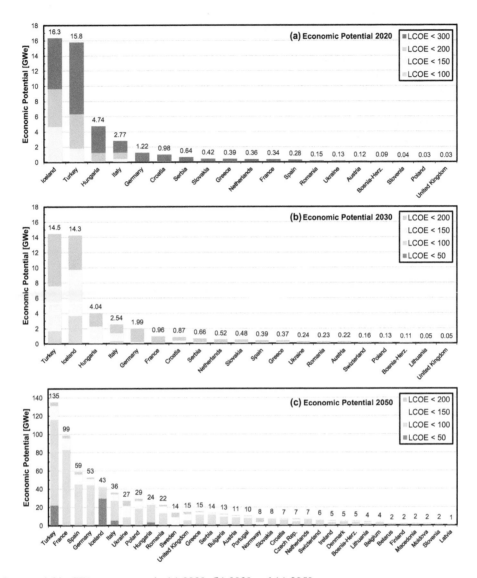

**Figure 7.** Economic potential in $GW_e$ per country in (**a**) 2020, (**b**) 2030 and (**c**) 2050.

ations of relevance to European geothermal potential estimates. However, for more detailed explorative studies, the incorporation of more detailed variations of thermal properties is key to unravel the temperature structure and prospective thermal anomalies (e.g., Bonté et al., 2012).

Furthermore, it is widely recognized that locally a conductive approximation for the temperature distribution may be oversimplified and models need to take into account the effects of convective fluid flow (e.g., Bonté et al., 2012; Guillou-Frottier et al., 2013; Calcagno et al., 2014).

Improvements to the quality of the temperature model could be attained by adopting data assimilation to borehole measurements of temperature, consistent with the constitutive equations for heat transfer and fluid flow. The successful implementation of the described improvements for all of Europe can only be achieved when the quality, quantity and

accessibility of geological information in Europe improves drastically.

One of the most important assumptions from a geological perspective is that the model uses a fixed flow rate. Since flow rate is one of the most sensitive parameters for the technical and economic performance of a geothermal system (e.g., Frick et al., 2010), care must be taken with the interpretation of the results. An ideal situation would be the use of location-specific flow rates, taking into account favorable conditions for creating new reservoirs or enhancing existing ones, such as lithology, natural (fracture) permeability and the in situ stress.

Furthermore, no distinction has been made between national differences regarding the economic situation, legislation, regulation and stimulation. These effects could potentially be significant but it is not in the scope of this study to quantify these differences. Nevertheless, for future work the model can easily be adjusted to nation specific scenarios.

Comparing the future economic potential for Europe obtained in this study to the results of other large-scale resource assessments is problematic because of differences in methodologies and assumptions; however, the results in Table 3 appear to be in agreement with other estimations. (Stefansson, 2005; Bertani, 2010; Goldstein et al., 2011; Chamorro et al., 2014).

## Appendix A: Input variables cash-flow model

For the LCOE calculation, the following input variables are used depending on the specific application. Most of the input parameters use the default values as specified below, whilst the values of some variables, including the base temperature and the relative efficiency, depend on the temperatures derived from the temperature voxet.

Fluid and rock properties:

- $Cp_{water}$ $(J\,Kg^{-1}\,K^{-1}) = 4250$: the heat capacity of the geothermal fluid

- $\rho_{water}$ $(Kg\,m^{-3}) = 1078$: the density of the geothermal fluid

- $Cp_{rock}$ $(J\,Kg^{-1}\,K^{-1}) = 1000$: the heat capacity of the reservoir rock

- $\rho_{rock}$ $(Kg\,m^{-3}) = 2500$: the density of the reservoir rock.

Power conversion:

- $\eta_{th}$ (%) = variable: total conversion efficiency

- $\eta_{relative}$ (%) = 60: the relative efficiency

- $T_r$ (°C) = variable: $T_r = T_0 + 80\,°C$

- $T_i$ (°C) = 80: offset for $T_r$.

Reservoir:

- $Q$ $(ls^{-1}) = 100$: flow rate

- $z$ (m) = variable: along hole depth of a single well

- $T_0$ (°C) = variable: surface temperature

- $T_x$ (°C) = variable: production temperature

- $t$ (years) = 30: economic lifetime.

Subsurface:

- scaling factor for ThermoGIS well cost model = 1.5

- well costs (EUR $10^6$ per well) = variable

- stimulation and other costs (EUR $10^6$ per well) = 10

- pump investment (EUR $10^6$ per pump) = 0.6

- number of wells = 2: depends on application

- subsurface CAPEX (EUR $10^6$) = variable

- maximum drilling depth (m) = 7000.

Subsurface parasitic:

- COP $(MW_{th}\,MW_e^{-1}) = 20$: coefficient of performance to drive the pumps

- electricity price for driving the pumps (EUR $MWh_e^{-1}$) = 140

- variable OPEX (EUR $MWh_{th}^{-1}$) = variable.

Power temperature range used:

- outlet temperature power plant (°C) = variable.

Power surface facilities:

- thermal power for electricity $(MW_{th})$ = variable

- electric power $(MW_e)$ = variable

- power load time (hours per year) = 8000

- power plant investment costs (EUR $10^6$ $MW_e^{-1}$) = 3

- power distance to grid (m) = 5000

- power grid investment (EUR $kW_e^{-1}$) = 80

- power grid connection variable (EUR $m^{-1}$) = 100

- power plant CAPEX (EUR $10^6$) = variable

- power fixed OPEX rate (%) = 1

- power fixed OPEX (EUR $10^3$ $MW_e^{-1}$) = variable

- power variable OPEX (EUR $MWh_e^{-1}$) = variable.

Fiscal stimulus:

- fiscal stimulus on lowering equity before tax (true or false) = false

- percentage of CAPEX for fiscal stimulus (%) = 0

- legal max in allowed tax deduction (EUR $10^6$) = 0

- NPV (net present value) of benefit to project (EUR $10^6$) = variable.

Economics:

- inflation or discount rate $r$ (%) = 0%

- loan rate (%) = 6 %

- required return on equity (%) = 15 %

- equity share in investment (%) = 20 % (100 % minus debt share in investment)

- debt share in investment (%) = 80 % (100 % minus equity share in investment)

- tax (%) = 25.5 %

- term loan (years) = 30

- depreciation period (years) = 30 .

**Acknowledgements.** The research leading to these results has
received funding from the Intelligent Energy Europe Programme
under grant agreement no. IEE/10/321 Project GeoElec and from
the European Community's Seventh Framework Programme under
grant agreement no. 608553 (Project IMAGE). This paper is largely
based on the master's thesis of J. Limberger that described the work
carried out during an internship at the Dutch Geological Survey
(TNO) from April 2012 to March 2013. We thank the editor and
two anonymous reviewers for their constructive comments, which
have helped us to improve the manuscript.

Edited by: G. Beardsmore
Reviewed by: two anonymous referees

# References

Agemar, T., Schellschmidt, R., and Schulz, R.: Subsurface temperature distribution in Germany, Geothermics, 44, 65–77, 2012.

Amante, C. and Eakins, B. W.: ETOPO1 1 Arc-Minute Global Relief Model: Procedures, Data Sources and Analysis, NOAA Technical Memorandum NESDIS NGDC-24, http://www.ngdc.noaa.gov/mgg/global/relief/ETOPO1/docs/ETOPO1.pdf, 2009.

Artemieva, I. M.: The Lithosphere: An Interdisciplinary Approach, Cambridge University Press, United Kingdom, ISBN 978-0-521-84396-6, 2011.

Astolfi, M., Romano, M. C., Bombarda, P., and Macchi, E.: Binary ORC (organic Rankine cycles) power plants for the exploitation of medium-low temperature geothermal sources – Part A: Thermodynamic optimization, Energy, 66, 423–434, 2014a.

Astolfi, M., Romano, M. C., Bombarda, P., and Macchi, E.: Binary ORC (organic Rankine cycles) power plants for the exploitation of medium-low temperature geothermal sources – Part B: Techno-economic optimization, Energy, 66, 435–446, 2014b.

Augustine, C. R.: Hydrothermal Spallation Drilling and Advanced Energy Conversion Technologies for Engineered Geothermal Systems, Ph.D. thesis, Massachusetts Institute of Technology, http://hdl.handle.net/1721.1/51671, 2009.

Augustine, C. R.: Updated U.S. Geothermal Supply Characterization and Representation for Market Penetration Model Input, U.S. Geological Survey, http://www.nrel.gov/docs/fy12osti/47459.pdf, 2011.

Augustine, C. R., Young, K. R., and Anderson, A.: Updated U.S. Geothermal Supply Curve, in: Proceedings Thirty-Fifth Workshop on Geothermal Reservoir Engineering, Stanford University, Stanford, California, 2010.

Bassin, C., Laske, G., and Masters, G.: The Current Limits of Resolution for Surface Wave Tomography in North America, Eos, Trans. Amer. Geophys. Union, 81, 2000.

Baumgärtner, J.: Insheim and Landau – recent experiences with EGS technology in the Upper Rhine Graben, Oral presentation presented at ICEGS 25 May 2012, Freiburg, 2012.

Beardsmore, G. R., Rybach, L., Blackwell, D., and Baron, C.: A Protocol for Estimating and Mapping Global EGS Potential, Geoth. Res. T., 34, 301–312, 2010.

Bertani, R.: Geothermal Power Generation in the World 2005–2010 Update Report, in: Proceedings World Geothermal Congress 2010, The International Geothermal Association, Bali, Indonesia, 2010.

Blackwell, D. D., Negraru, P. T., and Richards, M. C.: Assessment of the Enhanced Geothermal System Resource Base of the United States, Nat. Resour. Res., 15, 283–308, 2007.

Bonté, D., Guillou-Frottier, L., Garibaldi, C., Bourgine, B., Lopez, S., Bouchot, V., and Lucazeau, F.: Subsurface temperature maps in French sedimentary basins: new data compilation and interpolation, B. Soc. Geol. Fr., 181, 377–390, 2010.

Bonté, D., Van Wees, J.-D., and Verweij, J. M.: Subsurface temperature of the onshore Netherlands: new temperature dataset and modelling, Geol. Mijnbouw-N. J. G., 91, 491–515, 2012.

Breede, K., Dzebisashvili, K., Liu, X., and Falcone, G.: A systematic review of enhanced (or engineered) geothermal systems: past, present and future, Geothermal Energy, 1, 1–27, 2013.

Busby, J., Kingdon, A., and Williams, J.: The measured shallow temperature field in Britain, Q. J. Eng. Geol. Hydroge., 44, 373–387, 2011.

Calcagno, P., Baujard, C., Guillou-Frottier, L., and Genter, A. D. A.: Estimation of the deep geothermal potential within the Tertiary Limagne basin (French Massif Central): An integrated 3D geological and thermal approach, Geothermics, 51, 496–508, 2014.

Chamorro, C. R., García-Cuesta, J. L., Mondéjar, M. E., and Pérez-Madrazo, A.: Enhanced geothermal systems in Europe: An estimation and comparison of the technical and sustainable potentials, Energy, 65, 250–263, 2014.

Clauser, C.: Radiogenic Heat Production of Rocks, in: Encyclopedia of Solid Earth Geophysics – second edition, edited by Gupta, H., chap. 15, pp. 1018–1024, Springer, Dordrecht, The Netherlands, ISBN 978-90-481-8701-0, 2011.

Cloetingh, S., Van Wees, J. D., Ziegler, P., Lenkey, L., Beekman, F., Tesauro, M., Förster, A., Norden, B., Kaban, M., Hardebol, N., Bonté, D., Genter, A., Guillou-Frottier, L., Voorde, M. T., Sokoutis, D., Willingshofer, E., Cornu, T., and Worum, G.: Lithosphere tectonics and thermo-mechanical properties: An integrated modelling approach for Enhanced Geothermal Systems exploration in Europe, Earth-Sci. Rev., 102, 159–206, 2010.

Coskun, A., Bolatturk, A., and Kanoglu, M.: Thermodynamic and economic analysis and optimization of power cycles for a medium temperature geothermal resource, Energ. Convers. Manage., 78, 39–49, 2014.

DiPippo, R.: Ideal thermal efficiency for geothermal binary plants, Geothermics, 36, 276–285, 2007.

Dumas, P., Van Wees, J.-D., Manzella, A., Nardini, I., Angelino, L., Latham, A., and Simeonova, D.: GEOELEC Report, http://www.geoelec.eu/wp-content/uploads/2014/01/GEOELEC-report-web.pdf, 2013.

Exxon Production Research Company: Tectonic map of the world, American Association of Petroleum Geologists Foundation, 1995.

Frick, S., Kaltschmitt, M., and Schröder, G.: Life cycle assessment of geothermal binary power plants using enhanced low-temperature reservoirs, Energy, 35, 2281–2294, 2010.

Gérard, A., Genter, A., Kohl, T., Lutz, P., Rose, P., and Rummel, .: The deep EGS (Enhanced Geothermal System) project at Soultz-sous-Forêts (Alsace, France), Geothermics, 35, 473–483, 2006.

Goldstein, B., Hiriart, G., Bertani, R., Bromley, C., Gutiérrez-Negrín, L., Huenges, E., Muraoka, H., Ragnarsson, A., Tester, J., and Zui, V.: in: IPCC Special Report on Renewable Energy Sources and Climate Change Mitigation, edited by Edenhofer, O., Pichs-Madruga, R., Sokona, Y., Seyboth, K., Matschoss, P., Kadner, S., Zwickel, T., Hansen, P. E. G., Schlömer, S., and von Stechow, C., chap. Geothermal Energy, pp. 401–436, Cambridge University Press, Cambridge, United Kingdom and New York, NY, USA, ISBN 978-1-107-02340-6, 2011.

Goodman, R., Jones, G. L. I., Kelly, J., Slowey, E., and O'Neill, N.: Geothermal Energy Exploitation in Ireland – Review of Current Status and Proposals for Optimising Future Utilisation, CSA report 3085/02.04, http://www.seai.ie/Archive1/Files_Misc/FinalReport.pdf, 2004.

Guillou-Frottier, L., Carré, C., Bourgine, B., Bouchot, V., and Genter, A.: Structure of hydrothermal convection in the Upper Rhine Graben as inferred from corrected temperature data and basin-scale numerical models, J. Volcanol. Geoth. Res., 256, 29–49, 2013.

Hantschel, T. and Kauerauf, A. I.: Fundamentals of Basin and Petroleum Systems Modeling, Springer Berlin Heidelberg, Germany, ISBN 978-3540723172, 2009.

Hasterok, D. and Chapman, D. S.: Heat production and geotherms for the continental lithosphere, Earth. Planet. Sc. Lett., 307, 59–70, 2011.

Hijmans, R. J., Cameron, S. E., Parra, J. L., Jones, P. G., and Jarvis, A.: Very high resolution interpolated climate surfaces for global land areas, Int. J. Climatol., 25, 1965–1978, 2005.

Huenges, E., ed.: Geothermal Energy Systems – Eploration, Development, and Utilization, Wiley-VCH Verlag GmbH & Co. KGaA, Weinheim, Germany, ISBN 978-3-527-40831-3, 2010.

Hurter, S. J. and Haenel, R., eds.: Atlas of Geothermal Resources in Europe (EUR 17811), Office for Official Publications of the European Communities, Luxemburg, ISBN 92-828-0999-4, 2002.

Jones, A. G., Plomerova, J., Korja, T., Sodoudi, F., and Spakman, W.: Europe from the bottom up: A statistical examination of the central and northern European lithosphere-asthenosphere boundary from comparing seismological and electromagnetic observations, Lithos, 120, 14–29, 2010.

Kost, C., Mayer, J. N., Thomsen, J., Hartmann, N., Senkpiel, C., Philipps, S., Nold, S., Lude, S., Saad, N., and Schlegl, T.: Levelized Cost of Electricity Renewable Energy Technologies, Fraunhofer Institute for Solar Energy Systems ISE, 2013.

Kramers, L., Wees, J. D. V., Pluymaekers, M. P. D., Kronimus, A., and Boxem, T.: Direct heat resource assessment and subsurface information systems for geothermal aquifers; the Dutch perspective, Geol. Mijnbouw-N. J. G., 91, 637–649, 2012.

Lako, P., Luxembourg, S. L., Ruiter, A. J., and in 't Groen, B.: Geothermal energy and SDE – Inventarisatie van de kosten van geothermische energie bij opname in de SDE, ECN-E–11-022, http://www.ecn.nl/docs/library/report/2011/e11022.pdf, in Dutch, 2011.

Laske, G. and Masters, G.: A Global Digital Map of Sediment Thickness, EOS Transactions, American Geophysical Union, 78, 1997.

Limberger, J: European geothermal resource assessment for electricity production (GEOELEC), master's thesis, VU University Amsterdam, 2013.

Pharaoh, T. C.: Palaeozoic terranes and their lithospheric boundaries within the Trans-European Suture Zone (TESZ): a review, Tectonophysics, 314, 17–41, 1999.

Pollack, H. N. and Chapman, D. S.: On the regional variation of heat flow, geotherms, and lithospheric thickness, Tectonophysics, 38, 279–296, 1977.

REN21: Renewables 2014 Global Status Report, REN21 Secretariat, Paris, France, ISBN 978-3-9815934-2-6, 2014.

Rybach, L.: "The Future of Geothermal Energy" and Its Challenges, in: Proceedings World Geothermal Congress 2010, The International Geothermal Association, Bali, Indonesia, 2010.

Sanyal, S. K.: Sustainability and Renewability of Geothermal Power Capacity, in: Proceedings World Geothermal Congress 2005, The International Geothermal Association, Antalya, Turkey, 2005.

Sanyal, S. K., Morrow, J. W., and Butler, S. J.: Net power capacity of geothermal wells versus reservoir temperature – a practical perspective, in: Proceedings Thirty-Second Workshop on Geothermal Reservoir Engineering, Stanford University, Stanford, California, 2007.

Stefansson, V.: World Geothermal Assessment, in: Proceedings World Geothermal Congress 2005, The International Geothermal Association, Antalya, Turkey 2005, 2005.

Tesauro, M., Kaban, M. K., and Cloetingh, S. A. P. L.: EuCRUST-07: A new reference model for the European crust, Geophys. Res. Lett., 35, 1–5, 2008.

Tester, J. W., ed.: The Future of Geothermal Energy – Impact of Enhanced Geothermal Systems (EGS) on the United States in the 21st Century – An assessment by an MIT-led interdisciplinary panel, Massachusetts Institute of Technology, Cambridge, USA, ISBN 0-615-13438-6, 2006.

U.S. Energy Information Administration: Annual Energy Outlook 2012 with Projections to 2035, http://www.eia.gov/forecasts/aeo/pdf/0383(2012).pdf, 2012.

U.S. Energy Information Administration: Levelized Cost and Levelized Avoided Cost of New Generation Resources in the Annual Energy Outlook 2014, http://www.eia.gov/forecasts/aeo/pdf/electricity_generation.pdf, 2014.

Van Oversteeg, K., Lipsey, L., Pluymaekers, M., Van Wees, J.-D., Fokker, P. A., and Spiers, C.: Fracture Permeability Assessment in Deeply Buried Carbonates and Implications for Enhanced Geothermal Systems: Inferences from a Detailed Well Study at Luttelgeest-01, The Netherlands, in: Proceedings Thirty-Eighth Workshop on Geothermal Reservoir Engineering, Stanford University, Stanford, California, 2014.

Van Wees, J.-D., Bergen, F. V., David, P., Nepveu, M., Beekman, F., Cloetingh, S., and Bonté, D.: Probabilistic tectonic heat flow modeling for basin maturation: Assessment method and applications, Mar. and Petrol. Geol., 26, 536–551, 2009.

Van Wees, J.-D., Kronimus, A., van putten, M., Pluymaekers, M. P. D., Mijnlieff, H., van Hooff, P., obdam, A., and Kramers, L.: Geothermal aquifer performance assessment for direct heat production – Methodology and application to Rotliegend aquifers, Geol. Mijnbouw-N. J. G., 91, 651–665, 2012.

Vilà, M., Fernández, M., and Jiménez-Munt, I.: Radiogenic heat production variability of some common lithological groups and its significance to lithospheric thermal modeling, Tectonophysics, 490, 152–164, 2010.

Williams, C. F., Reed, M. J., and Mariner, R. H.: A Review of Methods Applied by the U.S. Geological Survey in the Assessment of Identified Geothermal Resources, U.S. Geological Survey, http://pubs.usgs.gov/of/2008/1296/pdf/of2008-1296.pdf, 2008.

# 11

# Influence of major fault zones on 3-D coupled fluid and heat transport for the Brandenburg region (NE German Basin)

**Y. Cherubini**[1,2], **M. Cacace**[2], **M. Scheck-Wenderoth**[2], and **V. Noack**[2,*]

[1]University of Potsdam, Institute of Earth and Environmental Science, Potsdam, Germany
[2]Helmholtz Centre Potsdam – GFZ German Research Centre for Geosciences, Potsdam, Germany
[*]now at: Federal Institute for Geosciences and Natural Resources, Berlin, Germany

*Correspondence to:* Y. Cherubini (yvonne.cherubini@gfz-potsdam.de)

**Abstract.** To quantify the influence of major fault zones on the groundwater and thermal field, 3-D finite-element simulations are carried out. Two fault zones – the Gardelegen and Lausitz escarpments – have been integrated into an existing 3-D structure of the Brandenburg region in northeastern Germany. Different geological scenarios in terms of modelled fault permeability have been considered, of which two end-member models are discussed in detail. In addition, results from these end-member simulations are compared to a reference case in which no faults are considered.

The study provides interesting results with respect to the interaction between faults and surrounding sediments and how it affects the regional groundwater circulation system and thermal field.

Impermeable fault zones seem to induce no remarkable effects on the temperature distribution; that is, the thermal field is similar to the no-fault model. In addition, tight faults have only a local impact on the fluid circulation within a domain of limited spatial extent centred on the fault zone. Fluid flow from the surrounding aquifers is deviated in close proximity of the fault zones acting as hydraulic barriers that prevent lateral fluid inflow into the fault zones.

Permeable fault zones induce a pronounced thermal signature with alternating up- and downward flow along the same structures. Fluid flow along the plane of the faults is principally driven by existing hydraulic head gradients, but may be further enhanced by buoyancy forces. Within recharge domains, fluid advection induces a strong cooling in the fault zones. Discharge domains at shallow depth levels ($\sim < -450\,\mathrm{m}$) are instead characterized by the presence of rising warm fluids, which results in a local increase of temperatures which are up to 15 °C higher than in the no-fault case.

This study is the first attempt to investigate the impact of major fault zones on a 3-D basin scale for the coupled fluid and heat transport in the Brandenburg region. The approach enables a quantification of mechanisms controlling fluid flow and temperature distribution both within surrounding sediments and fault zones as well as how they dynamically interact. Therefore, the results from the modelling provide useful indications for geothermal energy exploration.

## 1 Introduction

Faults can significantly influence physical processes that control heat transfer and fluid motion in the subsurface. Faults provide permeable pathways for fluids at a variety of scales, from great depth in the crust to flow through fractured groundwater, geothermal and hydrocarbon reservoirs (Barton et al., 1995). Faults are also important because they may offset porous aquifer rocks against shales, rendering permeable rocks a dead end in terms of fluid flow (Bjørlykke, 2010). To understand the role of faults on the fluid and thermal field is also important for geothermal applications, as they may modify the overall reservoir permeability structure and therefore change its flow dynamics. Numerical simulations provide a useful tool for analysing heat and fluid transport processes in complex sedimentary basin systems integrating fault zones.

The aim of this study is to investigate the impact of major fault zones on the fluid and heat transport by 3-D numerical simulations. Our study is based on a recently published structural model of the Brandenburg area in the southeastern part of the Northeast German Basin (NEGB) (Noack et al., 2010). This refined 3-D structural model was constructed by integrating different types of data sets, including depth as well as thickness maps, data from previous models of the NEGB (Scheck and Bayer, 1999) and the Central European Basin System (Scheck-Wenderoth and Lamarche, 2005; Maystrenko et al., 2010), and well data (Noack et al., 2010). Subsequently, newly available data were integrated additionally to structurally refine the Tertiary unit (Noack et al., 2013). Figure 1a and b outline the location of the study area and its present-day topographic elevation. The dominantly clastic sedimentary succession of the NEGB resolved in the model ranges from the Permian to Cenozoic and reaches up to 8000 m thick in the central part of the basin (Fig. 2a). In response to variations in lithologies, four aquitards of regional extent subdivide the sedimentary succession into different aquifer systems (Fig. 2b). These aquitard layers are from bottom to top, the Permian basement forming the lowermost impermeable layer in the model (Fig. 3a), the Upper Permian Zechstein salt (Fig. 3b), the Middle Triassic Muschelkalk limestones (Fig. 3c) and the Tertiary Rupelian clays (Fig. 3d). Model detailed information about the hydrogeological configuration of sedimentary layers of interest is given in Sect. 2.1.

Along the southern margin the basin is dissected by two major fault zones, the Gardelegen and Lausitz escarpments (Fig. 1b), which vertically offset the pre-Permian basement by several km. As a result, the basement is uplifted by about 5 km coming close to the surface south of the Gardelegen Fault (Scheck-Wenderoth et al., 2008) (see also Fig. 3a).

The conductive thermal field of the Brandenburg region was first calculated by Noack et al. (2010, 2012). A comparison of the model results with published temperature measurements of 52 wells showed that the model predictions are largely consistent with the observations, indicating predominantly conductive heat transport (Noack et al., 2010, 2012). Local deviations between observations and model results were interpreted to be the result of additional fluid-related processes. Indeed, recent 3-D coupled fluid and heat transport simulations have revealed that the shallow thermal field is influenced by forced convective processes due to hydraulic gradients (Noack et al., 2013). Another aspect that could be responsible for the deviations between observed and predicted temperatures are faults, which may provide pathways for moving fluids and which have been not included in the model (Noack et al., 2012, 2013).

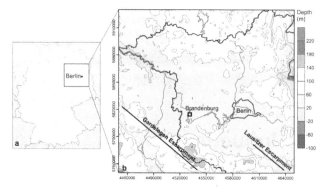

**Figure 1. (a)** Map of Germany showing the outlines of the study area located in the federal state of Brandenburg. The study area (red rectangle) covers a surface of 180 km in the N–S direction and of 200 km in the E–W. **(b)** Topography map (top of Quaternary) of the model area in UTM zone 33° N (ETOPO1, after Amante and Eakins, 2009) with the borderline of Brandenburg (black solid line), main rivers (blue lines) and the location of the Gardelegen and Lausitz escarpments, which are part of the larger Elbe fault system. The approximated traces of these two major fault zones are given by the straight black solid lines.

These previous studies have provided deeper insights into the present-day thermal structure of the Brandenburg area. However, the impact that major existing fault zones may have on the groundwater system and thermal field has not been investigated so far. Previous 2-D numerical studies applied to different geological settings showed that faults may significantly influence the hydrothermal field (e.g. Bense et al., 2008; Garven et al., 2001; Lampe and Person, 2002; Magri et al., 2010; Simms and Garven, 2004; Yang et al., 2004a, b). These investigations demonstrated that along-fault convection may be an important heat transport mechanism in permeable faults and may give rise to significant variations of the thermal field. Results from 3-D studies seem to confirm these conclusions (Alt-Epping and Zhao, 2010; Bächler et al., 2003; Baietto et al., 2008; Cacace et al., 2013; Cherubini et al., 2013; López and Smith, 1995, 1996; Yang, 2006). However, differences between 2-D and 3-D studies have been found, due to the fact that the longitudinal fluid flow and heat transport along the strike of the faults are ignored in 2-D studies (Yang, 2006).

With the study area, previous attempts to investigate the influence of the Gardelegen fault zone have relied on 2-D coupled fluid and heat transport simulations. The study (Pommer, 2012) revealed a hydraulic interaction between shallow and deep aquifers in which upward convective and downward advective flow through the permeable fault coexist. The present study aims to upgrade the results obtained so far by assessing the influence of major fault zones on the 3-D basin scale. In the following, the results of the first 3-D hydrogeological model of the Brandenburg area that integrates two major fault zones and that couples transient fluid and heat

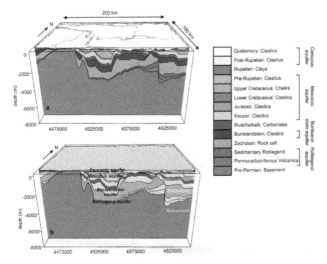

**Figure 2.** (a) Three-dimensional geological model of the study area with the stratigraphic layers resolved (vertical exaggeration: 7 : 1). Note the exposed pre-Permian basement coming close to the surface at the southeastern margin and the Permian Zechstein diapirs controlling the structural configuration of the overburden sediments. At the southern basin margin, the Gardelegen and Lausitz escarpments vertically offset the pre-Permian basement against the Permian to Cenozoic basin fill by several kilometres. (b) Distribution of aquitards and aquifers in the 3-D geological model (vertical exaggeration: 7 : 1). Four main layers act as regional barriers to fluid flow comprising the lowermost pre-Permian basement, the Permian Zechstein, the Middle Triassic Muschelkalk and the Tertiary Rupelian in the shallower part of the model. These aquitards hydraulically decouple the sedimentary succession into four main aquifer systems: from bottom to top, the Rotliegend aquifer, the Lower Triassic Buntsandstein aquifer, the Mesozoic aquifer and the Cenozoic aquifer.

transport in finite-element simulations are presented and discussed.

Due to variations in the regional stress field, different geological scenarios with an idealized fault zone representation in terms of their hydraulic behaviour have been tested, of which three end-member models are presented: (1) no fault (model 1), (2) tight fault zones (model 2) and (3) highly permeable fault zones (model 3). Model 1 is the reference case. It describes the "undisturbed" system with respect to the regional groundwater and thermal field. Thus, the regional thermal field is investigated by considering the interaction of different fluid and heat transport processes with respect to the hydrogeological setting of the study area. In model 2, a very low permeability is assigned for the faults zones, making them effectively impermeable for fluid flow. By contrast, in model 3, fault zones have a high permeability and are therefore supposed to act as hydraulically conductive structures. By means of these two end-member models, the influence of major fault zones on the coupled fluid and heat transport is

quantified and the interaction between fault zones and surrounding sediments is addressed.

To assess the respective impact of the different faults on the thermal field and fluid circulation, the results of all three models are compared with each other and the outcomes are discussed.

## 2 Data and method

### 2.1 Hydrogeological model

The 3-D structural model covers an area of 200 km in E–W and 180 km in N–S direction reaching down to −8000 m depth with a horizontal resolution of 1000 m × 1000 m, corresponding to 210 × 180 grid points. The model integrates 14 geological layers ranging from the pre-Permian basement at the bottom to the Quaternary at the top (Fig. 2a). In Table 1 all stratigraphic units are listed with predominant lithologies and corresponding physical properties adopted for the numerical simulations.

As summarized in Fig. 2b, the sedimentary succession is hydraulically decoupled by four regional aquitards (pre-Permian basement, Permian Zechstein, Middle Triassic Muschelkalk, Tertiary Rupelian) into four main aquifer systems (sedimentary Rotliegend, Lower Triassic Buntsandstein, Mesozoic aquifer, Cenozoic aquifer). The corresponding thickness maps for all aquitard layers are given in Fig. 3.

The lowermost, highly compacted pre-Permian represents the basement layer, which is assumed to be hydraulically impermeable due to its burial depth acting as the fourth major aquitard in the model (Fig. 2b and Table 1). The basement thickness gradually increases from northwest to southeast revealing up to ∼ 7000–8000 m thickness across the southern margin ("Flechtingen High") (Fig. 3a). It is followed upward by the Permo-Carboniferous volcanics and the clastic sedimentary Rotliegend, representing a prominent target horizon for geothermal exploration (e.g. Zimmermann et al., 2007). Evaporites, composed mainly of rock salt, form the overlying Permian Zechstein layer. The Zechstein salt layer represents the third regional aquitard in the model (Fig. 2b). Apart from its quasi-impermeable behaviour, rock salt has a relatively high thermal conductivity with respect to values for common sedimentary rocks (Table 1). Due to its thermal and fluid properties as well as its high level of geological structuration with up to 4500 thick salt diapirs locally piercing the overburden (Fig. 3b), this layer exerts a primary role in controlling the deeper groundwater circulation patterns and the thermal field. Above, the Lower Triassic Buntsandstein aquifer mainly consists of clastic sediments. Though limestones are the predominant lithology of the overlying Middle Triassic Muschelkalk, this layer acts as a second hydraulic barrier in the sedimentary succession, due to alternating anhydrite sequences (Fig. 2b and Table 1). In the central and northern parts of the study area, the Muschelkalk aquitard reaches down to −4000 m depth, whereas it

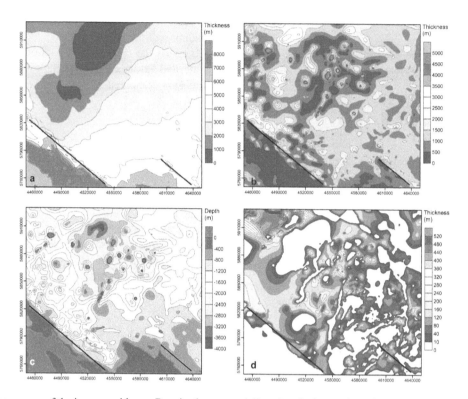

**Figure 3. (a)** Thickness map of the impermeable pre-Permian basement delineating the inverted southern margin where up to ~ 8000 m thick basement comes close to the surface ("Flechtingen High"). The basement thickness thins abruptly near the large Gardelegen and Lausitz escarpments and gradually decreases north of these faults towards the basin centre. Location of the Gardelegen and Lausitz escarpments are given by the solid black lines (also in the following subfigures). **(b)** Map showing the highly variable thickness distribution of the Permian Zechstein, characterized by numerous salt pillows that can reach locally thicknesses of up to 4500 m. Around the salt structures (salt rim synclines), reduced thicknesses indicate areas where the overburden Mesozoic and Cenozoic clastics reach their maximum thicknesses. **(c)** Depth of the top Middle Triassic Muschelkalk. This layer represents the lower limit of the Mesozoic aquifer. **(d)** Thickness map of the Tertiary Rupelian aquitard revealing large geological windows where this clay-rich layer was not deposited or has been eroded.

comes close to the surface at the southern margin (Fig. 3c). The Muschelkalk aquitard is followed upward by the clastic Upper Triassic Keuper, Jurassic, and Lower and Upper Cretaceous layers, together forming the Mesozoic aquifer complex. The clastic Tertiary overlays the Mesozoic aquifer complex. The Tertiary has been resolved into post-Rupelian, Rupelian and pre-Rupelian sub-units. Of these, the clay-rich Rupelian unit is characterized by a very low permeability (Table 1), thus representing the shallowest hydraulic barrier in the sedimentary succession separating the Quaternary and Tertiary units from the deeper aquifers (Figs. 2b and 3d) (cf. Hebig et al., 2012). Over most part of the study area, the thickness of the Rupelian clay varies between ~ 0 and 280 m; only the northwestern part is characterized by a higher thickness up to ~ 520 m (Fig. 3d). Areas where the Rupelian clay has not been deposited or eroded, "Rupelian windows", hydraulically connect the Mesozoic aquifer with the shallow Cenozoic aquifer complex. The latter is formed by the post-Rupelian and the uppermost Quaternary layer consisting of unconsolidated, permeable sediments.

## 2.2 Fault system

The Gardelegen and Lausitz escarpments are two major WNW–ESE-striking structures of the larger Elbe fault system (EFS). The EFS encompasses an approximately 800 km long, WNW–ESE-striking zone extending from the southeastern North Sea to southwestern Poland along the present southern margin of the North German and Polish basins (Scheck et al., 2002). During a Late Cretaceous–early Cenozoic compressional event induced by the Alpine convergence and by the opening of the North Atlantic Ocean, individual faults of the EFS in northern Germany were reactivated to a certain extent as thrust or transpressional faults. Along these faults vertical offsets may reach several kilometres (Scheck et al., 2002). As part of these reactivated structures, the Gardelegen and Lausitz fault zones sub-vertically cut through all sedimentary layers in the structural model and offset the uplifted pre-Permian basement (Flechtingen High) across the southern margin by up to 4 km against the Permian to Cenozoic sedimentary succession.

Faults' orientation within the present-day stress field is a primary factor controlling their hydraulic behaviour with

**Table 1.** Stratigraphic units with predominant lithologies and corresponding physical properties used for the numerical simulations. Hydrogeological barriers separating the stratigraphic succession into different aquifer systems are highlighted (bold).
Porosity and heat capacity after Magri (2005). Permeability values assigned for the Cenozoic (post-Rupelian, Rupelian, pre-Rupelian and Quaternary) after Noack et al. (2013), and for the Mesozoic and Palaeozoic after Magri (2005).
Thermal conductivity and radiogenic heat production used for numerical simulations of the thermal field for the Brandenburg area: thermal conductivities and radiogenic heat production after Bayer et al. (1997), thermal properties for post-Rupelian, Rupelian and pre-Rupelian after Noack et al. (2013) and radiogenic heat production of Rupelian after Balling et al. (1981).

| Stratigraphic unit (predominant lithologies) | Permeability $\kappa$ [m$^2$] | Porosity $\varepsilon$ [%] | Rock heat capacity $c_s$ [MJ m$^{-3}$ K$^{-1}$] | Bulk thermal conductivity $\lambda$ [W m$^{-1}$ K$^{-1}$] | Radiogenic heat production $Q_T$ [$10^{-7}$ W m$^{-3}$] |
|---|---|---|---|---|---|
| Quaternary (sand, silt, clay) | $1.0 \times 10^{-13}$ | 23 | 3.15 | 1.5 | 7 |
| Post-Rupelian (sand, silt, clay) | $1.0 \times 10^{-14}$ | 23 | 3.15 | 1.5 | 7 |
| **Rupelian (clay)** | 1.0E-16 | 20 | 3.3 | 1.0 | 4.5 |
| Pre-Rupelian (sand, silt, clay) | $1.0 \times 10^{-14}$ | 10 | 2.4 | 1.9 | 3 |
| Upper Cretaceous (chalk) | 1.0E-13 | 10 | 2.4 | 1.9 | 3 |
| Lower Cretaceous (clays with sand and silt) | $1.0 \times 10^{-13}$ | 13 | 3.19 | 2.0 | 14 |
| Jurassic (clays with sand, silt, marl) | $1.0 \times 10^{-13}$ | 13 | 3.19 | 2.0 | 14 |
| Keuper (clays with marl and gypsum) | $1.0 \times 10^{-14}$ | 6 | 3.19 | 2.3 | 14 |
| **Muschelkalk (limestone)** | 1.0E-18 | 0.1 | 2.4 | 1.85 | 3 |
| Buntsandstein (silts with sand, clay, evaporite) | $1.0 \times 10^{-14}$ | 4 | 3.15 | 2.0 | 10 |
| **Zechstein (evaporites)** | Impermeable $\sim 0$ | $\sim 0$ | 1.81 | 3.5 | 0.9 |
| Sedimentary Rotliegend (clay, silt, sandstone) | $1.0 \times 10^{-14}$ | 0.3 | 2.67 | 2.16 | 10 |
| Permo-Carboniferous volcanics (rhyolite and andesite) | 1.0E-14 | 0.3 | 2.67 | 2.5 | 20 |
| **Pre-Permian basement** | Impermeable $\sim 0$ | $\sim 0$ | 2.46 | 2.65 | 15 |

critically stressed faults acting as highly permeable hydraulic conduits and non-critically stressed faults acting as hydraulic barriers (Barton et al., 1995). Therefore, a characterization of hydraulic properties of the faults integrated in the model could be achieved via a previous knowledge of the present-day in situ stress field.

The present-day regional stress field in Germany and surrounding regions as summarized by the most recent release of the World Stress Map shows a broad-scale NW–SE direction of the maximum horizontal stress $S_{Hmax}$ in the western part with a rotation toward NE–SW in the easternmost part of the region (World Stress Map, 2000, and references therein). Borehole stress data show a different orientation of $S_{Hmax}$

in the suprasalt and subsalt complexes (Roth and Flecken-stein, 2001). Subsalt in situ stress data show a more uniform $S_{Hmax}$ orientation with an approximate N–S direction open-ing to a more NE–SW direction in the eastern parts of the basin (Marotta et al., 2002; Cacace et al., 2008). In contrast, the stress field in the suprasalt is highly inhomogeneous and shows no dominant direction of the maximum component (Röckel and Lempp, 2003).

From what has been stated above, no simple relationship between in situ stress field and fault pattern can be obtained. In order to represent the different possible stress states, vari-ous geological scenarios have been tested, of which two end-member models are represented: (1) fault zones are consid-ered as barriers to fluid flow because they are non-critically stressed (model 2), and (2) fault zones are supposed to act as hydraulically conductive structures because they are crit-ically stressed (model 3). In addition, a simulation in which no fault is included is also presented (model 1). Simulation results of the two fault models (model 2 and model 3) are then compared to the no-fault case (model 1) to quantify their respective impact on the coupled fluid and heat transport.

## 2.3  Set-up of the numerical model

Three-dimensional coupled fluid and heat transport simula-tions are carried out with the numerical simulator FEFLOW® (Diersch, 2002). This commercial software package is based on the finite-element method (FEM) and enables the mod-elling of coupled fluid flow and transport processes in vari-ably saturated porous media. The governing equations of density coupled thermal convection in saturated porous me-dia are given in Appendix A.

Within coupled simulations, different fluid and heat trans-fer processes are taken into account, including conduction, convection and advection.

Conductive heat transfer occurs due to an existing temper-ature gradient through rock molecules transmitting their ki-netic energy by collision (Turcotte and Schubert, 2002). The flow of heat is directly proportional to the existing tempera-ture gradient via the medium bulk thermal conductivity (see Eq. 3).

Heat transport by moving fluids includes convection and advection. Convective heat transport is a form of buoyant flow due to differences in fluid temperature (or salinity), whereas advection is triggered by gradients in the hydraulic head inducing flow from higher to lower hydraulic potentials (Bjørlykke, 2010). For the present study, only temperature-induced density changes of the fluid are considered and the influence of salinity of the fluid is neglected within the sim-ulations. Mixed convection is the result of all these different fluid and heat transfer processes acting on the same geologi-cal system.

### 2.3.1  FEM model construction

In order to build the FEM, the outlines of the study area ($180\,km \times 200\,km$) need to be defined as a "super-element" in FEFLOW®. To integrate the faults into the model, two lines representing the approximated traces of the fault zones are implemented into the super-element. Within its frame, a two-dimensional unstructured triangle mesh is then gen-erated, referred to as a "reference slice". To build up a 3-D model, copies of the reference slice (all with the same horizontal mesh resolution) are vertically connected at each nodal point of the mesh. One geological layer is represented by the 3-D body between a top and a bottom slice. From the number of 14 geological layers, 15 slices are required to con-struct the 3-D model. To reproduce the geological subsurface structure, the geometry of the stratigraphic layers is derived from the structural model (Fig. 2a). The extracted $z$ coordi-nates (elevations) of each geological top and base surface are assigned to each node of the corresponding top and bottom slice in the numerical model. Therefore, the resulting layer thicknesses a priori determine the vertical resolution of the numerical model. To guarantee numerical stability, the verti-cal resolution of the model is refined by subdividing all layers into two sub-layers each. The model is closed along its base by inserting a planar slice at a constant depth of $-8000\,m$.

According to the main lithology of each geological unit, hydraulic and thermal rock properties are assigned to each corresponding layer in the numerical model (Tables 1). Each layer is considered homogenous and isotropic with respect to its physical properties.

Permeable fault zones are implemented as a combination of discrete feature elements and equivalent porous media. Discrete feature elements are finite elements of lower dimen-sionality, which can be inserted at element edges and faces (Diersch, 2002). We use vertical 2-D discrete elements and assume Darcy's law as governing law of fluid motion within the fault elements as well. A highly permeable (permeabil-ity equal to that of the discrete fault) domain extending for 500 m on either side of the fault trace has been additionally integrated. This domain surrounding the discrete fault ele-ments has been locally refined to ensure a stable calculation of the simulated physical processes. Mesh resolution within this domain is approximately 100 m. Impermeable faults are modelled as equivalent porous media having a lateral extent of 1 km, i.e. equal to the permeable fault zones case.

The final 3-D finite-element model consists of 28 layers (accordingly 29 slices) with approximately 3.5 million ele-ments (i.e. triangulated prisms).

Table 2 summarized the fault properties adopted in the two model realizations.

**Table 2.** Thermal and hydraulic properties used for the fault modelling.

| Properties | Faults – impermeable (model 2) | Faults – permeable (model 3) |
|---|---|---|
| Porosity $\varepsilon$ [%] | $\sim 0$ | $30^1$ |
| Rock heat capacity $c_s$ [M J m$^{-3}$ K$^{-1}$] | $2.6^2$ | $2.49^1$ |
| Bulk thermal conductivity $\lambda$ [W m$^{-1}$ K$^{-1}$] | $2.4^2$ | $2.63^1$ |
| Permeability $\kappa$ [m$^2$] | $\sim 0$ | $1 \times 10^{-12^1}$ |
| Radiogenic heat Production $Q_T$ [$10^{-7}$ W m$^{-3}$] | $0.9^2$ | $0.89^1$ |

[1] Values after Pommer (2012); [2] mean values of the geological layers averaged by their thicknesses.

### 2.3.2 Time setting

Though the present study aims to address the present-day state, transient coupled simulations for all models are run for 250 000 years. This is done to let the system equilibrate, thus obtaining pseudo-steady-state conditions. Given the goal of the study, results are shown only for the final simulation state at 250 000 years.

### 2.3.3 Boundary and initial conditions

A fixed hydraulic head equal to the topographic elevation is assigned at the top of the model. Due to this upper flow boundary condition, groundwater flow is predominantly controlled by gradients in the topography. No-flow boundary conditions are set along the bottom and lateral boundaries of the model.

As upper thermal boundary condition, we assume a fixed constant surface temperature of 8 °C, according to the average surface temperature in northeastern Germany (Diener et al., 1984). At the model base (−8000 m depth), a variable temperature distribution is defined which has been extracted from a lithosphere-scale conductive thermal model of Brandenburg taking into account the thermal effects of the underlying differentiated crust and lithosphere, down to a depth of −125 km (Noack et al., 2012). Figure 4 illustrates the variations imposed to the bottom thermal boundary. Highest temperatures ($\sim$ 260–285 °C) characterize the southern and eastern area. Furthermore, temperatures increase up to $\sim$ 275 °C along the western margin. By contrast, lowest temperatures are present at the southern margin (down to 220 °C) and in the northern and northwestern model area. The regional temperature field at this depth is predominantly controlled by the thickness distribution of the underlying upper crust (characterized by higher values of thermal conductivity and radiogenic heat production) and of the low conductive post-salt deposits (Noack et al., 2012).

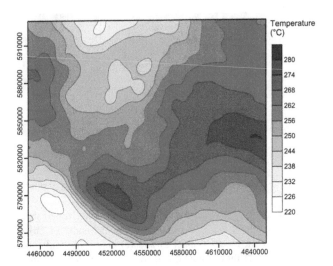

**Figure 4.** Laterally variable temperature distribution at −8000 m depth used as the lower thermal boundary condition for all numerical simulations presented in the manuscript. The distribution has been extracted from a lithosphere-scale conductive thermal model of Brandenburg by Noack et al. (2012).

Initial pressure and temperature conditions are derived from steady-state uncoupled flow and heat transport simulations, respectively.

## 3 Modelling results

### 3.1 Regional thermal field

A preliminary investigation of the regional thermal field at different depth levels enables a first assessment to be carried out of the temperature distribution with respect to the regional hydrogeological setting. This evaluation is necessary to quantify more completely the influence of the fault configurations on the fluid and thermal regime. Therefore, in the following Sects. 3.1.1 and 3.1.2, the results of model 1 in which no fault is included are first discussed and then subsequently compared to the simulation results of the two fault models (model 2 and model 3).

Figures 5a displays the horizontal temperature distribution at −1000 m depth for model 1 (no faults). The results for model 2 (impermeable fault zones) and model 3 (permeable fault zones) are shown in Fig. 5b and c, respectively. The depth level of −1000 m is located below the Tertiary Rupelian aquitard and cuts through the pre-Rupelian Mesozoic aquifer as well as the pre-Permian basement along the southern margin of the basin (Figs. 2b and 3d). The temperature maps at −3000 m depth of model 1 (Fig. 5d), model 2 (Fig. 5e) and model 3 (Fig. 5f) cut through the Mesozoic sediments, major Permian salt diapirs and the pre-Permian basement at the southern margin (cf. Fig. 3c and d).

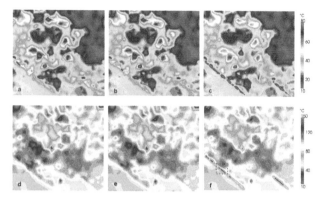

**Figure 5. (a–c)** Temperature distribution along a horizontal slice at −1000 m depth cutting the **(a)** no-fault model (model 1), **(b)** impermeable fault model (model 2) and **(c)** permeable fault model (model 3). **(d–f)** Temperature distribution along a horizontal slice at −3000 m depth cutting the **(d)** no-fault model (model 1), **(e)** impermeable fault model (model 2) and **(f)** permeable fault model (model 3). In **(f)** the black rectangle displays the position of the map view in Fig. 10b. All temperature maps encompass the complete model domain. Note the significant difference in the modelled temperatures in the proximity of the faults for model 3 compared to the very similar model 1 and model 2.

### 3.1.1   Horizontal temperature distribution: −1000 m depth

At −1000 m depth, temperatures range between ~30 and 50 °C in wide parts of the study area. Local spots of higher temperatures (up to 80 °C) are also visible in the western and central domains (Fig. 5a). Across the southern margin of the basin, temperatures are sensitively lower (~22–34 °C) and are even colder in larger parts of the north and northeastern model area (10–20 °C).

The regional temperature pattern can be correlated to the thickness of the overlying Rupelian aquitard (Fig. 3d). Locations where the Rupelian clay is missing (Rupelian windows) correspond to areas of strongly reduced temperatures. The reason for the observed cooling trend should be related to cold water inflow from the top surface (set to 8 °C) which penetrates unhampered into deeper parts of the model, thus leading to the modelled cooling within the Mesozoic aquifer (cf. Fig. 2b). Higher topographic elevations (recharge areas) above these Rupelian windows (cf. Fig. 1b) enhance cold water advection and hence the observed cooling.

However, local spots of distinctly higher temperatures in the western and central domains can be correlated with areas with an increased thickness of the Rupelian aquitard, which prevents cold water inflow from the surface to reach this depth levels (cf. Fig. 3d). As a consequence, significantly higher temperatures are generated below the Rupelian. At these locations, temperatures additionally increase in the vicinity of major Zechstein salt diapirs (Fig. 3b) due to the thermal blanketing effect by thick and low conductive Mesozoic sediments.

Across the southern margin, larger areas of lower temperatures correspond to domains in which the pre-Permian basement almost reaches the surface (Flechtingen High) (Fig. 3a). The thick impermeable and thermally highly conductive basement leads to very efficient conductive heat transport towards the surface. Due to the absence of overlying sediments, heat cannot be stored by thermal blanketing and gets lost at the surface, thus explaining the observed thermal trend along the margin of the basin (cf. Fig. 2a).

### 3.1.2   Horizontal temperature distribution: −3000 m depth

At −3000 m depth the thermal field shows a wider range of variations with respect to the shallower temperature distribution described above. At this depth level, modelled temperatures vary between minima of approximately 10 °C and maxima of up to 150 °C (Fig. 5d). Local spots of reduced temperatures (~44–52 °C) occur in the central and northwestern parts. As already observed at −1000 m depth, the southern margin is characterized by relatively lower temperatures with respect to the central domains (~80 °C). Higher temperatures of up to 150 °C, however, occur in the western, northwestern, eastern and central parts. These positive thermal anomalies show both short-wavelength ("spot-like anomalies") and long-wavelength ("elongated anomalies") characteristic geometries.

The occurrence of local negative thermal anomalies in the central and northwestern domains can be structurally linked to the topology of the Rupelian aquitard and of the Middle Triassic Muschelkalk aquitard, the latter limiting the Mesozoic aquifer at its base (Fig. 3c). Indeed, cooler temperatures are present in areas where the Mesozoic aquifer reaches deeper than −3000 m beneath major hydraulic windows in the Rupelian thickness distribution (cf. Fig. 3a and b). Therefore, cold water from the top surface can flow downward and penetrate to greater depths.

The long term spatial distribution of modelled temperature in the remaining part of the study area is mainly controlled by the geometry and thickness of the pre-Permian basement (Fig. 3a). Reduced temperatures are present across the southern margin, where the basement reaches its highest thickness and shallowest depths. Similar to the thermal pattern at −1000 m depth, heat loss is caused by the lack of insulating cover sediments. By contrast, domains of increased temperatures in the basin centre evolve beneath a thick basement overlain by a thick sequence of Mesozoic sediments that act as a thermal blanket.

Additionally, temperatures rise in the vicinity of major salt diapirs (spot-like anomalies). These are locations where low conductive Mesozoic and Cenozoic sediments reach their highest thicknesses and hence where their thermal blanketing effect is most effective.

From the results described above it can be concluded that the temperature distribution is the result of superposed thermal effects generated by the complex interaction between the hydrodynamics induced by the boundary settings and the hydrogeological configuration of the basin comprising different sediment types as (a) the permeable but thermally insulating Mesozoic sediments, (b) the intercalated but partially discontinuous aquitards (Rupelian and Middle Triassic Muschelkalk) (c) the impermeable but thermally conductive Zechstein salt and (d) the pre-Permian basement.

Modelled temperature distributions for model 2 (impermeable faults) show a striking resemblance with those obtained for model 1 (no faults) both at −1000 and −3000 m depth (Fig. 5b and e). In order to quantify the range of magnitudes of calculated differences between the two model realizations, as well as their spatial correlation, contour maps of temperature differences between model 1 and 2 are shown in Fig. 6a (−1000 m depth) and Fig. 6b (−3000 m depth). A close inspection of these two figures reveals only minor temperature differences (maximum about 10 K), which spatially correlate with the location and geometry of the two fault zones. The localized temperature differences and the overall similarity to the no-fault scenario indicate that the impermeable fault zones only have a minor influence on the thermal field. Adding to this first conclusion, it is worth emphasizing that the areas affected by the presence of the fault zones can be further used to spatially constrain what we from here on refer to as the "range of influence" of the two faults on the regional thermal pattern and groundwater dynamics. Such a range of influence is found to vary between the two fault zones (10 km for the Gardelegen Escarpment and 4 km for the Lausitzer Escarpment). This last aspect can be related to existing differences in geometry and spatial extension between the two faulted areas – an observation that clearly points toward the presence of a thermal and pressure field – which are structurally controlled by the local fault configuration cutting through the sedimentary sequence.

At −1000 m depth, model 3 (highly permeable faults) shows distinctly cooler temperatures (10–35 °C) along both fault zones with an alternating thermal signature (Fig. 5c). A temperature difference map between model 1 (no faults) and model 3 shows differences of ∼ 20 to 80 K at the permeable fault zones (Fig. 6c). As found for model 2, the range of influence varies between the two fault areas, at the Lausitzer fault zone ranging between ∼ 3.5 and 7.4 km and that at the Gardelegen fault zone between ∼ 2.4 and 8.8 km.

Spatial variations in the extent of the range of influence for a single fault zone are structurally related to the presence of differing sedimentary layers bordering the fault zone. The range of influence is decreased in areas where the fault zone is bordered the pre-Permian, Zechstein and Muschelkalk aquitards. Due to their hydraulically impermeable setting, heat is transferred only by diffusion within these layers, thus leading to a less effective propagation of thermal anomalies far from the fault domains. By contrast, the range of influence is increased in those areas where the fault zones are next to aquifers because the heated fluid can spread out into

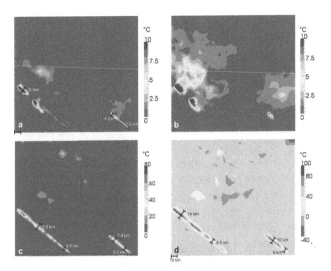

**Figure 6.** (a–b) Temperature difference maps between no-fault model 1 and impermeable fault model 2 at (a) −1000 m depth and (b) at −3000 m depth. (c–d) Temperature difference maps between no-fault model 1 and permeable fault model 3 at (c) −1000 m depth and (d) at −3000 m depth. Positive values indicate higher temperatures in the no-fault model and vice versa. Local temperature differences (maximum 10 K) indicate that the impermeable fault zones slightly affect the temperature distribution. By contrast, the permeable fault zones exert a significant influence on the thermal field, reflected by distinctly cooler temperatures (of maximum 100 K) that characterize the internal parts of the fault zones.

the permeable sediments, reaching greater distances within the layers.

Differences in the local thermal field can also be observed at −3000 m depth between model 3 and model 1 (Fig. 5f). Along the fault planes, isotherms are arranged in an alternating pattern, displaying values between 10 and 56 °C. At these locations temperatures are about 35–45 K (maximum 100 K) lower than those obtained for model 1 (no faults) (Fig. 6d). The total range of influence varies between ∼ 6 and 10 km for the Lausitz fault zone and between ∼ 6.5 km and 14 km for the Gardelegen fault zone.

Moving away from the fault zones, the three models show the same regional trend in the resulting thermal field. These similarities found in the regional thermal configuration should be considered as an additional indication of the sub-regional influence exerted by the two fault zones. Indeed, considering a different fault configuration does not seem to alter the main physical processes driving thermal and groundwater transport in the sedimentary compartments away from the faults' range of influence proper.

By looking at Fig. 6 it can be noticed that despite a very similar regional thermal trend between all three models, differences are found in computed magnitudes at greater distance from the fault zones.

In the case of model 1 (no faults) minus model 2 (impermeable fault zones), these differences range between 1 and maximum 8 K and are spatially constrained to spots of relatively small extent in the centre of the model domain north of the two fault zones (Fig. 6a and b). In the case of model 1 (no faults) minus model 3 (permeable fault zones), spots of temperature differences are confined to the central and northern model domain at −1000 m depth (Fig. 6c). At −3000 m depth, temperatures locally differ from the no-fault model in the central domain (−20 to 20 K) as well as at the NE edge (up to −40 K) (Fig. 6d).

In this regard it is worth mentioning that the three models differ only with respect to the adopted parameterization considered for representing the two fault zones. Accordingly, the observed local variations in magnitudes result from having different model configurations. Indeed, all simulations are performed considering the system thermodynamically closed, which in turn requires an overall conservation of the internal energy of the system. As the boundary conditions are the same in all three simulations, changes implemented for the domains of the fault zones will also cause differences for the remaining model domain in response to mass- and energy conservation. Therefore, local changes in magnitudes of the temperature field as observed in these areas far away from the fault zones are the consequence of the differences in model configuration. These differences can be considered as a physical sound consequence of variations in the groundwater dynamics and related thermal field due to different hydrogeological characteristics for the three models. Following this reasoning, it naturally follows that the greatest differences in magnitude of modelled temperatures are to be expected for those simulations which differ the most. Indeed, the largest difference is between model 1 and model 3 (Fig. 6c and d).

### 3.2 Temperature distribution in the fault zones

To understand the pronounced thermal signature within both permeable faults (model 3), a more detailed analysis of the thermal state along the entire fault zones is carried out. Figure 7a displays the temperature distribution along the Gardelegen and Lausitz fault zones. Temperatures range between 8 and 75 °C along both faults. Convex upward-shaped isotherms (corresponding to temperatures as high as 35–75 °C) alternate with isotherms that are bent convexly downward (indicating temperatures as low as 8–30 °C). This alternating temperature pattern reflects thermal anomalies already described in the regional thermal field at −1000 and −3000 m depth along the permeable fault zones (Fig. 5c and f).

A combination plot of isotherms and fluid velocity vectors along the Gardelegen Fault (similar conclusions are also valid for the Lausitz fault zone) shows that convex downward-shaped isotherms (cold domains) correspond to relatively fast downward groundwater flow ($\sim 1 \times 10^{-2}$–$3 \times 10^{-2}$ m d$^{-1}$) (Fig. 7b). By contrast, convex upward-bent isotherms conform to mainly upward groundwater flow of lower fluid velocities ($\sim 1 \times 10^{-4}$–$1 \times 10^{-3}$ m d$^{-1}$).

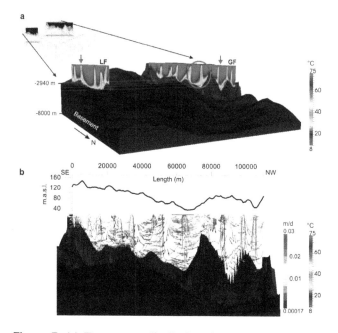

**Figure 7.** (a) Temperature distribution along the fault planes in strike direction of the permeable Gardelegen (GF) and Lausitz faults (LF). Alternating hotter and colder domains characterize the thermal state within both fault zones. The red arrows indicate the location of two vertical cross sections shown in Fig. 8. The cross sections cut through the faults zones parallel to the N–S axis. The red circle shows the location of the close-up of Fig. 10 (vertical exaggeration: 7 : 1). (b) Combination plot of fluid velocity vectors (length of vectors are not scaled) and isotherms along the entire length of the Gardelegen Fault assumed permeable (model 3). At the top, the topographic elevation used as a proxy for hydraulic head boundary condition at the surface is delineated along the trace of the fault zone. Down- and upward-oriented flow are initiated by hydraulic head gradients but may be enhanced by a convective fluid flow component.

The temperature distribution and corresponding fluid dynamics within both faults can be spatially correlated with the topographic elevation along the fault zones (Fig. 7b). At areas of high topographic elevation, recharge areas, cold water (8 °C) is forced to enter from the surface of the model due to the hydraulic boundary conditions imposed. According to Darcy's law (see Eq. 2 in Appendix A), steep hydraulic head gradients result in higher flow velocities of infiltrating water. Due to the higher permeability of the fault zone (Table 2), fluid can easily flow downwards, being fast enough to result in the observed cooling. Reaching deeper parts of the fault zone, the fluid is heated by thermal equilibration with the surrounding matrix system (its velocity diminishes with depth) and will rise upwards at specific areas following the distribution of the hydraulic head gradients. These hotter domains within the fault mainly correspond to areas of lower topographic elevation, discharge areas.

In conclusion, fluid motion within the fault is driven principally by fluid advection due to hydraulic head gradients

Influence of major fault zones on 3-D coupled fluid and heat transport for the Brandenburg region...

125

imposed along the top boundary. Depending on the pressure and temperature conditions at depth, local upward movement of the fluid may be locally enhanced by buoyancy forces. However, thermally buoyant forces exert only a secondary contribution to the fluid movement, thus having a little impact on the resulting thermal field.

## 3.3 Interaction between fault zones and surrounding sediments

To analyse the temperature field and fluid behaviour across the fault zones, and to understand the interaction between fault zones and surrounding sediments, two representative vertical profiles are chosen. The first profile cuts through a recharge zone at the Gardelegen Fault (GF), whereas the second profile dissects a discharge area at the Lausitz fault zone (LF) (Fig. 7a). Based on these two profiles, the results are visualized by focusing only on the fault zone areas. For both sections, both temperature distribution and flow field are illustrated for models 1, 2 and 3 (Figs. 8a–f and 9a–f). These profiles should be considered as representative for the other parts of the fault zones characterized by similar hydraulic conditions.

### 3.3.1 Recharge area – Gardelegen fault zone

#### Model 1

An initial consideration of the temperature distribution for model 1 reveals relatively flat isotherms in the area of the pre-Permian basement, displaying somewhat cooler temperatures in the south (Fig. 8a). Convex upward-shaped isotherms characterize the central part of the profile in the narrow transition zone between the deep and the shallow Rotliegend aquifer within a transition region extending between the pre-Permian basement and the Permian Zechstein. Within the Permian Zechstein layer, the isotherms distinctively bend upwards. This thermal anomaly indicates locally higher temperatures in the salt decreasing in the surrounding sediments.

Flat isotherms in the area of the pre-Permian basement reflect conductive heat transfer through this hydraulically impermeable layer. In addition, its higher thermal conductivity leads to an efficient heat transfer towards the surface. In the south, the basement is nearly exposed at the surface and it is not covered by insulating sediments. Due to its special geological setting, reduced temperatures are observed. This cooling trend reflects the thermal signature already observed in the regional thermal field along the southern margin of the study area (Fig. 5a and d).

The observed thermal anomaly throughout the Permian Zechstein salt is caused by thermal refraction, triggered by the sharp contrast between the thermally highly conductive salt and the low thermal conductivity of the surrounding sediments. Because the salt acts as a heat chimney, conductively transferring the heat towards the surface, it results in higher temperatures in the salt than within the surrounding less conductive sediments.

So far, the results demonstrate that conductive heat transfer decisively shapes the local thermal field through the thick pre-Permian basement and the Permian Zechstein, both impermeable to fluid flow and thermally more conductive than the surrounding sediments (Table 1).

In the shallow Cenozoic, Mesozoic and Buntsandstein aquifers, fluid velocity vectors indicate a diverging regional flow at the vertical offset in the central part of the profile (marking the position of the Gardelegen Fault) (Fig. 8b). This flow pattern results from the location of the profile beneath a major recharge area where cold water enters into the model and flows into the shallow aquifer sediments following the regional groundwater flow.

In the Rotliegend aquifer, fluid velocity vectors indicate regional flow from north to south. They also display a hydraulic connection of reduced velocities ($\sim 1 \times 10^{-6}$ m d$^{-1}$) within the narrow transition domain between the deep and shallow Rotliegend aquifer. Due to the adjacent impermeable basement, fluid flows upward following the geometry of the basement and flanking Zechstein salt layer. Upward flow of heated fluid leads to the observed upward-bent isotherms across this transition domain (Fig. 8a). By entering the shallow Rotliegend aquifer, fluid locally mixes with the regional flow.

#### Model 2

The thermal field for model 2 shows only weak disturbances in the fault zone area with respect to the case previously described (Fig. 8c). Compared to the thermal field of the no-fault model 1 (Fig. 8a), only a slight temperature increase (max. 6 °C) can be distinguished in the domain of the hydraulic connection between the shallow and deep Rotliegend aquifer. Moving to the fluid velocity results (Fig. 8d), the results nicely show that fluid cannot enter the fault zone. Instead, fluid flow is redirected along the impermeable structure with very low fluid velocities ($\sim 1 \times 10^{-10}$ m d$^{-1}$) and flows laterally into the aquifers with increased velocities ($\sim 1 \times 10^{-8}$–$1 \times 10^{-6}$ m d$^{-1}$). Generally, fluid flow direction and fluid velocities are similar to the flow field of the no-fault model 1 within the aquifer systems (cf. Fig. 8b).

The presence of a tight fault has only a minor influence on the flow dynamics. Indeed, neither the fluid velocity in the surrounding sedimentary layers nor the fluid direction appears to change in the vicinity of the fault zone when compared to the no-fault model 1 (Fig. 8b). Very low fluid velocities along the fault-matrix boundary and spatially limited influence of the fault zone result in no remarkable influence on the thermal field. A minor temperature difference between the two models results from the locally disturbed fluid flow at the fault offset which marks the transition between shallow and deep Rotliegend aquifer.

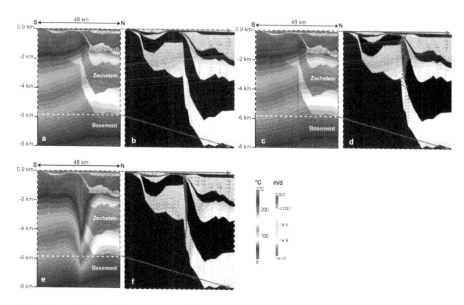

**Figure 8.** N–S cross section through the northwestern part of the Gardelegen fault (GF) plane (the location of the cross section is outlined in Fig. 7a by the red arrow on the right-hand side). In the combination plots of temperature distribution and fluid velocity vectors (**b, d, f**), the lengths of the fluid velocity vectors are non-scaled and the temperature distribution is shown with reduced intensity in the background. (**a**) Temperature distribution for model 1 in which no faults are integrated. The almost flat character of the isotherms reflects the diffusive nature of conductive heat transfer throughout the impermeable pre-Permian basement. Isotherm deflection is only present where the conductive Zechstein salt is thick. (**b**) Fluid velocity vectors and temperature distribution for the no-fault model 1 indicating horizontal flow in the upper aquifers. Predominantly upward directed flow only occurs where a hydraulic connection exists between the shallow and the deep Rotliegend aquifer. (**c**) Temperature distribution for the model 2 in which the fault zones are integrated as impermeable structures. The position of the fault zone is displayed by the dotted red line. Only a slight temperature increase is observed at the offset between shallow and deep Rotliegend aquifer when compared to the no-fault model. (**d**) Fluid velocity vectors and temperature distribution for the impermeable fault model 2. Very low fluid velocities are evident along the fault, and the communication of shallow and deep Rotliegend aquifers is inhibited. Apart from that, within the surrounding aquifer sediments, fluid direction and fluid velocities are similar to the no-fault model 1. (**e**) Temperature distribution for the permeable fault model 3. A strong cooling is induced by the permeable fault zone cutting through the central part of the profile (indicated by the dotted red line). (**f**) Fluid velocity vectors and temperature distribution for fault model 3 displaying a fast downward-oriented flow inside the permeable fault zone and a lateral outflow into the surrounding aquifer sediments.

## Model 3

Considering the fault zone as more permeable than the neighbouring sedimentary layers significantly changes both the thermal field (Fig. 8e) and the fluid circulation pattern (Fig. 8f) compared to the no-fault case (Fig. 8a and b). A negative thermal anomaly characterizes the thermal state in which the isotherms are sharply bent downward, indicating a strong cooling in the central domain of the fault plane. This pattern continues to greater depth but weakens towards the pre-Permian basement. Along both sides of the fault zone, a step-wise increase in modelled temperatures is visible at a distance of approximately 3 km from the fault plane. The corresponding fluid velocity vectors display a fast downward-oriented flow (of up to $0.003\,\mathrm{m\,d^{-1}}$) in the central part of the fault zone (Fig. 8f). Along the fault flanks, velocity vectors indicate a lateral fluid outflow from the fault zone into the permeable sediments of the different aquifer systems. While this lateral flow direction is maintained throughout the aquifer systems, the fluid velocity gradually decreases with increasing distance from the fault zone.

The observed fast downward-oriented flow in the fault zone explains the strong cooling observed at the same location. The relatively strong temperature drop is induced by the surface water inflow ($8\,^{\circ}\mathrm{C}$) from the upper tip of the fault zone located beneath a major recharge area (cf. Fig. 7b). The higher permeability of the fault zone translates to high fluid velocities and downward-oriented flow through the fault zone. Reaching the lower tip of the fault, fluid cannot penetrate the impermeable pre-Permian basement. Within this layer, heat is transferred only by conduction. This transition between heat transfer processes is displayed by the weakening of downward-bent isotherms throughout the pre-Permian basement layer.

As a possible further consequence of the fast downward flow in the fault zone, fluid penetration occurs from the fault zone into the surrounding aquifers. The gradual decrease of the fluid velocities with increasing distance from the fault mainly results from the differences in the permeability of the fault and of the surrounding aquifers, and the permeability decreases from the fault zone to the sedimentary layers

Influence of major fault zones on 3-D coupled fluid and heat transport for the Brandenburg region...

127

(Tables 1 and 2). Accordingly, the velocity of modelled lateral flow may vary within each aquifer.

The induced lateral outflow from the fault zones into the surrounding aquifers is also reflected by an increased range of influence of the permeable fault zones in the regional temperature distribution at −1000 and −3000 m depth as described previously (cf. Fig. 5c and f).

The shallow and deep Rotliegend aquifers are still hydraulically connected through offset by the presence of the permeable fault, which considerably changes fluid pathways and fluid velocities. Most significantly, the direction of fluid flow is reversed along the offset, changing from slowly upward flow in the no-fault case (Fig. 8b) to fast downward-oriented flow in the permeable fault zone (Fig. 8f). As a further consequence of the downward flow in the fault, fluid flow direction is inverted within the deep Rotliegend aquifer sediments.

In summary, when located beneath a major recharge area, the highly permeable fault is characterized by downward-oriented fluid flow which induces net cooling in the permeable fault zone. In the adjacent aquifers, fluid flow is directed away from the fault zone and varies in magnitudes in relation to the specific permeability values of the individual sedimentary layers.

### 3.3.2 Discharge area – Lausitz fault zone

#### Model 1

In the no-fault model 1, flat and almost horizontal parallel isotherms characterize the area of the thick pre-Permian basement, whereas slightly disturbed isotherms can be observed within the Zechstein salt (Fig. 9a). In the Mesozoic aquifer, the isotherms indicate slightly decreased temperatures below the Rupelian windows in the northern part of the profile.

The geometric distribution of the isotherms within the basement points to conductive heat transfer through this impermeable layer. Thermal refraction is reflected by the pattern of modelled isotherms throughout the Zechstein salt and surrounding sediments. Cooler temperatures in the Cenozoic and Mesozoic aquifers are induced by the inflow of cold surface water through the Rupelian windows. Beneath the latter hydraulic window, fluid velocity vectors indicate vigorous fluid flow (Fig. 9b). In the deeper Buntsandstein and Rotliegend aquifers, a predominant lateral fluid flow direction from north to south is displayed by the fluid velocity vectors. This regional trend is only disturbed at the narrow transition between deep and shallow Buntsandstein and Rotliegend aquifers (marking the location of the Lausitz fault zone) where the basement and Zechstein salt are closest.

The regional flow direction from north to south is induced by the location of the profile at a discharge area, adjacent to a major recharge zone in the north (cf. Fig. 1b). Due the steep hydraulic head gradient (connecting the discharge with the recharge area), enhanced fluid inflow occurs from the north and results in the regional north–south-directed flow pattern. At the narrow hydraulic connection between deep and shallow Buntsandstein and Rotliegend aquifers, fluid flows upwards by following the relief of the impermeable basement and Zechstein salt layers. By reaching the shallow parts of the Buntsandstein and Rotliegend aquifers, fluid mixes with the shallow aquifer fluids.

#### Model 2

The isotherm pattern for model 2 (Fig. 9c) is very similar to the thermal field of the no-fault model 1 (Fig. 9a) and confirms previous conclusions made on the regional temperature field (Fig. 5b, e) and on the temperature distribution around the Gardelegen fault zone (Fig. 8c). In the shallow Cenozoic and Mesozoic aquifers, fluid velocity vectors resemble the regional trend as observed in no-fault model 1 (Fig. 9d). Along the fault zone, fluid is characterized by very low fluid velocities ($\sim 1 \times 10^{-10}$ m d$^{-1}$). Fluid flow is locally disturbed and redirected north of the fault zone within the deeper part of the Buntsandstein aquifer. Throughout the deeper part of the Rotliegend aquifer, the regional north–south flow pattern is dominant. Compared to the no-fault model 1, fluid flows from the shallow Buntsandstein and Rotliegend sediments towards the fault zone and no fluid mixing can be observed at the narrow transition.

Along the fault zone, fluid flow is characterized by very low fluid velocities due to impermeable conditions of the latter. Because fluid cannot discharge into the fault zone, it is redirected in the adjacent aquifer sediments. No hydraulic connection exists between the shallow and deep Buntsandstein and Rotliegend aquifer domains. Therefore, no mixing occurs between the shallow and deep aquifer. But fluid flows laterally from the shallow Buntsandstein and Rotliegend sediments towards the impermeable fault zone.

In conclusion, the results for model 2 confirm a spatially limited impact of a tight fault zone which leads to an almost unchanged temperature field when compared to the no-fault scenario. The effects of the impermeable fault zone are only local and result in a disturbance of the flow field near the fault zone which inhibits any hydraulic communication between shallow and deep aquifers which are offset at depths by the fault (Buntsandstein and Rotliegend).

#### Model 3

In the case of a permeable fault (model 3), significant differences can be observed in the thermal field compared to the models 1 and 2 (Fig. 9e). In the central part of the fault zone, the isotherms are bent upwards, revealing locally increased temperatures, which gradually decrease within the underlying pre-Permian basement. At the fault–sediment interface, temperatures are decreased (e.g. 75 °C at −3500 m

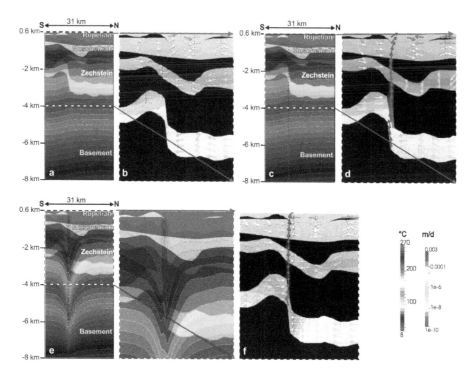

**Figure 9.** N–S cross section through the southeastern part of the Lausitz fault (LF) plane (the location of the cross section is outlined in Fig. 7a by the red arrow on the left-hand side). In the combination plots of temperature distribution and fluid velocity vectors (**b, d, f**), the lengths of the fluid velocity vectors are non-scaled and the temperature distribution is shown with reduced intensity in the background. (**a**) Temperature distribution for model 1 in which no faults are implemented. The isotherms indicate a net cooling effect in the shallow Cenozoic and Mesozoic aquifers, induced by unhampered cold water inflow through the Rupelian windows in the northern part of the profile. (**b**) Fluid velocity vectors and temperature distribution for the no-fault model 1. The regional flow direction is from north to south. At the offset, where the basement, Zechstein and Buntsandstein layers are in close contact, fluid mixing occurs within the aquifers. (**c**) Temperature distribution for the impermeable fault model 2. The position of the fault zone is displayed by the dotted red line. The isotherm pattern closely resembles the thermal field for the no-fault model 2. (**d**) Fluid velocity vectors and temperature distribution for model 2. Because the fault zone acts as a fluid flow barrier, no hydraulic connection exists between deep and shallow aquifers at the offset. (**e**) Temperature distribution for the permeable fault model 3. The temperature zoom on the right shows that the isotherms are bent upwards inside the fault zone, indicating a temperature increase down to a depth of $\sim -4000$ m. (**f**) Fluid velocity vectors and temperature distribution for the permeable fault model 3. Fast upward-oriented flow characterizes the central part of the fault zone, whereas lateral fluid advection is observed from surrounding sediments towards the fault with reduced fluid velocities.

depth) compared to the temperatures within the surrounding aquifer sediments ($\sim 110\,°C$ at $-3500$ depth).

Fluid velocity vectors display a fast upward directed flow in the central part of the fault zone (max. $1 \times 10^{-4}$ m d$^{-1}$) (Fig. 9f). Similar to model 1, fluid velocity vectors display vigorous fluid flow below the Rupelian windows within the shallow Cenozoic and Mesozoic aquifers and a regional flow from north to south in the deeper Buntsandstein and Rotliegend aquifers north of the permeable fault zone. In general, fluid inflow occurs from all surrounding aquifer sediments into the fault zone ($\sim 1 \times 10^{-6}$–$1 \times 10^{-4}$ m d$^{-1}$).

Fast upward-oriented flow reflects a local temperature increase within the fault zone (Fig. 9e). Within the pre-Permian basement, fluid flow is impeded and the heat is transferred by conduction only. Within the fault zone, fluid flows upwards following the hydraulic head potential (cf. Fig. 7b).

The temperature drop next to the fault zone is induced by an inflow of colder fluid ($8$–$10\,°C$), which has been initially transported downwards the fault zone below a recharge area (cf. Figs. 7b, 8f), laterally impressed into the aquifer sediments (Fig. 8f), and from there has been finally redirected towards the fault zone. This lateral inflow from the aquifer into the fault zone is induced the by high permeability contrast between fault and surrounding sediments (Tables 1 and 2). At the fault–sediment interface thermal equilibration takes place, thus resulting in the observed overall temperature gradient between the fault and the sedimentary covers. The main result of this complex dynamics is reflected in the presence of a cone of depression in the temperature centred at the upflow domain within the flow, the extent of which increases with depth due to larger temperature gradients between the fault and the sediments reaching greater depth levels.

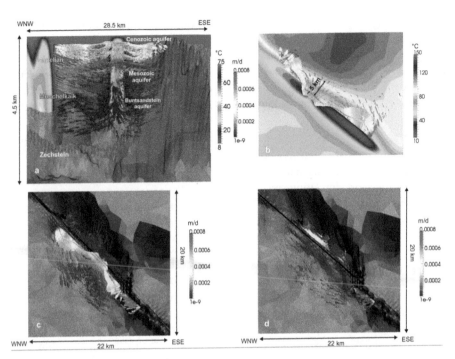

**Figure 10. (a)** Close-up of the 3-D flow field around the permeable fault zone (model 3) above the Permian Zechstein layer (grey-shaded). The temperature distribution along the strike of the Gardelegen fault zone is displayed in the background. Location of the zoom is delineated in Fig. 7a by the red circle. The velocity vectors clearly indicate flow from the surrounding aquifers towards the permeable fault zone. The lengths of the fluid velocity vectors are non-scaled. **(b)** Top view of Fig. 10a zooming in on the temperature distribution at −3000 m depth (position is outlined in Fig. 5f by a black rectangle) indicating the range of influence of the fault on the flow and thermal field. **(c)** Top view on the 3-D flow field around the permeable fault zone (model 3) for the same location as in Fig. 10a and b. Fluid advection towards the fault zone is displayed by the vectors, and is induced by the high permeability contrast between fault and neighbouring sedimentary layers. **(d)** Top view on the 3-D flow field around the impermeable fault zone (model 2) for the same location as in the previous subfigures. Fluid flow in the surrounding sediments is redirected parallel to the fault zone due to the impermeable nature of the fault.

In conclusion, upward-oriented fluid flow induces higher temperatures in the permeable fault zone. Fluid inflow is observed from the less permeable surrounding aquifers into the permeable Lausitz fault zone located beneath a major discharge area.

### 3.3.3   3-D sections

To get a final overview of the fluid behaviour around both fault zones, 3-D close-up views are shown for the two fault models (model 2 and model 3).

Figure 10a shows a close-up plot of the flow field in the sedimentary layers above the Zechstein salt layer around the permeable Gardelegen fault zone (model 3). The temperature field along the strike direction of the Gardelegen fault zone is shown in the background.

Velocity vectors indicate increased fluid inflow from the top of the section into the Cenozoic aquifer with highest velocities ($4 \times 10^{-4}$–$8 \times 10^{-4}$ m d$^{-1}$) down to the depth level of the Rupelian aquitard. Throughout the Mesozoic aquifer, fluid velocities are reduced ($\sim 2 \times 10^{-4}$–$4 \times 10^{-4}$ m d$^{-1}$) compared to the shallow Cenozoic aquifer. Below the Muschel-

kalk aquitard, fluid velocities further decrease down to $\sim 1 \times 10^{-9}$ m d$^{-1}$ within the Buntsandstein aquifer.

While fluid velocities gradually drop with depth due to decreasing permeabilities of the aquifers (Table 1) and aquitards, modelled fluid flow shows a consistent direction within the whole aquifer systems. A downward-oriented flow is displayed in the shallow Cenozoic aquifer. Downward flow occurs in areas where cooler temperatures ($\sim 8$–$25\,°C$) characterize the internal part of the fault zone (in the background, Fig. 10a). Fluid is then redirected towards the fault zone in all aquifers due to the high permeability of the latter. Finally, upward-oriented flow can be seen in areas where higher temperatures ($\sim 25$–$50\,°C$) are present in the fault zone.

Looking from top of the 3-D image in map view, we can conclude that the fault zone influences the fluid movement and the temperature field in the surrounding sediments at −3000 m depth within an influence radius of $\sim 5$ km (Fig. 10b; cf. Fig. 5c and f).

Figure 10c and d show top views for the same 3-D block section as in Fig. 10a for the permeable fault model 3 and for the impermeable fault model 2, respectively. Fluid velocity vectors clearly reflect enhanced fluid advection from the surrounding sediments towards the permeable fault zone

(Fig. 10c). By contrast, fluid cannot enter the impermeable fault zone (Fig. 10d). Furthermore, overall reduced fluid velocities characterize the flow in close proximity to the fault zone ($\sim 1 \times 10^{-9}$–$4 \times 10^{-4}$ m d$^{-1}$) where the fluid flow is deviated along the same structure. The range of influence is restricted to a distance of $\sim 1$ km on each side of the fault.

## 4 Discussion and conclusions

The investigation of fault zones with varying permeability by 3-D coupled fluid and heat transport simulations reveals a distinct local influence on the regional thermal field and fluid system for the different permeability configurations considered.

### 4.1 Regional thermal field without faults

The regional temperature distribution at different depth levels is the result of superposed thermal effects generated by the complex interaction between aquifers and aquitards of varying thickness and different fluid and heat transport processes, i.e. advection, buoyant flow and heat conduction.

Advective processes strongly affect the shallow thermal field in the Cenozoic and Mesozoic aquifers down to a maximum penetration depth of approximately $-3000$ m. Where the Rupelian clay is missing, inflow of cold water from the top surface, as triggered by the upper boundary conditions, induces a pronounced cooling in the Mesozoic aquifer. Three-dimensional models of coupled fluid and heat transport of the NEGB have already shown that pressure forces triggered by local topographic gradients may be strong enough to induce a net cooling on the shallow aquifer system, whereby higher permeability of the corresponding layers promotes greater penetration depths of cold water (Kaiser et al., 2011).

Indications for convective flow in the shallower Mesozoic aquifer play an additional, though secondary, role in areas confined by a thick sequence of Rupelian clay, preventing inflow of cold water from the Cenozoic aquifer and shielding the influence of superficial hydraulic head gradient on the underlying Mesozoic aquifer (Noack et al., 2013).

Heat conduction through the thick pre-Permian basement and the Permian Zechstein (both impermeable to fluid flow but thermally higher conductive than the surrounding sediments), decisively shapes the local thermal field across the southern margin and in the vicinity of salt diapirs in the central basin.

### 4.2 Influence of impermeable fault zones

When implementing fault zones as low permeable structures, the temperature distribution resembles the regional thermal field modelled when no faults are considered. The influence of impermeable faults on the flow field is local, limited to the fault zone itself and its close proximity. Fluid flow, both

the direction and magnitudes of fluid velocity, within the surrounding sediments is unaffected by the presence of a tight fault. At the fault zone, fluid flow is deviated from the sediments with very low velocities. Acting as hydraulic barriers, fault zones prevent a lateral fluid inflow. Fluid cannot be transmitted through the fault zones resulting in very low fluid velocities, thus leaving conduction as the predominant heat transport process. Due to low fluid velocities and a spatially limited influence of the fault, no remarkable influence on the thermal field can be observed.

### 4.3 Influence of permeable fault zones

Highly permeable fault zones (model 3) may locally exert a considerable influence on the thermal field. Along both permeable fault zones distinctly cooler temperatures than in its surroundings characterize the thermal field, expressed by an alternating thermal signature. The range of influence of the permeable fault zones extends over a distance of $\sim 2.4$–$8.8$ km in $-1000$ m depth and $\sim 6$–$14$ km in $-3000$ m depth. Inside the fault zones, a net cooling effect is induced by relatively fast downward flow. Triggered by fluid advection due to hydraulic head gradients and enhanced by the high permeability of the fault zones, cold water can easily flow downwards, generating the cooling.

The observed cooling alternates with higher thermal anomalies through upward directed flow below discharge areas, which may be locally enhanced by buoyancy forces having a secondary effect on both the hydrodynamics and thermal field.

### 4.4 Interaction between fault zones and surrounding sediments

The fluid behaviour in the sediments surrounding both fault zones is principally controlled by existing hydraulic head gradients and by the permeability of the fault zones. Modelled fluid flow and thermal field within the surrounding sediments are locally influenced by the presence of the permeable fault zones along their strike direction. Across these fault zones, the thermal field and fluid flow are affected by the thickness and permeability of the sedimentary layers adjacent to the faults.

Below recharge areas and adjacent steep hydraulic gradients, downward-oriented flow inside the fault zones in turn affects the temperature and flow field in the surrounding sediments. Fast downward flow impresses a lateral fluid discharge into the surrounding aquifers. Below discharge areas, the regional flow pattern induced by topographic gradients combined with the higher fault permeability leads to lateral fluid inflow from the aquifers towards the fault zones. Both the lateral outflow from faults into the aquifers below recharge areas and the lateral inflow from aquifers into the faults below discharge areas clearly demonstrate the dynamic interaction between surrounding sediments and fault zones.

Outside the range of influence of the fault zones, the temperature distribution is controlled by different heat transport processes closely linked to the distribution of the aquifers and aquitards.

## 4.5  Inferences for geothermal applications

The fault model outcomes provide valuable inferences on fault zone behaviour and its impact on the surrounding groundwater circulation and thermal field, which might be useful for geothermal energy exploration.

The study has highlighted two major implications for geothermal applications.

1. Impermeable fault zones have little effect for the thermal field and very locally deviate the flow field in the sediments next to the faults. Therefore, this setting would be an unfavourable place to drill a geothermal well.

2. Drilling a geothermal well into or in close proximity to a permeable fault zone would be more prospective in the shallow part of the model domain (up to ∼ 450 m depth), where rising warm fluids are in concert with high permeability of the faults. These spots of rising warm water are locally restricted but temperatures are increased (up to 15 °C compared to the thermal field without faults). By contrast, domains of colder temperatures next to permeable sediments are the most unfavourable areas for geothermal utilization.

   In the deeper Buntsandstein and Rotliegend aquifers the temperature distribution along permeable fault zones is affected by the net cooling effect propagating downward below major recharge areas and steep hydraulic head gradients. This net cooling effect is enhanced by the high permeability of the fault zones. Below adjacent discharge areas, a temperature increase in the fault zones is observed down to a depth of ∼ −4000 m due to upward-oriented flow.

   In summary, the best place to drill a prospective geothermal well would need to be chosen with care, as two conditions should be fulfilled: (1) enhanced inflow of warm and deep water into the fault and (2) a rising branch of heated fluid within the fault zone.

## 4.6  Model limitations and outlook for future studies

Some limitations of the study presented include (1) the structural resolution of the model, (2) the physical fault zone representation and (3) the choice of hydraulic boundary conditions adopted along the top boundary.

1. The structural resolution of the model in particular concerns the distribution of sedimentary layers in the proximity of the fault zones, as it has been concluded that the fluid flow in the surroundings may locally contribute to the fluid behaviour of the permeable fault. As the current study revealed that fault zones locally influence the thermal and fluid system, in a subsequent step, a higher resolution of the model could be achieved by a structural refinement and by focussing only on the areas around the fault zones by decreasing the model size considerably. This step, however, would also assume better constraints on physical rock properties for both the sedimentary and the fault zones (see also point 2 below).

2. As the two fault zones are represented as idealized, homogenously permeable/impermeable zones of a finite width, the modelled thermal and hydraulic pattern in the fault zones are first-order effects. In a real-case scenario, these effects might be more complex. Thus, the models may locally under/overestimate the absolute amplitude of the signal, which is nonetheless present. Further studies may consider a more heterogeneous composition of the fault zone, possibly consisting of a damage zone and a fault core. However, more detailed data would be required on the structure and composition of the Gardelegen and Lausitz fault zones. These could be beneficial for future studies aiming at a more quantitative assessment of their impact on the thermal and groundwater field. Local reservoir-scale models integrating more highly resolved structural data, information derived from boreholes, and in situ measurements may provide better constraints for the characterization of fault zones (e.g. Cacace et al., 2013; Blöcher et al., 2010). Better constraints on the hydraulic behaviour of fault zones may be provided by prospective integrated geophysical methods (e.g. magnetotellurics) that could help to discriminate between different end-member models. At present, a direct assessment of the hydraulic behaviour of faults by the determination of their orientation within the present-day stress field is not possible. This was reflected in the modelling approach followed, in which different end-member stress states have been tested. Given the state of the art of available information, this study represents the best quantitative approach so far to characterize the thermal and hydraulic behaviour of major fault zones in a basin-scale model.

3. The cooling observed in the permeable fault zone and in the Mesozoic aquifer may overestimate the influence of forced convection processes due to the crude boundary setting adopted. Future studies are needed that integrate more realistic information on recharge rates and may be additionally improved by a dynamic coupling with surface water transport modelling.

This study is the first attempt to investigate the impact of major fault zones on a 3-D basin scale for the coupled fluid and heat transport in the Brandenburg region. The simulation outcomes provide new insights into the dynamic mechanisms that control the fluid behaviour and thermal evolution of fault zones with varying permeability in the context of a complex hydrogeological setting. While the lateral influence of a 1 km wide fault zone is rather limited with respect to the neighbouring sediments, temperature variations within the fault zone may be significant, thus having interesting application for issues related to geothermal exploration.

**Acknowledgements.** This work is part of project GeoEn and has been funded by the German Federal Ministry of Education and Research within the programme "Spitzenforschung in den neuen Ländern" (BMBF grant 03G0671A/B/C).

Björn Lewerenz and Björn Kaiser are thanked for helpful computational support and Peter Klitzke for his contributions to visualization. We thank Michael Schneider, the anonymous reviewer and Rüdiger Schulz for their helpful and constructive comments that helped to improve the quality of the manuscript. All numerical results are illustrated by ParaView, an open-source, multi-platform and visualization application.

The service charges for this open access publication have been covered by a Research Centre of the Helmholtz Association.

# References

Alt-Epping, P. and Zhao, C.: Reactive mass transport modelling of a three-dimensional vertical fault zone with a finger-like convective flow regime, J. Geochem. Explor., 106, 8–23, 2010.

Amante, C. and Eakins, B. W.: ETOPO1 1 Arc-Minute Global Relief Model: Procedures, Data Sources and Analysis, NOAA Technical Memorandum NESDIS NGDC-24, 19 pp., 2009.

Bächler, D., Kohl, T., and Rybach, L.: Impact of graben-parallel faults on hydrothermal convection – Rhine Graben case study, Phys. Chem. Earth, 28, 431–441, 2003.

Baietto, A., Cadoppi, P., Martinotti, G., Perello, P., Perrochet, P., and Vuataz, F.-D.: Assessment of thermal circulations in strike-slip fault systems: the Terme di Valdieri case (Italian western Alps), Geol. Soc., London, Special Publications 299, 317–339, doi:10.1144/SP299.19, 2008.

Balling, N., Kristiansen, J. I., Breiner, N., Poulsen, K. D., Rasmussen, R., and Saxov, S.: Geothermal measurements and subsurface temperature modelling in Denmark, GeoSrifter, Department of Geology Aarhus University, 16, 1981.

Barton, C. A., Zoback, M. D., and Moos, D.: Fluid flow along potentially active faults in crystalline rock, Geology 23, 683–686, 1995.

Bayer, U., Scheck, M., and Koehler, M.: Modeling of the 3-D thermal field in the northeast German Basin, Geol. Rundsch., 86, 241–251, 1997.

Bense, V. F., Person, M. A., Chaudhary, K., You, Y., Cremer, N., and Simon, S.: Thermal anomalies indicate preferential flow along faults in unconsolidated sedimentary aquifers, Geophys. Res. Lett., 35, L24406, doi:10.1029/2008GL036017, 2008.

Bjørlykke, K.: Subsurface water and fluid flow in sedimentary basins, Petroleum Geoscience: From Sedimentary Environments to Rock Physics, 259–279, Springer, Heidelberg, doi:10.1007/978-3-642-02332-3_10, 2010.

Blöcher, M. G., Zimmermann, G., Moeck, I., Brandt, W., Hassanzadegan, A., and Magri, F.: 3D Numerical Modeling of Hydrothermal Processes during the Lifetime of a Deep Geothermal Reservoir, Geofluids, 10, 406–421, 2010.

Cacace, M., Bayer, U., and Marotta, A. M.: Strain localization due to structural in-homogeneities in the Central European Basin, Int. J. Earth Sci. (Geol. Rundsch.), 97, 899–913, 2008.

Cacace, M., Blöcher, G., Watanabe, N., Moeck, I., Börsing, N., Scheck-Wenderoth, M., Kolditz, O., and Huenges, E.: Modelling of fractured carbonate reservoirs-outline of a novel technique via a case study from the Molasse Basin, southern Bavaria (Germany), Environ. Earth Sci., 70, 3585–3602, doi:10.1007/s12665-013-2402-3, 2013.

Cherubini, Y., Cacace, M., Scheck-Wenderoth, M., Moeck, I., and Lewerenz, B.: Controls on the deep thermal field – implications from 3-D numerical simulations for the geothermal research site Groß Schönebeck, Environ. Earth Sci., 70, 3619–3642, doi:10.1007/s12665-013-2519-4, 2013.

Diener, J., Katzung, G., and Kühn, P. et al.: Geothermie-Atlas der Deutschen Demokratischen Republik, Zentrales Geologisches Institut-ZGI-Berlin, 1984.

Diersch, H.-J. G.: FEFLOW Finite-Element Subsurface Flow and Transport Simulation System, User's Manual/Reference Manual/White Papers, Release 5.0. WASY GmbH, Berlin, 2002.

Garven, G., Bull, S. W., and Large, R. R.: Hydrothermal fluid flow models of stratiform ore genesis in the McArthur Basin, Northern Territory, Australia, Geofluids, 1, 289–311, 2001.

Hebig, K. H., Ito, N., Scheytt, T., and Marui, A.: Review: Deep groundwater research with focus on Germany, Hydrogeol. J., 20, 227–243, doi:210.1007/s10040-10011-10815-10041, 2012.

Kaiser, B. O., Cacace, M., Scheck-Wenderoth, M., and Lewerenz, B.: Characterization of main heat transport processes in the Northeast German Basin: Constraints from 3-D numerical models, Geochem. Geophy. Geosy., 12, Q07011, doi:10.1029/2011GC003535, 2011.

Lampe, C. and Person, M.: Advective cooling wihtin sedimentary rift basins – application to the Upper Rhinegraben (Germany), Mar. Pertol. Geol., 19, 361–375, 2002.

López, D. L. and Smith, L.: Fluid flow in fault zones: Analysis of the interplay of convective circulation and topographically driven groundwater flow, Water Resour. Res., 31, 1489–1503, 1995.

López, D. L. and Smith, L.: Fluid flow in fault zones: Influence of hydraulic anisotropy and heterogeneity on the fluid flow and heat transfer regime, Water Resour. Res., 32, 3227–3235, 1996.

Magri, F.: Derivation of the coefficients of thermal expansion and compressibility for use in FEFLOW (implementation code in C++), White Pap., III, pp. 13–23, WASY GmbH Inst. for Water Resour. Plann. And Syst. Res., Berlin, 2004.

Magri, F.: Mechanismus und Fluiddynamik der Salzwasserzirkulation im Norddeutschen Becken: Ergebnisse thermohaliner numerischer Simulationen (Dissertation Thesis, Freie Universität Berlin), Scientific Technical Report STR05/12, GeoForschungsZentrum Potsdam, 131 pp., 2005.

Magri, F., Akar, T., Gemici, U., and Pekdeker, A.: Deep geothermal groundwater flow in the Seferihisar-Balçova area, Turkey: results from transient numerical simulations of coupled fluid flow and heat transport processes, Geofluids, 10, 388–405, 2010.

Marotta, A. M., Bayer, U., Thybo, H., and Scheck, M.: Origin of the regional stress in the North German basin: results from numerical modeling, Tectonophysics, 360, 245–264, 2002.

Maystrenko, Y., Bayer, U., and Scheck-Wenderoth, M.: Structure and Evolution of the Central European Basin System according to 3D modeling, DGMK Research Report, 577-2/2-1, 2010.

Noack, V., Cherubini, Y., Scheck-Wenderoth, M., Lewerenz, B., Höding, T., Simon, A., and Moeck, I.: Assessment of the present-day thermal field (NE German Basin) – Inferences from 3-D modeling, Chemie der Erde, 70, 47–62, 2010.

Noack, V., Scheck-Wenderoth, M., and Cacace, M.: Sensitivity of 3-D thermal models to the choice of boundary conditions and thermal properties – a case study for the area of Brandenburg (NE German Basin), Environ. Earth Sci., 67, 1695–1711, 2012.

Noack, V., Scheck-Wenderoth, M., Cacace, M., and Schneider, M.: Influence of moving fluids on the regional thermal field: results from 3-D numerical modelling for the area of Brandenburg (North German Basin), Environ. Earth Sci., 70, 3523–3544, 2013.

Pommer, H.: The control of faults on the thermal field in Northern Germany – Constraints from 2-D coupled numerical simulations, Department of Earth Sciences, Freie Universität Berlin, 69 pp., 2012.

Röckel, T. and Lempp, C.: Spannungszustand im Norddeutschen Becken, Erdöl Erdgas Kohle, 119, 73–80, 2003.

Roth, F. and Fleckenstein, P.: Stress orientations found in the NE Germany differ from the West-European trend, Terra Nova, 13, 289–296, 2001.

Scheck, M. and Bayer, U.: Evolution of the Northeast German Basin – inferences from a 3D structural model and subsidence analysis, Tectonophysics, 313, 145–169, 1999.

Scheck, M., Bayer, U., Otto, V., Lamarche, J., Banka, D., and Pharaoh, T.: The Elbe Fault System in North Central Europe – a basement controlled zone of crustal weakness, Tectonophysics, 360, 281–299, 2002.

Scheck-Wenderoth, M. and Lamarche, J.: Crustal memory and basin evolution in the Central European Basin System – new insights from a 3D structural model, Tectonophysics, 397, 132–165, 2005.

Scheck-Wenderoth, M., Krzywiec, P., Zühlke, R., Maystrenko, Y., and Froitzheim, N.: Permian to Cretaceous tectonics, in: The geology of Central Europe, T. McCann (Ed.), London, Geological society of London, 2, Mesozoic and Cenozoic, 999–1030, 2008.

Simms, M. A. and Garven, G.: Thermal convection in faulted extensional sedimentary basins: theoretical results from finite-element modeling, Geofluids, 4, 109–130, 2004.

Turcotte, D. L. and Schubert, G.: Geodynamics, Cambridge University Press, New York, 2002.

World Stress Map, www.world-stress-map.org/, 2000.

Yang, J.: Full 3-D numerical simulation of hydrothermal field flow in faulted sedimentary basins: Example of the McArthur Basin, Northern Australia, J. Geochem. Explor., 89, 440–444, 2006.

Yang, J., Large, R., and Bull, S.: Factors controlling free thermal convection in faults in sedimentary basins: implications for the formation of zinc–lead mineral deposits, Geofluids, 4, 1–11, 2004a.

Yang, J., Bull, S., and Large, R.: Numerical investigation of salinity in controlling ore-forming fluid transport in sedimentary basins: example of the HYC deposit, Northern Australia, Miner. Deposita, 39, 622–631, 2004b.

Zimmermann, G., Reinicke, A., Blöcher, G., Milsch, H., Gehrke, D., Holl, H.-G., Moeck, I., Brandt, W., Saadat, A., and Huenges, E.: Well path design and stimulation treatments at the geothermal research well GtGrSk4/05 in Groß Schönebeck, PROCEEDINGS, Thirty-Second Workshop on Geothermal Reservoir Engineering, Stanford University, Stanford, California, 22–24 January 2007, SGP-TR-183, 2007.

## Appendix A

The system of flow equations with variable fluid density $\rho^f$ and viscosity $\mu^f$ is given by the mass conservation of the fluid –

$$\frac{\partial(\varepsilon\rho^f)}{\partial t} + \nabla\cdot\left(\rho^f q^f\right) = \varepsilon Q_\rho, \tag{A1}$$

where $\varepsilon$ is the porosity, $\rho^f$ the mass density of the fluid, $q^f$ the specific discharge (Darcy's velocity), and $Q_\rho$ the sink–source mass term – as well as by the generalized Darcy law:

$$q^f = -K\left(\nabla h + \frac{\rho^f - \rho_0^f}{\rho_0^f}\frac{g}{|g|}\right), \tag{A2}$$

where $K$ is the hydraulic conductivity tensor of the porous media given by $K = \frac{\rho_0^f g}{\mu^f}k$, with $k$ being the permeability tensor and $g$ the gravity acceleration.

Equation (2) is written in terms of hydraulic head rather than pressure as the primary variable to conform to the mathematical formulation used in FEFLOW®. Assuming thermal equilibrium between the porous medium and the fluid (i.e. $Tf = T = Ts$) and if density gradients are neglected, applying the law of energy conservation the following heat transfer equation results:

$$(\rho c)_{fs}\frac{\partial T}{\partial t} + \rho^f c^f \nabla\cdot\left(q^f T\right) - \nabla\cdot(\lambda\nabla T) = Q_T, \tag{A3}$$

with $(\rho c)_{fs}$ being the bulk specific heat capacity of the fluid ($f$) plus solid ($s$) phase system, defined as

$$(\rho c)_{fs} = \left\lfloor \varepsilon\rho^f c^f + (1-\varepsilon)\rho^s c^s \right\rfloor. \tag{A4}$$

$Q_T$ is a heat source–sink function. $\lambda$ is the equivalent thermal conductivity tensor of the fluid and the porous medium; it takes both conductive (Fourier) and thermodispersive (mixing) effects into account. Accordingly, the equivalent thermal conductivity may be subdivided into its two components as

$$\lambda = \lambda_{DISP} + \lambda_{COND} \tag{A5}$$

where the first term on the right-hand side of Eq. (5) is the thermodispersive term

$$\lambda_{DISP} = \rho^f c^f \left[\alpha_T\sqrt{(q_i^f q_i^f)}\mathbf{I} + (\alpha_L - \alpha_T)\frac{q_i^f q_i^f}{\sqrt{(q_i^f q_i^f)}}\right] \tag{A6}$$

and the second term is the conductive one:

$$\lambda_{COND} = \varepsilon\lambda^f + (1-\varepsilon)\lambda^s \tag{A7}$$

In Eq. (6), $\alpha_L$ and $\alpha_T$ are respectively the longitudinal and transversal dispersion lengths, while $\lambda^f$ and $\lambda^s$ appearing in Eq. (7) represent the thermal conductivity of the fluid and solid phase and $\mathbf{I}$ is the unit matrix.

The balance Eqs. (2) and (3) are coupled by the fluid density according to an equation of state after Magri (2004) and Blöcher et al. (2010).

# 12

# Geothermal heat pump system assisted by geothermal hot spring

**M. Nakagawa**[1] **and Y. Koizumi**[1,a]

[1]Colorado School of Mines, Golden, Colorado, USA
[a]now at: Kajima Corporation, Tokyo, Japan

*Correspondence to:* M. Nakagawa (mnakagaw@mines.edu)

**Abstract.** The authors propose a hybrid geothermal heat pump system that could cool buildings in summer and melt snow on the pedestrian sidewalks in winter, utilizing cold mine water and hot spring water. In the proposed system, mine water would be used as cold thermal energy storage, and the heat from the hot spring after its commercial use would be used to melt snow for a certain section of sidewalks. Neither of these sources is viable for direct use application of geothermal resources, however, they become contributing energy factors without producing any greenhouse gases. To assess the feasibility of the proposed system, a series of temperature measurements in the Edgar Mine (Colorado School of Mines' experimental mine) in Idaho Springs, Colorado, were first conducted, and heat/mass transfer analyses of geothermal hot spring water was carried out. The result of the temperature measurements proved that the temperature of Edgar Mine would be low enough to store cold groundwater for use in summer. The heat loss of the hot spring water during its transportation was also calculated, and the heat requirement for snow melt was compared with the heat available from the hot spring water. It was concluded that the heat supply in the proposed usage of hot spring water was insufficient to melt the snow for the entire area that was initially proposed. This feasibility study should serve as an example of "local consumption of locally available energy". If communities start harnessing economically viable local energy in a responsible manner, there will be a foundation upon which to build a sustainable community.

## 1   Introduction

Geothermal energy is a safe, 24/7 renewable energy source with a unique ability for cascading usage, thus being well suited for use in the development of a sustainable community. For example, Hachijojima is a volcanic island located 300 km south of Tokyo, Japan with a population of about 9500 according to Yamashita et al. (2000). The Hachijojima Geothermal Power Station, operated by the Tokyo Electronic Power Company (TEPCO) since 1999, has the rated output of 3300 kW of electricity to meet about 30 % of the electricity demand for this isolated island community. Before commercial operation of the geothermal power plant, all electricity demand was met by a diesel power station. In addition to the power generation, the 43 °C condensed vapor is utilized for heating greenhouses. This is a good example of a sustainable energy utilization system within a community, with local consumption of locally produced energy through a cascading use of geothermal resources.

Although there are no geothermal power stations in Colorado, USA, the state is blessed with low-enthalpy geothermal resources. Local residents and tourists enjoy natural hot springs in many places. The Geo-Heat Center Quarterly Bulletin has described six Colorado hot springs and direct use applications for those springs. One of the side effects of Colorado's booming mining economy during the middle of the 19th century was that there were many mines in the state that were left abandoned when the mining boom ceased.

Abandoned mines are usually environmental and safety liabilities for communities in which they are located, but their potential use as thermal storage resources should not be overlooked. Pingjia and Ning (2011) studied three different usages of abandoned mines, and one of the usages identified in their study is the "Thermal Accumulator". In this

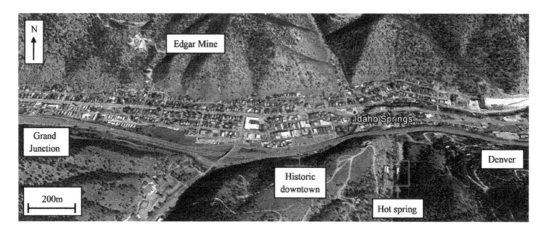

**Figure 1.** Map of Idaho Springs, Colorado.

usage, two underground mines, a cold mine and a hot mine, would be identified and utilized. Cold water stored in the cold mine would be used in summer, and warm water stored in the hot mine would be used in winter to improve geothermal heat pump efficiency. In a different study, Rodriguez and Diaz (2009) considered a deep underground mine in Spain to be used as a geothermal heat exchanger. According to their measurement, the temperature of the rock mass in the underground mine was a constant 27 °C. Therefore, the authors proposed that cold water be pumped into the mine to run through 1000 m of galleries in order to gain heat. The heated water would be used to improve the heat pump efficiency of buildings in winter. Furthermore, the authors also conducted an economic analysis and proved that the proposed system would economically be viable.

## 2 Geothermal heat pump system assisted by a geothermal hot spring

### 2.1 General Information about Idaho Springs, Colorado, USA

The city of Idaho Springs, Colorado, USA, is located about 50 km west of Denver in the foothills of the Rockies. The town was founded in 1859 by mining prospectors and flourished as a mining community throughout the 1860s. Today, the town has a population of about 1900 and attracts tourists to its historic downtown, hot spring, and the experimental Edgar Mine located just north of the town. Figure 1 shows the map of the city of Idaho Springs and the locations of the historic downtown, the hot spring, and the Edgar Mine.

### 2.2 Proposed hybrid geothermal heat pump system

The proposed heating/cooling geothermal heat pump system for the city of Idaho Springs can be visualized in Fig. 2. It was assumed that the rock mass temperature in the Edgar Mine would be relatively low since the mine was located at a shallow depth. Therefore, the authors envisioned that the collected cold groundwater would be stored in a closed sec-

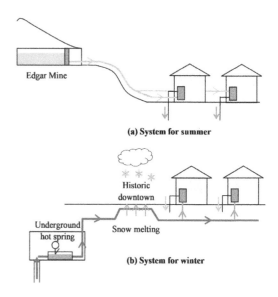

**Figure 2.** Proposed system for Idaho Springs.

tion of the Edgar Mine in winter and used to cool down the condensers of heat pumps in summer as shown in Fig. 3a. In this way, the efficiency of heat pumps would be improved. In winter, geothermal hot spring water used for commercial bathing would be transported to the historic downtown and used to melt the snow on the pedestrian sidewalks. Furthermore, any residual heat from the hot spring water after snow melting could be used to heat the heat pump evaporators, which would improve the efficiency of the heat pumps (Fig. 3b).

As reviewed in the previous section, the use of a mine to improve heat pump efficiency is not a new idea. For example, Shiba (2008) also reported a case study in which hot spring water consumed in a public bathing facility was re-used to improve heat pump efficiency of buildings in Japan. However, to our best knowledge, we know of no case study in which a mine and hot spring are jointly utilized to assist in

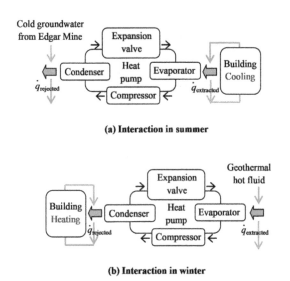

**(a) Interaction in summer**

**(b) Interaction in winter**

**Figure 3.** Interaction between proposed system, heat pump, and building.

improving the efficiency of a geothermal heat pump system. The close proximity of a hot spring and available mine also makes the location of Idaho Springs and its potential distribution of a geothermal resource unique.

We also point out that there are acceptable flow and temperature range with heat pump source water. The extended range water–water heat pumps prefer a source and load flow rate of 2.5 to 3.0 gpm per nominal ton of capacity, i.e., 0.16 to $0.19\,\mathrm{L\,s^{-1}}$ per 3.5 kW. For most heat pumps (water–water or water–air) using R410a refrigerant for heating, 27 °C entering water temperature on the source side is the theoretical maximum for reliable operation to produce a leaving water temperature on the load side of 49 °C.

## 3  Temperature measurement in the Edgar Mine

The Edgar Mine produced high-grade silver, gold, lead and copper in the mid 19th century. Colorado School of Mines (CSM) acquired the Edgar Mine in 1921 when a bankrupt mining company agreed to lease it to the school, and CSM has been using the mine for education and research ever since. As an example of its use by the school, junior students in the Mining Engineering Department take a course entitled "Mining Engineering Laboratory" at the Edgar Mine where they receive practical training in operating jackleg drills, jumbo drills, load–haul–dump machines, etc. In other classes, students gain hands-on experience in underground mine surveying, geological mapping, mine ventilation field studies, mine safety, and so on. Photo 1 shows the entrance of the Edgar Mine (Miami Tunnel), and Photo 2 shows a classroom inside the mine. Research is conducted at the Edgar Mine by numerous academic, government, and industry groups including the CSM Mining Engineering Department, the National Institute for Occupational

Safety and Health (NIOSH), the US Army, the US Department of Energy and others. Research topics cover tunnel detection, blasting, rock mechanics, development of a model circulation system for geothermal study, and development of new mining equipment and methods. For more information, the CSM website is available using the following link: http://inside.mines.edu/Mining-Edgar-Mine.

### 3.1  Temperature measurement

In order to assess the thermal capacity of the Edgar Mine, the temperature field inside the Mine was mapped. Temperature measurements were carried out on three different dates; 17 September, 24 October, and 25 November of 2013. Figure 4 shows the data for temperature, precipitation, and snowfall in Idaho Springs in 2013. The data were obtained from the website of AccuWeather.com. The ambient temperature decreased significantly during the measurement period. Measurements taken include the surface temperature of the rock mass, ambient temperature, and humidity at the 24 locations shown in Fig. 5. There are two main tunnels in the Edgar Mine. The eastern tunnel is called the Miami Tunnel, and the western tunnel the Army Tunnel. The height and width of the Miami Tunnel are about 2 and 2 m, and those of the Army Tunnel are about 4 and 5 m. An infrared thermometer (Fluke, Model: 62Max) was used to measure the rock surface temperature as shown in Photo 3. The temperature of groundwater accumulated at location 21 was also measured. The area between the entrance of the Army Tunnel (Location 24) and the location 19 was wet during this period.

### 3.2  The temperature mapping in the Edgar Mine

The measured temperatures of the rock mass surface at the 24 locations are shown in Fig. 5. The temperature measured near the entrance of the two tunnels decreased during the three measurements due to the influence of the ambient temperature. On the other hand, the temperature inside the mine was stable. The highest surface temperature was always measured at location 9 (U.S. Geological Survey (USGS) classroom), and it was about 12.5 °C. The second highest temperature of 12 °C was measured at location 6.

Compared to the surface temperature of 27 °C of the rock mass that was reported in the paper by Rodriguez and Diaz (2009), the surface temperature of the rock mass in the Edgar Mine is much lower; however, this makes sense because the measured areas in the Edgar Mine has been excavated horizontally on just one level so that the typical temperature gradient $(20–30\,\mathrm{°C\,km^{-1}})$ in depth is not expected.

Figure 6a and b show how the measured rock surface and the ambient temperatures change with the distance from the entrance. More specifically, the temperatures measured at locations 1, 2, 3, 4, 5, and 6 are shown in Fig. 6a, and the temperatures measured at locations 18, 19, 21, 23, and 24 are shown in Fig. 6b. Comparing the surface temperatures

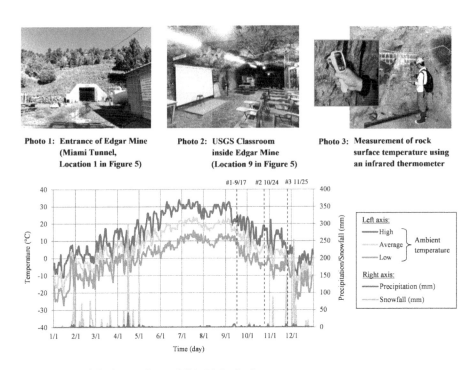

Photo 1: **Entrance of Edgar Mine** (Miami Tunnel, Location 1 in Figure 5)

Photo 2: **USGS Classroom** inside Edgar Mine (Location 9 in Figure 5)

Photo 3: **Measurement of rock** surface temperature using an infrared thermometer

**Figure 4.** Ambient temperature, precipitation, and snowfall in Idaho Springs.

for locations near the entrance to each of the two tunnels, it is found that the surface temperature increases with the increasing distance from the entrance in the Miami Tunnel (Fig. 6a), and the surface temperatures at locations 4, 5, and 6 show no significant difference between the three measurement dates even though the ambient temperature decreases. Figure 6b, on the other hand, indicates that the surface temperature does not increase as the distance becomes greater in the Army Tunnel. Additionally, the surface temperatures shown in Fig. 6b are significantly different for each of the three measurements.

In order to understand the difference in the temperature profiles between the Miami Tunnel and the Army Tunnel, the relationship between the cover that is defined as the difference between the elevation of ground surface and that of the tunnel and the distance from the entrance is shown in Fig. 7. It can be seen that the cover above the Army Tunnel does not increase as much as that for the Miami Tunnel. The heat is transferred through the rock mass by conduction since no significant wind blows inside the underground mine. Therefore, as the cover becomes greater, the surface temperature of the rock mass is less influenced by the ambient temperature outside the Edgar Mine. The rock surface temperatures at locations 4, 5, and 6, with the cover greater than 150 m, is considered to be independent of the ambient temperature outside the mine. Figure 7 also shows the different topography above the Miami Tunnel and the Army Tunnel, which should explain why the Miami Tunnel is dry and the Army Tunnel is wetter. The thin and relatively flat cover above the

Army Tunnel provide more opportunity for the surface water to permeate through to the tunnel.

In conclusion, it is found that the surface temperature of the rock mass in the Edgar Mine is relatively low due to its shallow depth. Therefore, the mine would be suitable for thermal energy storage in which cold groundwater would be stored in winter and used in summer as proposed in Fig. 2. Our recommendation is to utilize the groundwater that is naturally flowing in the mine in a designated storage area.

## 4 Temperature loss of transported hot spring water

In the proposed system, hot spring water will be transported from the hot spring to the east end of historic downtown area through a pipe buried in the ground. In this section, the temperature loss of the transported hot spring water is estimated.

### 4.1 Temperature profile of the ground at various depths

In order to understand the heat transfer between the flowing hot spring water and the ground, the ground temperature $T$ (°C) is first obtained following Kasuda and Achenbach (1965).

$$T = T_{mean} - T_{amp} \times \exp\left[-D\left(\frac{\pi}{365\alpha}\right)\right]$$
$$\times \cos\left\{\frac{2\pi}{365}\left[t - t_{shift} - \frac{D}{2}\left(\frac{365}{\pi\alpha}\right)\right]\right\}, \quad (1)$$

where $T_{mean}$, $T_{amp}$, $D$, $\alpha$, $t$, $t_{shift}$ are the mean surface temperature, the amplitude of surface temperature, the depth below the surface, thermal diffusivity of the ground, time (day),

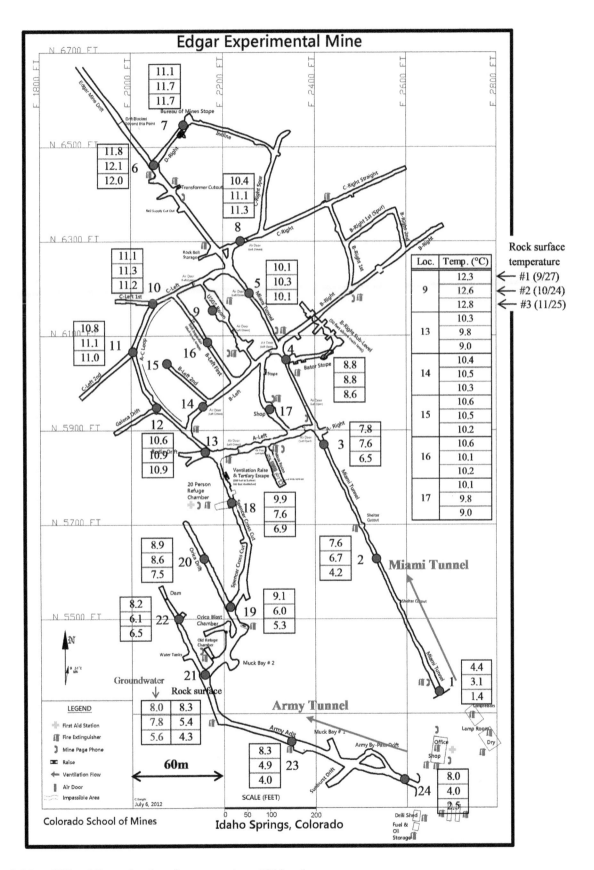

**Figure 5.** Map of Edgar Mine and rock surface temperature at 24 locations.

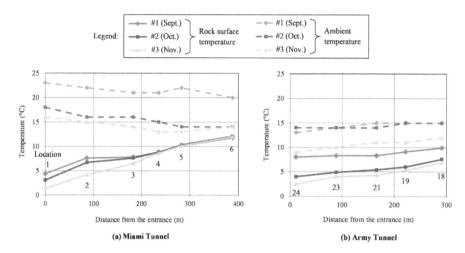

**Figure 6.** Relationship between temperature and distance from the entrance.

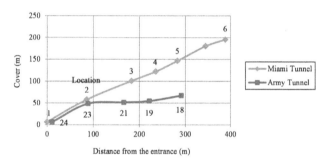

**Figure 7.** Relationship between cover and distance from the entrance.

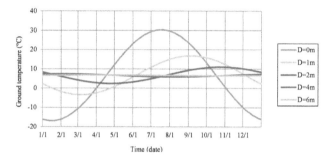

**Figure 8.** Ground temperature as a function of time and depth.

and day of the year of the minimum surface temperature, respectively. We set $T_{mean}$ to be 6.7 °C defined by the annual average air temperature; $T_{amp}$ to be 23.7 °C defined as the average between the maximum, 29.9 °C, and the minimum, −17.5 °C, air temperature in July and January, respectively; $D$ to be the depth in meters; $\alpha$ to be $1.19 \times 10^{-2}$ m² day⁻¹ estimated value of thermal diffusivity of the ground; $t$ to be the number of days for the present calculation; $t_{shift}$ to be 15 days in January defined by the number of days of the year with the minimum surface temperature.

Figure 8 shows the temperature profile in the ground at various depths and indicates that the ground temperature becomes more or less a constant value of 6.7 °C at the depth of approximately 6 m. Considering potentially significant excavation costs, we assume that the pipe is buried at the depth of 2 m, at which the ground temperature is not exactly constant but allows us to assume a constant temperature of 6.7 °C to simplify the calculation hereafter.

## 4.2 Temperature loss estimation

Figure 9 shows the schematic representation of the hot spring water flowing to the historic downtown. According to Rep-

plier et al. (1982), the flow rate of the geothermal well used by the hot spring is 136 L min⁻¹. Considering 25 % loss of the flow, the flow rate is assumed to be 100 L min⁻¹. Under the conditions shown in the figure, the temperature loss was calculated as follows.

The Reynolds number of the internal flow through a pipe, $Re_D$ is

$$Re_D = \frac{\rho u_m D_i}{\mu} = \frac{995.0 \times 0.212 \times 0.1}{7.69 \times 10^{-4}} = 27\,430, \quad (2)$$

where $\rho$, $\mu$, $u_m$ are density (kg m⁻³) and viscosity ($N \times$ s m⁻²) of the fluid, and internal flow rate (m s⁻¹).

$Re_D$ is larger than 2300 indicating that the flow is turbulent. As the temperature of the fluid is higher than that of the surrounding ground temperature, the following equation can be applied to calculate the Nusselt number, $Nu_D$:

$$Nu_D = 0.023 \times Re^{0.8} \times Pr^{0.3} = 0.023 \times 27430^{0.8} \times 5.2^{0.3} = 134.0, \quad (3)$$

where $Pr$ is Prandtl number.

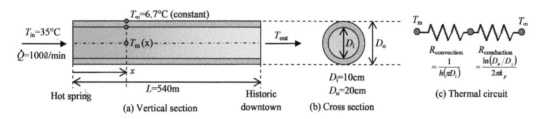

**Figure 9.** Schematic representation of the geothermal hot fluid flowing from hot spring to historic downtown.

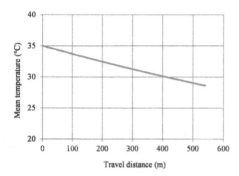

**Figure 10.** Relationship between mean temperature of geothermal hot fluid and travel distance.

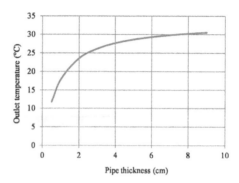

**Figure 11.** Relationship between outlet temperature of geothermal hot fluid and pipe thickness.

Convection heat transfer coefficient, $h$ (W m$^{-2} \times$ K) is

$$h = \frac{Nu_D \times k}{D_i} = \frac{134.0 \times 0.62}{0.1} = 830.8, \tag{4}$$

where $k$ is the thermal conductivity of the fluid (W m$^{-2} \times$ K).

Total thermal resistance per unit length, $R_{tot}$ (m $\times$ K W$^{-1}$) is

$$R_{tot} = R_{conv} + R_{cond} = \frac{1}{h(\pi D_i)} + \frac{\ln(D_o/D_i)}{2\pi k_p}$$

$$= \frac{1}{830.8 \times (\pi \times 0.1)} + \frac{\ln(0.2/0.1)}{2 \times \pi \times 0.16} = 0.306, \tag{5}$$

where $k_p$ is the thermal conductivity of a pipe (W m$^{-1} \times$ K).

$$\frac{T_\infty - T_x}{T_\infty - T_{in}} = \exp\left(-\frac{x}{\dot{m}c_p R_{tot}}\right) T_x$$

$$= 6.7 - \exp\left(-\frac{x}{1.66 \times 4178 \times 0.306}\right)$$

$$\times (6.7 - 35), \tag{6}$$

where $T_x, T_\infty, T_{in}, m, c_p$ are mean temperature at $x = x$ (°C), constant surface temperature (°C), inlet temperature (°C), mass flow rate (kg s$^{-1}$), and specific heat of the fluid (J kg$^{-1} \times$ K).

Figure 10 shows the relationship between the mean temperature of the geothermal hot spring water and its traveled distance. It is found that the temperature would decrease from 35 to 28.6 °C while it is transported to the historic

downtown over the distance of 540 m. In this calculation, the thickness of a pipe was assumed to be 5 cm. Figure 11 shows the result of a parametric study when the thickness of the pipe is varied from 1 to 9 cm. Figure 11 indicates that the outlet temperature monotonically increases with increasing thickness, but beyond the thickness of 5 cm, the increase of the temperature becomes very modest. Considering its high cost and the difficulties working with thick pipes, 5 cm thickness assumed in the calculation above is the most reasonable.

## 5 Energy balance analysis for snow melting

The exiting temperature 28.6 °C is used to melt snow and to improve the heat pump efficiency of the buildings in the historic downtown. In this section, the feasibility of the designed snow melting system is discussed by comparing heat supply and heat requirement for snow melting.

### 5.1 Heat supply from the hot spring water to pedestrian sidewalks

Figure 12 shows the design of the proposed snow melting system for the sidewalks of the historic downtown. The hot water transported from the hot spring source flows through the pipe system in the north and south pedestrian sidewalks. The depth, the spacing, and the size of buried pipes are appropriately adapted after having reviewed two previous case studies in which similar snow melting systems were installed. One case study covers the system in Kla-

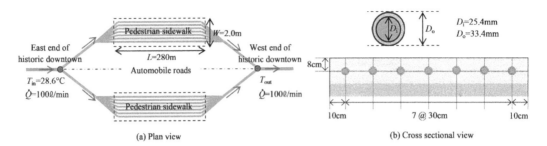

**Figure 12.** Schematic representation of the geothermal hot fluid flowing from hot spring to historic downtown.

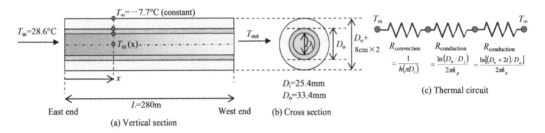

**Figure 13.** Schematic representation of geothermal hot fluid flowing from hot spring to historic downtown.

math Falls, Oregon, reported by Lund (1999) and the other covers the system in Sapporo, Japan, reported by Sato and Sekioka (1979).

In order to calculate the heat supply from the hot spring water to the pedestrian sidewalks, the temperature decrease of the flow through the pipes buried in the sidewalks was calculated in the same way as the calculation of the temperature loss in Sect. 4. However, the buried depth of the pipes shown in Fig. 12 is only 8 cm, while that of the pipe in the previous section was assumed to be 2 m. Therefore, the geothermal hot water would significantly be influenced by cold ambient temperature. Thus, it is assumed that the flow is cooled down by the constant temperature, $-7.7\,^\circ$C, as shown in Fig. 13. The average temperature of the ground surface in 2013 was $-7.7\,^\circ$C between 1 November and 30 April, the average months of snowfall in Idaho Springs as shown in Fig. 4.

The outlet temperature $T_{\text{out}}$ was found to be $3.1\,^\circ$C. Therefore, the heat supply from the geothermal hot fluid to the surrounding ground $q$ (W) is

$$q = \dot{m} c_p (T_{\text{in}} - T_{\text{out}}) = 1.66 \times 4184 \times (28.6 - 3.1)$$
$$= 177.108\,\text{W} = 177.1\,\text{kW}. \tag{7}$$

According to Adlam (1950), 70 % of the heat is usefully consumed for snow melting, because approximately 8–10 % is emitted to the atmosphere and approximately 20–22 % is lost downwards to the ground. Therefore, the available heat for snow melting is

$$q_{\text{available}} = 0.7 \times q = 0.7 \times 177.1 = 124.0\,\text{kW}. \tag{8}$$

The outlet temperature of the geothermal fluid ($3.1\,^\circ$C) is so cold that it could not be used to heat up an evaporator

of a heat pump. Therefore, the proposed system shown in Fig. 2b is not realistic and should be modified. The geothermal hot fluid flows 8 cm under the pedestrian sidewalks only on snowy days. Otherwise, it is transported to a heat pump of each building directly so that the heat can be used more effectively.

### 5.2  Heat flux requirement for snow melting

Chapman and Katunich (1956) estimated the required total heat flux for snow melting, $q_0$ (W m$^{-2}$) as follows:

$$q_0 = q_s + q_m + A_r(q_h + q_e), \tag{9}$$

where $q_s$, $q_m$, $A_r$, $q_h$, $q_e$ are sensible heat flux (W m$^{-2}$), latent heat flux (W m$^{-2}$), snow-free area ratio, convective and radiative heat flux from snow-free surface (W m$^{-2}$), and heat flux of evaporation (W m$^{-2}$), respectively. According to Chapman and Katunich (1956), when $A_r$ is 1.0, the system melts snow rapidly enough that no accumulation of snow occurs. However, this condition requires the maximum energy supply. When $A_r$ is 0, the surface is covered with snow of sufficient thickness. This condition is not desirable but could be tolerable when time for snow clearance is not critical. Chapman and Katunich (1956) recommended that an intermediate value of $A_r = 0.5$ would be used for many non-critical situations. Thus, in this calculation, three required total heat fluxes for snow melting were calculated when $A_r$ is 0, 0.5 and 1.0, and the required heat with $A_r = 0.5$ is considered to be the most important value in practice. The document prepared by ASHRAE (American Society of Heating, Refrigerating and Air-Conditioning Engineers) (2011) is used to calculate each heat flux as shown below.

Sensible heat flux, $q_s$ (W m$^{-2}$) is

$$q_s = \rho_{water} s \left[ c_{p,\,ice}(t_s - t_a) + c_{p,\,water}(t_f - t_s) \right] / c_1$$
$$= 1000 \times \frac{2.36}{3.6 \times 10^6}$$
$$[2100 \times (0 + 6.2) + 4290 \times (0.56 - 0)] = 10.1, \quad (10)$$

where $\rho_{water}$, $s$, $c_{p,\,ice}$, $c_{p,\,water}$, $t_a$, $t_f$, $t_s$, $c_1$ are density of water (kg m$^{-3}$), snowfall rate water equivalent (mm h$^{-1}$), specific heat of ice (J kg$^{-1} \times$ K), specific heat of water (J kg$^{-1} \times$ K), ambient temperature coincident with snowfall (°C), liquid film temperature (°C), melting temperature (°C), and conversion factor. We find that it snowed in Idaho Springs for a total of 34 days in 2013. Therefore, the ambient temperature during the snowfall is assumed to be $-6.2$ °C, which is the average of ambient temperature of the 34 snowy days. Snowfall rate water equivalent is assumed 2.36 mm h$^{-1}$ as the average snowfall rate per day is 56.6 mm day$^{-1}$ in Idaho Springs.

Latent heat flux, $q_m$ (W m$^{-2}$) is

$$q_m = \rho_{water} s h_{if}/c_1 = 1000 \times \frac{2.36}{3.6 \times 10^6} \times 334000 = 219.0, \quad (11)$$

where $h_{if}$ is heat of fusion of snow (J kg$^{-1}$).

Convective and radiative heat flux from a snow-free surface, $q_h$ (W m$^{-2}$), is

$$q_h = h_c(t_s - t_a) + \sigma \varepsilon_s \left( T_f^4 - T_a^4 \right) = 12.3 \times (0.56 + 6.2)$$
$$+ \left( 5.67 \times 10^{-8} \right)(0.9)\left( 273.7^4 - 264.9^4 \right) = 110.7, \quad (12)$$

$$h_c = 0.037 \left( \frac{k_{air}}{L} \right) Re_L^{0.8} Pr^{\frac{1}{3}}$$
$$= 0.037 \left( \frac{k_{air}}{L} \right) \left( \frac{VL}{\nu_{air}} c_2 \right)^{0.8} Pr^{\frac{1}{3}}$$
$$= 0.037 \left( \frac{0.0235}{6.1} \right) \left( \frac{10.0 \times 2.0 \times 0.278}{1.3 \times 10^{-5}} \right)^{0.8}$$
$$(0.7)^{\frac{1}{3}} = 12.3, \quad (13)$$

where $h_c$, $T_f$, $T_{MR}$, $\sigma$, $\varepsilon_s$, $k_{air}$, $L$, $Re_L$, $Pr$, $V$, $\nu_{air}$, $c_2$ are convection heat transfer coefficient for turbulent flow (W m$^{-2}$), liquid film temperature (K), mean radiant temperature of surroundings (K), Stefan–Boltzmann constant ($= 5.67 \times 10^{-8}$ W m$^{-2} \times$ K$^4$), emittance of surface ($= 0.7$ assumed), thermal conductivity of air at $t_a$ (W m$^{-1} \times$ K), characteristic length of slab in direction of wind (m), Reynolds number based on characteristic length $L$, and Prandtl number for air ($= 0.7$), design wind speed near slab surface ($= 10$ km h$^{-1}$ assumed), kinematic viscosity of air (m$^2$ s$^{-1}$), and conversion factor ($= 0.278$), respectively.

Evaporation heat flux, $q_e$ (W m$^{-2}$) is

$$q_e = \rho_{dry\,air} h_m (W_f - W_a) h_{fg} = 1.33 \times 0.0102$$
$$\times (0.00393 - 0.00160) \times 2499 \times 10^3 = 79.2, \quad (14)$$

$$h_m = \left( \frac{Pr}{Sc} \right)^{\frac{2}{3}} \frac{h_c}{\rho_{dry\,air} c_{p,\,air}} = \left( \frac{0.7}{0.6} \right)^{\frac{2}{3}} \frac{12.3}{1.33 \times 1005} = 0.0102, \quad (15)$$

where $h_m$, $W_a$, $W_f$, $h_{fg}$, $\rho_{dry\,air}$, $Sc$ are mass transfer coefficient (m s$^{-1}$), humidity ratio of ambient air, humidity ratio of saturated air at film surface temperature, heat of vaporization (J kg$^{-1}$), density of dry air (kg m$^{-3}$), and Schmidt number ($= 0.6$).

As a result, the required total heat flux for snow melting is 229.1, 324.0, and 418.9 W m$^{-2}$ when $A_r$ is 0, 0.5, and 1.0, respectively. The area of the pedestrian sidewalks is 1120 m$^2$ ($= 280$ m $\times$ 2 m$\times$ 2); therefore, the required total heat is 256.6, 362.9, and 469.2 kW when $A_r$ is 0, 0.5, and 1.0.

Comparing the heat supply from the geothermal hot fluid (124.0 kW) with the required total heat shown above, it is found that the heat supply would be insufficient to melt snow even when $A_r$ is 0 for the entire section of sidewalks. As shown in Eq. (7), the heat supply depends on mass flow rate and the difference between inlet and outlet temperature. As it is difficult to make the temperature difference larger, the required heat would be supplemented by drilling roughly two additional geothermal wells to satisfy the required total heat with $A_r = 0.5$, 362.9 kW. Or the snow melting area should be limited to approximately one-third of the total area of the pedestrian sidewalks.

## 6  Conclusion

In this study, the authors proposed a hybrid geothermal heat pump system that is coupled with mine water and hot spring water. We mapped the temperature profile in the Edgar Mine, assessed its thermal capacity, and analyzed the heat/mass transfer of the geothermal hot spring water.

The temperature measurements showed that the temperature of rock surface was approximately 12 °C at maximum in the Edgar Mine, and the mine should be designed as a cold thermal energy storage. As the Army Tunnel was wetter and colder than the Miami Tunnel, the Army Tunnel is a more suitable mine tunnel to store cold groundwater from winter to summer. Unlike the Miami Tunnel, however, the rock surface temperature in the Army Tunnel was not constant because of its thinner cover. Therefore, temperature measurement in the Army Tunnel should be continued for the rest of the year to design a cold thermal storage system.

The heat/mass transfer analyses showed that the temperature of the geothermal hot water decreased from 35 to 28.6 °C when the thickness of a pipe was assumed to be 5 cm during a 540 m transportation. In order to minimize this temperature loss, the thickness of a pipe should be much larger than

5 cm. However, as the correlation between the pipe thickness and the temperature loss is non-linear, the pipe thickness has to be chosen appropriately considering the cost and ease of construction work.

The energy balance analyses showed that the proposed system would not melt snow-covered pedestrian sidewalks effectively. In order to satisfy the heat requirement with $A_r = 0.5$, the heat supply has to be increased approximately three times, which could be achieved by drilling additional geothermal wells if appropriate. Alternatively, additional thermal insulation measures can be taken to reduce the temperature drop. It should be acknowledged that underlayment insulation is very effective on snowmelt slabs to maximize snowmelt performance. With good underlayment insulation and optimized hydronic pipe spacing it may be expected to increase the use of the provided flow and temperature of hot water by 30 % or more, and the snowmelt slab will react faster to melt snow/ice.

**Acknowledgements.** The authors would like to thank Terry Proffer for his critical review of the manuscript.

## References

AccuWeather.com: Local weather record of Idaho Springs, CO, USA, available at: http://www.accuweather.com/en/us/idaho-springs-co/80452/december-weather/337419?monyr=12/1/2013&view=table, last access: 1 January 2014.

Adlam, T. N.: Snow Melting, The Industrial Press, New York, USA, 1950.

ASHRAE: Snow Melting and Freeze Protection, 2011 ASHRAE Handbook – HVAC Applications, 51, 51.1–51.20, 2011.

Chapman, W. P. and Katunich, S.: Heat Requirements of Snow Melting Systems, ASHRAE Tran., 62, 359–372, 1956.

Kusuda, T. and Achenbach, P. P.: Earh Temperature and Thermal Diffusivity at Selected Stations in the United States, Summary of Research Report 8972, National Bureau of Standard, 1965.

Lund, J. W.: Reconstruction of a Pavement Geothermal Deicing System, Geo-Heat Center Quarterly Bulletin, 20, 14–17, 1999.

Pingjia, L. and Ning, C.: Abandoned Coal Mine Tunnels: Future Heating/power Supply Centers, Mining Science and Technology (China), 21, 637–640, 2011.

Repplier, F. N., Zacharakis, T. G., and Ringrose, C. D.: Geothermal Resource Assessment of Idaho Springs, Colorado, Colorado Geological Survey, Golden, Colorado, 1982.

Rodriguez, R. and Diaz, M. B.: Analysis of the Utilization of Mine Galleries as Geothermal Heat Exchangers by Means a Semi-empirical Prediction Method, Renew. Energ., 34, 1716–1725, 2009.

Sato, M. and Sekioka, M.: Geothermal Snow Melting at Sapporo, Japan, Geo-Heat Center Quarterly Bulletin, 4, 16–18, 1979.

Shiba, Y.: Ground-Source Heat Pump System, Building Utilities and Piping Work, 2, 20–24, 2008 (in Japanese).

Yamashita, M., Majima, T., Tsujita, M., and Matsuyama, K.: Geothermal Development in Hachijojima, Proceedings, World Geothermal Congress 2000, 28 May–10 June 2000, Kyusyu-Tohoku, Japan, 2000.

# 13

# Geothermal resources and reserves in Indonesia: an updated revision

**A. Fauzi**

PT. Geo Power Indonesia, Menara Palma #15-02A-B Kuningan, 12950 Jakarta, Indonesia

*Correspondence to:* A. Fauzi (fauziamir2000@yahoo.com)

**Abstract.** More than 300 high- to low-enthalpy geothermal sources have been identified throughout Indonesia. From the early 1980s until the late 1990s, the geothermal potential for power production in Indonesia was estimated to be about 20 000 MWe. The most recent estimate exceeds 29 000 MWe derived from the 300 sites (Geological Agency, December 2013).

This resource estimate has been obtained by adding all of the estimated geothermal potential resources and reserves classified as "speculative", "hypothetical", "possible", "probable", and "proven" from all sites where such information is available. However, this approach to estimating the geothermal potential is flawed because it includes double counting of some reserve estimates as resource estimates, thus giving an inflated figure for the total national geothermal potential.

This paper describes an updated revision of the geothermal resource estimate in Indonesia using a more realistic methodology. The methodology proposes that the preliminary "Speculative Resource" category should cover the full potential of a geothermal area and form the base reference figure for the resource of the area. Further investigation of this resource may improve the level of confidence of the category of reserves but will not necessarily increase the figure of the "preliminary resource estimate" as a whole, unless the result of the investigation is higher.

A previous paper (Fauzi, 2013a, b) redefined and revised the geothermal resource estimate for Indonesia. The methodology, adopted from Fauzi (2013a, b), will be fully described in this paper.

As a result of using the revised methodology, the potential geothermal resources and reserves for Indonesia are estimated to be about 24 000 MWe, some 5000 MWe less than the 2013 national estimate.

## 1 Introduction

Much of the Indonesian Archipelago is situated on the boundary of the Indo-Australian and the Eurasian tectonic plates. Numerous active volcanoes are associated with this plate boundary, stretching from Sumatera to Java, Bali and Maluku up to Sangihe Island, resulting in many high-enthalpy geothermal resources in these areas.

There are more than 177 volcanic centres in the Archipelago, of which about 88 bear evidence of fumarolic and solfataric activity (Radja, 1990). Reconnaissance studies carried out since the mid-1960s have found in excess of 300 sites as potential high- to low-enthalpy geothermal sources.

The first national geothermal energy potential estimate was reported to be about 16 000 MWe (Radja, 1990). This was later revised by Pertamina to be about 20 000 MWe (Fauzi et al., 2000; Pertamina, 2005). A short time later, at the 2000 World Geothermal Congress, the government of Indonesia released an estimate of the national geothermal energy potential to be approximately 27 000 MWe. The most recent estimate of geothermal potential exceeds 29 000 MWe (Geological Agency and Ebtke, December 2013).

This paper presents the results of an updated revision of the geothermal energy potential of Indonesia based on the most updated data (Geological Agency, December 2013). The revision uses a new summation methodology which was described in Fauzi (2013a, b) by using the 2008 geothermal potential data. The new summation methodology will be fully described in this paper.

## 2  Definitions

In general, geothermal resources are defined as thermal energy beneath the earth at depths shallow enough to be tapped/extracted economically and legally by drilling at a specific time. Geothermal reserves are defined as that portion of the geothermal resource that is commercially extractable. Below are examples of specific definitions.

Geothermal resources: the estimates of geothermal potential determined on the basis of limited data and yet to be proven as potential reserves (SNI, 1998).

Geothermal resource is a term used to define a quantity of geothermal heat that is likely available if certain technologic and economic conditions are found in the future (specified) time (Pertamina, 2005).

A geothermal resource is a geothermal play which exists in such a form, quality and quantity that there are reasonable prospects for eventual economic extraction. Resources are known, estimated or interpreted from specific geological evidence and knowledge. Geothermal resources are subdivided, in order of increasing geological confidence, into categories of inferred, indicated and measured (Australian Geothermal Code, 2010).

Speculative resources: the estimate based on the presence of geothermal surface manifestations and other signs of heat flow (modified from Pertamina, 2005). (Note: speculation is about something that cannot be definitely proven, i.e. theoretical.)

Hypothetical resources: determined from regional geologic surveys and geochemistry of thermal features. Stored-heat calculations and estimates are used for resource sizing (modified from Pertamina, 2005). (Note: in science, a hypothetical conclusion is drawn before all the facts have been discovered and adopted for the time being as a guide to further investigation.)

Geothermal reserves: total heat content stored in the subsurface that is economically recoverable and estimated using the geosciences survey as a tool (SNI, 1998).

Geothermal reserves consist of identified economic resources that are recoverable at a cost that is competitive now with other commercially developed energy sources. Reserve estimates are calculated based on well data and or geoscientific data (modified from Pertamina, 2005).

A geothermal reserve is that portion of an indicated or measured geothermal resource that is deemed to be economically recoverable after the consideration of both the geothermal resource parameters and modifying factors. These assessments demonstrate at the time of reporting that energy extraction could reasonably be economically and technically justified (Australian Geothermal Code, 2010).

Possible reserves: the geothermal energy potential in the subsurface is estimated by using integrated surface geoscientific survey data (modified from Pertamina, 2005).

Probable reserves: the geothermal energy potential in the subsurface is estimated by using integrated surface geoscien-

tific survey data and the result of one discovery well (modified from Pertamina, 2005).

A probable geothermal reserve is the economically recoverable part of an indicated or, in some circumstances, a measured geothermal resource (Australian Geothermal Code, 2010).

Proven reserves: the geothermal energy potential in the subsurface is estimated by using integrated surface geoscientific survey works with reasonable certainty and includes one discovery well and two producing delineation wells to obtain data on the subsurface parts of the system (modified from Pertamina, 2005).

A proven geothermal reserve is the economically recoverable part of a measured geothermal resource. It includes a drilled well, the deliverability of which has been demonstrated in commercial production for the assumed lifetime of the project with a high degree of confidence (Australian Geothermal Code, 2010).

Cut-off temperature: minimum temperature required for wells to self-discharge in convective geothermal development. For pumped flows, the minimum economic reservoir fluid temperature for commercial energy extraction is based on the estimated temperatures from surface geochemistry (Australian Geothermal Code, 2010).

Note that cut-off temperature is the minimum temperature below which it is not economic to include the reservoir volume as part of the resource. The cut-off temperature typically used by geothermal scientists and engineers is $\sim 180\,°C$ for the purpose of generating electricity using a "conventional steam turbine". This figure has also been discussed in Australian Geothermal Code, 2010 – p. 18–19.

The definitions above that are quoted from several sources are only very slightly different between themselves. The difference showed in the terminology used; for example in Australian Geothermal Code the categories "inferred", "indicated" and "measured" were used, and in others the terms "speculative", "hypothetical", "possible", "probable" and "proven" were used. This sequence is given in order of increasing geological confidence. Furthermore, there is a reasonable degree of consistency between them with respect to the definition of "resources" versus "reserves".

## 3  History of resources estimation

The estimates of resource and reserve potential of geothermal energy in Indonesia has gone through three phases.

In the First Phase, from the early 1980s until the late 1990s, estimates of resources and reserves were predominantly carried out by the Directorate of Geology & Mineral Resources (in Suryantoro, 2000, 2005) and Pertamina (2005). In this phase, the attention was paid on the high enthalpy resources.

In the Second Phase, the estimates were made by the Director General of Geology and Mineral Resources (in

**Table 1.** Indonesia geothermal resources and reserves: December 2013 status.

| Location | Resources (MWe) | | Reserves (MWe) | | | Installed (MWe) |
| --- | --- | --- | --- | --- | --- | --- |
| | I | II | III | IV | V | |
| Sumatera | 3122 | 2451 | 7102 | 29 | 380 | 122 |
| Java | 1657 | 1826 | 3769 | 658 | 1815 | 1134 |
| Bali | 70 | 22 | 262 | – | – | – |
| N. Tenggara | 407.5 | 359 | 797 | – | 15 | 5 |
| Kalimantan | 145 | – | – | – | – | – |
| Sulawesi | 1434 | 241 | 1345 | 150 | 78 | 80 |
| Maluku | 545 | 76 | 397 | – | – | – |
| Papua | 75 | – | – | – | – | – |
| Total | 7455.5 | 4975 | 13 672 | 837 | 2288 | 1341 |
| | 12 430.5 | | | 16 797 | | |
| | | | 29 227.5 | | | |

Geological Agency & Ebtke (December 2013)
I – speculative; II – hypothetical; III – possible; IV – probable; V – proven.

Suryantoro, 2000, 2005), and first released at the 2000 World Geothermal Congress.

For the third phase, the estimates of resources and reserves were carried out by the Geological Agency of Indonesia, and published in API NEWS (2013) and in Geological Agency (2013).

Within the last two decades, all the prospects/resources from very low to high enthalpy have become the focus of further investigation. The results are shown in Table 1.

Volumetric method resource estimation, which is commonly used within the geothermal industry, has been applied to have the result as shown in Table 1 (Munandar, personal communication, 2004). It is considered that the volumetric estimation is the only one of the methods consistently applicable for resource estimation at an early stage in Indonesia.

The most recent geothermal resource figures were obtained by adding all the numbers from all of the categories speculative, hypothetical, possible, probable and proven as shown in the published information (Table 1). In the case of no new summation methodology being adopted, as proposed in Fauzi (2013a, b), it is almost certain that there will be no more significant new prospects inventory; it is highly likely that the resource figures of 29 000 MWe (Table 1) will be constant with time in the future.

## 4  Reassessment of the resources and reserves

Currently, the geothermal potential classification process includes estimating the geothermal potential of a prospect area in one of five categories defined above from a "speculative resource" (the least certain category of geothermal potential) to a "proven reserve" (the most reliable estimate of an area's geothermal potential).

The preliminary speculative resources category of resources should cover the full potential of the area and form the base reference figure for the resource of the potential

geothermal field. Eventually, further investigation of this resource may allow parts of the speculative resource to be upgraded to proven reserve. However, the total "resource capacity" does not increase in this upgrade (unless, of course, the estimated area of the geothermal field resource is increased).

Where detailed geological, geochemical and geophysical surveys are conducted over the resource area, then that part of the geothermal resource may be upgraded to the category of "possible reserve". Further studies and surveys including at least one exploration well must be able to prove the existence of a high-temperature fluid that can be used for producing electric power. This will allow that part of the resource to be categorized as "probable reserves". Again this does not make an addition to the possible reserves or the total resource. Additional produced exploration wells that can provide a three-dimensional model of the geothermal resource can result in the estimate being upgraded in the area of the drilling and model to proven reserves. Again, the estimate of the proven reserves does not increase the estimate of the geothermal resource as a whole as "geothermal reserves" are defined as that portion of the geothermal resource that is commercially extractable.

Thus, all activities (geoscientific surveys and drilling) that follow an initial resource estimate of the geothermal potential are tools to prove, with increasing certainty, the existence of geothermal energy stored in the reservoir rocks in the form of high-temperature fluid. Once a geothermal resource is established, any further estimates (that upgrade part of the resource to a hypothetical resource or a possible/probable/proven reserve) should not be added to the total resource capacity.

The previous practice of "summation" as depicted in Table 1 does not give a correct estimate as it leads to a greater estimate of the geothermal potential than is actually present. Based on this fact, the total summation methodology as a resource estimation tool needs to be improved.

**Table 2.** Geothermal resources and reserves: examples of proposed methodology using selected areas.

| Location | Resources (MWe) | | Reserves (MWe) | | | Installed |
|----------|------|-------|--------|-------|------|-----------|
|          | I (A) | II (B) | III (C) | IV (D) | V(E) | (Mwe) |
| Sibayak | – | 34 | 35 | – | 30 | 2 |
| Sibayak | 35 | 34 | 35 | – | 30 | 2 |
| Dolok Marawa | 100 | – | 40 | – | – | – |
| Dolok Marawa | 100 | – | 40 | – | – | – |
| Cubadak | 73 | – | 100 | – | – | – |
| Cubadak | 73/100 | – | 100 | – | – | – |
| Awibengkok | – | – | 110 | 110 | 375 | 375 |
| Awibengkok | –/375 | – | 110 | 110 | 375 | 375 |
| Kamojang | – | – | – | 300 | 260 | 140 |
| Kamojang | –/300 | – | – | 300 | 260 | 140 |
| Darajat | – | – | 160 | 150 | 300 | 255 |
| Darajat | –/300 | – | 160 | 150 | 300 | 255 |
| Total | 173/1210 | 34 | 445 | 560 | 965 | – |
|       |        |    | 2177/1210 | | | |

Original data from Geological Agency (2013). I – revised speculative resource as the whole resource; II – hypothetical; III – possible; IV – probable; V – proven.

**Table 3.** Revised total geothermal resources and reserves in Indonesia.

| Location | Resources (MWe) | | Reserves (MWe) | | | Installed (MWe) |
|----------|------|------|------|------|------|-----------------|
|          | I | II | III | IV | V | |
| Sumatera | 11 054 | 2451 | 7102 | 29 | 380 | 122 |
| Java | 7242 | 1826 | 3769 | 658 | 1815 | 1134 |
| Bali | 327 | 22 | 262 | – | – | – |
| N. Tenggara | 1399 | 359 | 797 | – | 15 | 5 |
| Kalimantan | 145 | – | – | – | – | – |
| Sulawesi | 2842 | 241 | 1345 | 150 | 78 | 80 |
| Maluku & Papua | 1011 | 76 | 397 | – | – | – |
| Total | 24 020 | 4975 | 13 672 | 837 | 2 288 | 1341 |
|       |        |      | 24 020 | | | |

Resummation speculative resource based on Geological Agency data (December 2013).
I – revised speculative resources as the whole resources; II – hypothetical; III – possible; IV – probable; V – proven.

In addition, the resource numbers shown in Table 1 are the result of using different cut-off temperatures. In the case of a country-wide resource assessment, like Indonesia, assumptions for electricity generation using a conventional steam turbine power plant should use a single cut-off temperature – typically $\sim 180\,°C$. In fact, the estimate of 29 000 MWe potential resource in Indonesia used variable cut-off temperatures, which ranged from 90 to 180 °C (SNI, 1999). This approach should be avoided.

## 5  Suggested approach

The initial estimation of the geothermal energy potential of an area is usually done through the study of literature and a site visit to the area. This estimation is classified as a resource and placed into the speculative and/or hypothetical category. Further study using geoscientific surveys and exploration drilling can improve the level of confidence to the category of reserves that consists of possible, probable and proven reserves.

The initial estimate that classifies the geothermal potential as a speculative resource, and the estimate using geoscientific surveys as a possible reserve, should be the main focus in terms of determining how large the geothermal energy resources are in a region. The initial estimates and the geoscientific survey will substitute one another unless the last two classifications, probable and proven reserves, prove higher than possible reserves and speculative resources. For example, see Table 2 for Darajat, Awibengkok and Kamojang fields. The proven reserves from the first two fields are the figures for the total resources of the fields. And the probable reserves from Kamojang are the total resources of field.

The estimate of a possible reserve may be lower or higher than a speculative resource. If the estimate of possible reserve remains lower than the estimated speculative resources, then the initial estimate remains being a benchmark for the calcu-

**Table 4.** Classification of geothermal resources in Indonesia: based on estimated subsurface temperature resource groups.

| Location | Resource classification (MWe) | |
| --- | --- | --- |
| | Low enthalpy < 100to 190 °C | High enthalpy > 190 °C |
| **Sumatera** | | |
| Aceh | 520 | 782 |
| North Sumatra | 170 | 2253 |
| West. Sumatra | 820 | 728 |
| Jambi + Riau | 224 | 778 |
| Bengkulu | 60 | 1005 |
| Bangka Belitung | 45 | – |
| South Sumatera | 555 | 987 |
| Lampung | 409 | 1718 |
| **Java** | | |
| Banten | 148 | 385 |
| West Java | 1131 | 3261 |
| Central Java | 350 | 954 |
| East Java & Bali | 182 | 1158 |
| Nusa Tenggara | 862 | 537 |
| Kalimantan | 145 | – |
| Sulawesi | 1696 | 1146 |
| Maluku & Papua | 569 | 442 |
| Total | 7886 | 16 134 |

lation of total resources for the region. On the other hand, if the estimate obtained from the results of further investigation is higher, this figure becomes the new reference number for the speculative resource (Table 2).

It is proposed that the resource calculation of a geothermal field not being estimated by summing all the numbers from speculative and hypothetical resources and from possible, probable and proven reserves. Thus the estimate for the calculation of total resources of an area is that estimate or figures recorded as the revised speculative resource (Table 2). As defined above, the hypothetical resources and all classes of reserves are that portion of the geothermal resources as resources in total (in Tables 2 and 3 as speculative resources).

In addition, the resources calculation should be treated differently for high- and low-temperature resources. For the high-temperature resources, a single cut-off temperature ($\sim 180\,°C$) should be assumed for conventional steam power plants. And for the low-temperature resources for non-conventional steam power plants (i.e. binary plant) or other uses, the different approaches should be applied.

Notes:

1. Sibayak: if (C) > (B) or (C) > (A), then the number in (C) should be transferred to (A) or (B) and be the same as the reference that is used in summation as a resource in total.

2. Dolok Marawa: if (C) < (A), then the number in (A) is the reference without any changes used in summation as a resource in total.

3. Cubadak: if (C) > (A), then the number in (C) should be transferred to (A) to be used in summation as a resource in total.

4. Awibengkok (Salak): if (E) > (A), then the number in (E) should be transferred to (A) to be used in summation as a resource in total. Similar for Darajat.

5. Kamojang: if (D) > (A), then the number of the potential in (D) should be transferred to (A) to be used in summation as a resource in total.

6. The resource calculation of a geothermal field is not estimated by summing all the numbers from speculative (A) and hypothetical (B) resources and from possible (C), probable (D) and proven (E) reserves.

7. Thus the total resource as a whole resource in this example is 1210 MWe.

8. The figure of the resource calculation using current methodology by summing all the numbers is 2177 MWe.

9. The dashes or slashes, especially in (A) entries, mean that the original estimate of total resources of each prospect/field is not available.

10. Pertamina (2005) proposed at the early stage that all the new resources have the speculative resource of 250 MWe. Assumed that each of them is considered to have an area of 20 km$^2$ and the recoverable resource has a power density at 12.5 MWe km$^{-2}$. In future, this proposal should be considered in order to avoid the dashes within the (A) entries.

11. The bold numbers are the revised estimates (see texts)

Using a more realistic methodology as described in Table 2, the geothermal resource potential of Indonesia is revised down to be approximately 24 000 MWe, some 5000 MWe less than the 2013 national estimate such as that summed up and concluded in Table 3. For example, the original total resource for Sumatera, by summing up all of column I to V in Table 1, is 13 084 MWe; after revision it is 11 054 MWe in Table 3. Thus, the figures in column I are the result after being revised from figures present in Table 1.

It is interesting to note that about 8000 MWe out of 24 000 MWe resources are classified as non-electrical-grade (< 100 °C) to low-temperature (150 to 190 °C) resources as defined by Sanyal (2005) (Table 4).

The cut-off temperature used by geothermal scientists and engineers typically is $\sim 180\,°C$. Based on this cut-off temperature, these non-electrical-grade (< 100 °C) to low-temperature (150 to 190 °C) resources are not efficiently converted into power (MWe). Consequently, the figure of 8000 MWe is flawed and should be reviewed. The real figure could be less than 8000 MWe and partly intended for direct use and/or electricity through binary technology or other.

## 6   Conclusions

The previous estimate of the geothermal resources of Indonesia included double counting of some reserves estimates as resources estimates, thus giving an inflated figure of the total national geothermal resource potential.

Using a more realistic methodology proposed in Fauzi (2013a, b), where this double counting is removed, the geothermal resource potential of Indonesia is revised down to approximately 24 000 MWe. This is some 5000 MWe less than the previous 2013 national geothermal resource estimate of 29 227.5 MWe.

**Acknowledgements.** The author thanks Kenneth B Alexander and Robert King for their valuable comments.

## References

API News-Sikumbang, Let's Investing Together on Geothermal; in API (Indonesian Geothermal Association) News No. 8 – October 2012, p.5., 2012.

Australian Geothermal Code Committee: Geothermal Lexicon For Resources and Reserves Definition and Reporting, Edition 2, Adelaide, 2010.

Directorate General of New Energy: Renewable and Conservation Energy (Ebtke), Profil Potensi Panas Bumi Indonesia, 2012.

Fauzi, A.: Geothermal development in Indonesia: An overview; Geothermia, Rev de Geoenergia, 14, 147–152, 1998.

Fauzi, A.: A proposed new methodology for estimating geothermal resources and reserves in Indonesia; in Petrominer Magazine No. 5, May 2013, 64–65, 2013a.

Fauzi, A.: Geothermal Resources and Reserves in Indonesia: A Revision; in: Proceedings, 13th Indonesia International Geothermal Convention & Exhibition 2013, Jakarta, Indonesia, 12–14 June, 2013b.

Fauzi, A., Bahri, S., and Akuanbatin, H.: Geothermal Development in Indonesia: An Overview of Industry Status and Future Growth; in Proc. World Geothermal Congress, 2000.

Geological Agency: (Badan Geologi Indonesia), Geothermal Area Distribution Map And Its Potential in Indonesia, Status Dec. 2013.

Pertamina: Pertamina Geothermal Development: Resources & Utilization: Publication of Pertamina Geothermal Division, 2005

Radja, V. T.: Review of the Status of Geothermal Development and Operation in Indonesia; in Proceedings of the Geothermal Resources Council Transaction 1990, Kona, Hawaii, 20–24 August, 14, 127–145, 1990,

Sanyal, S. K.: Technical Paper: Classification of Geothermal Systems – A Possible Scheme, United States Department of Energy (Washington, D.C.), 2005.

SN:, Klasifikasi Potensi Energi Panas Bumi di Indonesia; Standar Nasional Indonesia-SNI 13-5012-1998-ICS 73.020, 1998.

SNI: Metode Estimasi Potensi Energi Panas Bumi di Indonesia; Standar Nasional Indonesia-SNI 13-6171-1999 ICS 73.020, 1999.

Suari, S. and Fauzi, A.: Geothermal Prospects in Sumatera; in Proc. Indonesian Petroleum Association, 20th Annual Convention, Oct. 1991-IPA, 91-21.21, 1991.

Suryantoro, S.: Challenges of Geothermal Development in Indonesia; in Proc. World Geothermal Congress, 2000.

Suryantoro, S., Dwipa, S., Ariati, R., and Darma, S.: Geothermal Deregulation and Energy Policy in Indonesia; in Proc. World Geothermal Congress, 2005.

Wahyuningsih, R.: Potensi dan Wilayah Kerja Pertambangan Panas Bumi di Indonesia: Subdit Panas Bumi, Kolokium Hasil Lapangan-DIM, 2005.

# 14

# Rearrangement of stresses in fault zones – detecting major issues of coupled hydraulic–mechanical processes with relevance to geothermal applications

**G. Ziefle**[1,*]

[1]Leibniz Institute for Applied Geophysics (LIAG), Stilleweg 2, 30655 Hannover, Germany
[*]now at: Federal Institute for Geosciences and Natural Resources (BGR), Stilleweg 2, 30655 Hannover, Germany

*Correspondence to:* G. Ziefle (gesa.ziefle@bgr.de)

**Abstract.** The South German Molasse Basin provides favourable conditions for geothermal plants. Nevertheless, micro-seismic events occur in the vicinity of the geothermal Unterhaching Gt2 well and seem to be caused by the geothermal plant.

The injection and production are located in an existing fault system. The majority of seismic events takes place at a horizontal distance of 500 m or less of the borehole. However, none of the seismic events are located in the injection reservoir but in fact at a significantly greater depth. A deeper process understanding of the interacting thermal–hydraulic–mechanical effects in the vicinity of the well is desired.

This article presents a significantly simplified 2-D model, investigating interactions of the stress field in the vicinity of the geothermal well and movements in the fault system. This might be of special interest, as the operation of the geothermal plant might lead to changes in the material and fracture properties on the one hand and in the equilibrium state on the other. A detailed description of the model, as well as various parameter studies, is presented. It can be seen that boundary conditions such as direction of the stress field in relation to the fault system, geometry of the fault system and parameters of the fractures have a significant influence on stresses in the proximity of the geothermal well. A variation in the spatial stress field in some parts of the fault system is to be expected. For the chosen assumptions the dimension of this variation is about 25 % of the assumed stresses. Future work on this model might focus on the characteristics of the fault system, as well as on the influence of the coupled thermal–hydraulic–mechanical effects.

## 1 Introduction

The South German Molasse Basin provides favourable conditions for geothermal plants (BStWIVT, 2010). Consequently, about a dozen successfully working geothermal power plants are located in the greater Munich area (Ganz et al., 2013). However, micro-seismic events are observed in the proximity of the geothermal well Unterhaching Gt2 (Megies and Wassermann, 2014). The reason for these events – appearing at significantly greater depth than the well – is not yet known. A deeper understanding of the physical processes in this area is of great interest.

Generally, geophysical use of the underground, e.g. deep geothermal energy use, storage of radioactive waste, or the storage of $CO_2$ is characterized by various interacting physical processes. Numerical simulations are a good tool to provide an insight into the thermal–hydraulic–mechanical (THM) coupled processes in the subsurface. General information on the related topics can be found in, for example, Pollard and Fletcher (2005), Zoback (2007), Fossen (2010), and Gudmundsson (2011). Concerning geothermal plants, the injection and production of the water and the variations of temperature have an impact on the mechanical system in the target horizon of the geothermal wells. Taking into account the mechanical and fracture-mechanical effects in the

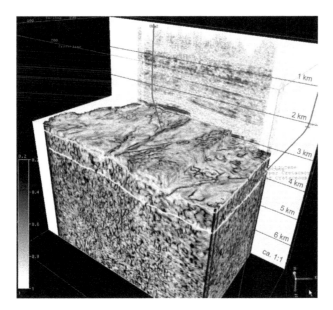

**Figure 1.** Seismic measurement in the area of the Unterhaching well. Here, the coherence (the similarity between two profiles) is presented. The Unterhaching Gt2 well (left) and the Kirchstockach well (right) are shown in red. The top of the Malm is shown by the blue line, and the bottom by the green line, according to Lüschen et al. (2011) and Lüschen et al. (2014).

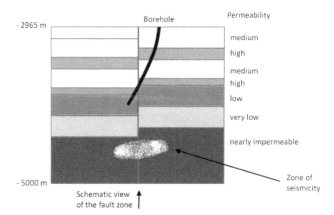

**Figure 2.** Schematic view of the geological situation in the near field of the Unterhaching geothermal well. Presented are the Malm (multicoloured) and the crystalline layer (in grey) in a vertical slice (after Schumacher, 2013).

deep underground, the existing structures and faults need to be mentioned. Complex, coupled THM problems have to be investigated, aiming for a better process understanding of the evolution of fault zones and their potential impact on seismic events. Utilizing numerical simulation tools to get a better understanding of the relevant effects is a demanding but promising challenge. However, on the one hand, there is a significant lack of information about the system and the specific properties and, on the other hand, the numerical simulation of such complex systems on a large scale is a major challenge even today. In the following, a simplified model approach is presented, focusing on the mechanical effects and the impact of the hydraulic properties in the surroundings of the Unterhaching Gt2 well.

The geological situation of the Unterhaching Gt2 well is given in Fig. 1 based on seismic results. The figure presents the coherence, which indicates fracture or fault zones. The Malm (Upper Jurassic) and the crystalline layer are given as 3-D structures. The Malm is characterized by highly fissured, karstic material. This area is the target horizon of the geothermal wells and is bordered here by the top and bottom lines. The layers above are only indicated by the vertical slice. The Unterhaching and the Kirchstockach wells are shown by red lines. A fault system is located in the near field of the Gt2 well, and will be the focus of the following sections.

Both the well and the seismic events are located close to the fault system, which is presented in Fig. 2 as a schematic view of a vertical slice. The target horizon, namely the Malm,

is characterized by several layers with different permeabilities. The well intersects the fault zone, which is indicated by the red line. The seismic events are located at a significant distance beneath the well in the crystalline.

## 2 Stress field analysis in the area of the geothermal well Unterhaching

### 2.1 Simplification of the model

The presented geophysical situation implies a three-dimensional model with thermal, hydraulic, mechanical and chemical effects. The following investigations concentrate on the mechanical processes and the interaction with some hydraulic effects, which are to be expected in this area. The focus is laid on the stress field in the vicinity of the fault system, which is influenced by elemental movements of the fracture zones. The following investigations give an idea of the impact of these movements – depending on the assumed parameters – on the stress field.

In context of mechanical processes, data describing the fracture-mechanical parameters, as well as information on the stress field and the mechanical parameters in the area of the micro-seismic events, are quite rare. Moreover, the numerical simulation of a three-dimensional geophysical problem incorporating fracture-mechanical processes is not yet possible. However, it is important to get a better understanding of the mechanical processes in and near the fault; therefore, a strictly simplified model is presented in the following section.

The simplification of the model is mainly characterized by the following steps:

1. 2-D instead of 3-D

   Naturally, the effects in the system are three-dimensional. However, today few numerical codes exist enabling a 3-D numerical simulation of a fully coupled THM problem incorporating fracture mechanics on the required scale. In addition, it is questionable whether such a numerical model would be meaningful due to the lack of input data. Nevertheless, in the simplified two-dimensional approach, it is not possible to picture all effects realistically. This simplification is based on the assumption that the main effects can be pictured by the two-dimensional approach and that this model results in a better understanding of the interacting effects.

2. Downscaling

   A downscaling from the large-scale system to a significantly downscaled model is carried out. The mechanic law of similarity as, for example, presented in Worthoff and Siemes (2012) must be kept in mind. It is applied in the modelling and by interpreting the results. The scale between model and reality is chosen to be 1 : 400.

3. Singular faults instead of fissured areas

   In reality, no singular faults are to be expected, as fault zones are characterized by fissured areas with a lateral extent of metres to tens of metres. To model this system it is necessary to reduce the fissured areas to single fracture surfaces due to restrictions of the boundary element (BEM) code FRACOD (Shen, 2002), which is used for this investigation. This simplification has a significant influence on the quantitative results, especially in the proximity of the fault zones. However, the present study focuses on identifying and understanding the dominant mechanical effects – a quantitative statement, though desirable, is beyond the scope of this paper. Nevertheless, even with a qualitative analysis of the results, principal effects can be pictured, future research can be specified and an increase in process understanding can be generated.

The model presented here enables investigations concerning various material parameters of the rock and the fault system. Furthermore, simplification strategies of the fault system, as well as boundary conditions such as stress field and direction of stresses in relation to the fault system, various borehole locations or injection pressures, can be investigated. The varying boundary conditions can describe various phases of the system's evolution and various depths or loading cases. The aim of this model is to increase process understanding and statements regarding the relevance of various input parameters and requirements in further investigations.

## 2.2 Model set-up

The geometry of the fault zones in the vicinity of the geothermal well Unterhaching Gt2 has a significant influence on mechanical and fracture-mechanical effects. To get a better idea of the impact of various properties, the mechanical effects were investigated with FRACOD. A manual for this

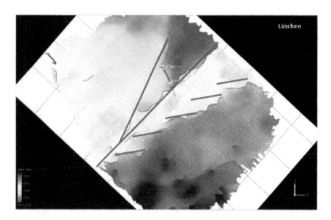

**Figure 3.** Simplified fault zones (blue lines), injection well (red) and depth of the top Malm horizon (colour coding from red to dark blue indicates from minimum to maximum depth; maximum depth difference ≈ 400 m) (Lüschen et al., 2014). Grid line interval is 1 km.

two-dimensional fracture initiation and propagation code is presented in Shen and Stephansson (1993b). Further information on this topic is given in Byerlee (1978), Ferrill and Morris (2003), Rinne (2008) and Shen et al. (2014). FRACOD utilizes the displacement discontinuity method (DDM) in combination with the Mohr–Coulomb failure criterion and a new fracture criterion (G criterion) proposed by Shen and Stephansson (1993a). FRACOD is designed to capture fundamental features of the fully coupled hydro-mechanical behaviour of elastic and isotropic rocks.

### 2.2.1 Geometry

Taking into account the seismic results given in Fig. 1, a significantly simplified model of the geometry of the fault zones is developed. Figure 3 depicts the depth to the top of the Malm and the simplified geometry of the fault zones in the simplified model.

The area given in Fig. 3 has an extension of about 8.0 km × 5.5 km. The zone of interest lies in the centre of the presented area and consists of the main fault (presenting the fault zone with an angle of about 45°), the minor fault (representing the fault zone with an angle of about 20°) and the en echelon fault system. The geometry of the downscaled model is depicted in Fig. 4. Due to the downscaling process (a factor of 1 / 400), the model area has an extension of 12 m × 12 m and depicts the zone of interest in the vicinity of the fault system.

### 2.2.2 Material properties

The material parameters can be split into material properties of rock and water and properties of the fractures. These are summarized in Table 1 for a so-called standard model, which will be the basis for all parameter studies. Some of

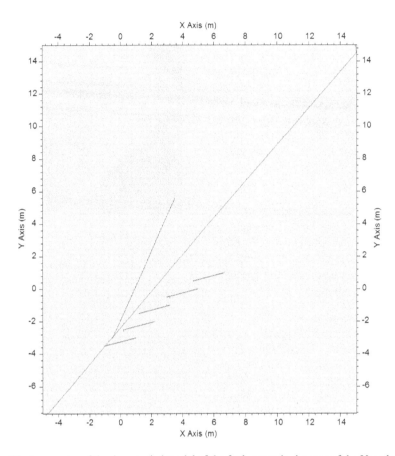

**Figure 4.** Significantly simplified geometry of the downscaled model of the fault zones in the area of the Unterhaching Gt2 geothermal well.

these parameters will be varied later for the purpose of parameter studies. The material parameters are calibrated based
on information taken from Gehle (2002), Mattle et al. (2003),
Rinne et al. (2003) and Moeck and Backers (2011). The calibration is carried out under the assumption that the system of
fault zones is nearly under equilibrium conditions. Finally, it
can be pointed out that the model is very sensitive to numerous parameters.

### 2.2.3  Simulation of the standard model

The model area is loaded by a complex stress field consisting of compressive and shear stresses. Simplifying this, the
boundary conditions in the model are related to the major
and minor horizontal stresses, which results in an anisotropic
compressive stress field. The direction of stresses in this area
is taken from Reinecker et al. (2010). Based on results taken
from Gehle (2002), they are assumed to be 120 MPa in the
north–south direction and 60 MPa in the east–west direction.
As the model is oriented north, the mean stresses are parallel
to the $y$ axis.

The resulting elastic displacements along the fault zones
are shown in Fig. 5 (left). Converted to realistic values according to the scaling assumption presented in Sect. 2.1,
these values are in the range of centimetres. More precisely,

the values vary between 2 and 22 cm, with lower values in
the en echelon system and maximums in the main fault.
Lastly, there exists a relative displacement along the main
fault. While the area located in the NW of the main fault
moves in the south-western direction, the area located in the
SE of the main fault moves in the NE direction. The fault displacements in the minor fault and in the system of en echelon
faults are significantly smaller than in the main fault.

Figure 5 (right) presents the deformations of the rock. The
scale used is equal to that of the displacements along the
faults given in Fig. 5 (left). The induced stresses result in
deformations of the order of decimetres. Due to the dynamic
in the system, some areas experience lower (blue) or higher
(red) deformations.

The resulting mean stresses are given in Fig. 6. The green
areas correspond to the initial compressive stress of 120 MPa.
In the red areas a decrease in the initial compressive stress
takes place. Here, the compressive stresses are reduced due
to a redistribution of stresses in the model area. The area
close to the en echelon system in particular is characterized by a decrease in the compressive stresses. In contrast
to that, the blue areas mark regions with an increased compressive stress. As might be expected, stress peaks arise at
the ends of the singular faults. Here, fracture processes are

**Table 1.** Material parameters of the standard case necessary for the FRACOD model

| Parameter | Symbol | Unit | Value |
|---|---|---|---|
| Rock | | | |
| Young's modulus | $E$ | GPa | 50 |
| Poisson's ratio | $\nu$ | – | 0.35 |
| Friction angle | $\phi$ | ° | 30 |
| Cohesion | $c$ | MPa | 0.25 |
| Tensile strength | $\sigma_Z$ | MPa | 12.1 |
| Mode I fracture toughness | $K_{IC}$ | MPa m$^{0.5}$ | 1.3 |
| Mode II fracture toughness | $K_{IIC}$ | MPa m$^{0.5}$ | 5.29 |
| Fractures | | | |
| Shear stiffness | $K_s$ | GPa m$^{-1}$ | 20 |
| Normal stiffness | $K_n$ | GPa m$^{-1}$ | 40 |
| Friction angle | $\phi$ | ° | 20 |
| Cohesion | $c$ | MPa | 1 |
| Cut-off level of fracture propagation | $f$ | – | 0.995 |
| Check-up level for elastic fracture growth | $e$ | – | 0.0 |
| Tolerance factor | $t$ | – | 1.0 |
| Hydraulic parameter | | | |
| Bulk modulus of the fluid | $K_w$ | GPa | 2.0 |
| Permeability | $k$ | m$^2$ | $1 \times 10^{-19}$ |
| Viscosity | $\mu$ | Pa s | $1 \times 10^{-3}$ |

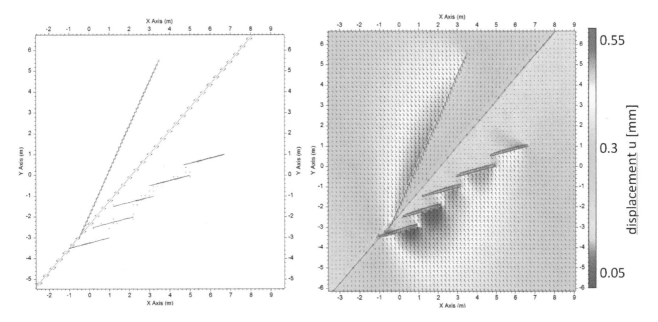

**Figure 5.** Displacements in the standard model due to the stress field applied (left: in the faults; right: in the rock). Converted to realistic values according to the assumptions in Sect. 2, the values are in the range of a few centimetres.

to be expected. However, these extreme changes in the stress field at the end of the faults do not fit well with reality – in actual fact they will be smaller because the existence of fissured areas instead of singular faults will dampen the changes.

Finally, it can be stated that a movement along the main fault can be observed due to the assumed boundary conditions and material parameters. This results in an area of reduced compressive stresses in the vicinity of the en echelon faults. This assumption coincides with the interpretation

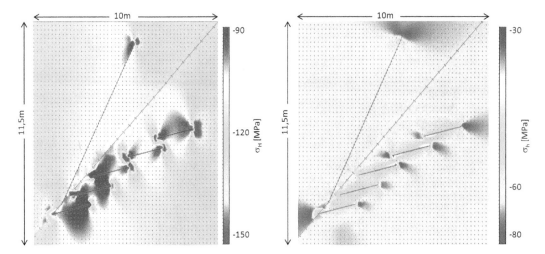

**Figure 6.** Variation in the major (left) and minor (right) stresses. In some areas of the fault system, variations of about 25 % can be observed in the large-scale stress field .

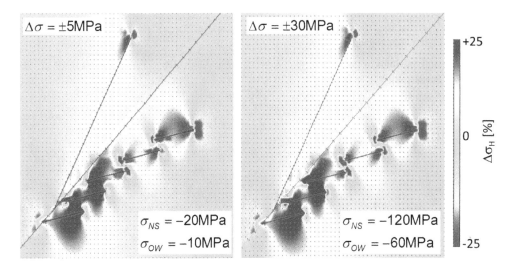

**Figure 7.** The variation in stresses in the near field of the fault system ranges by about 25 %. This relation is independent of the magnitude of the stress field.

based on the seismic measurements, which indicate areas of higher (in the vicinity of the minor fault) and lower (in the vicinity of the en echelon system) compressive stresses.

However, the presented stress field strongly depends on the boundary conditions of the system. Consequently, the influence of the stress field, as well as the material properties, will be investigated in the following sections. Some investigations concerning the influence of the borehole location are presented in Ziefle (2013). Depending on the depth, the location of the well changes with regard to the fault system. This might be of significant interest and could be a focus for further numerical investigations.

### 2.2.4 Influence of the stress field

The stress field in the Molasse Basin is discussed in Reinecker et al. (2010). It is assumed to be oriented in a N–S direction. Quantitative data are given by Seithel (2013). This information results from the investigation of borehole breakouts.

In this section the standard model, loaded with a stress field of 120 MPa in the N–S direction and 60 MPa in E–W direction, is compared with a similar model, loaded only by 20 MPa in the N–S direction and 10 MPa in E–W direction. The results are given in Fig. 7. The initial mean stress and the variation in the stress up to a level of 25 % are presented. It becomes clear that the initial value of 120 or 20 MPa remains constant in the far field (shown in green) while changes can be observed in the vicinity of the fault system. As expected,

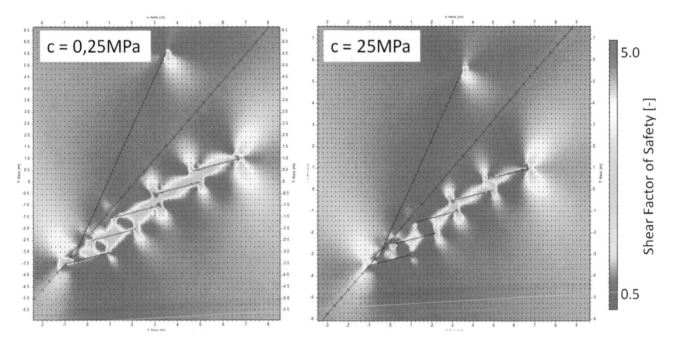

**Figure 8.** The variation in the cohesion has a moderate influence on the stress field. Pictured here is the safety factor depending on the relation of shear stress and shear stiffness.

the compressive stresses increase at the end points of the assumed singular faults (shown in blue). Moreover, a trend of a decreased compression field in the vicinity of the en echelon system is detected (shown in red). The variation in the stress reaches a level of about 25 %. Due to the difference in the initial stress, this variation has a value of 30 MPa in the standard case and a value of 5 MPa in the second case.

Finally, it can be stated that the variation in stresses for both cases remains, as a percentage, of the same order of magnitude. In the vicinity of the faults, variations of about 25 % related to the initial stress field have to be expected.

### 2.2.5   Influence of the material properties

The material properties consist of properties of the rock and the faults. Due to the significant simplification of the fissured areas as singular faults, the definition of fault properties is problematic. Parameter studies indicate a strong influence of fault properties on the results, while the influence of the material properties is comparatively low. In this section, the impact of some properties is summarized.

Figure 8 presents a comparison of two models with varying properties for the cohesion of the rock. While the assumed cohesion in the standard case remains 0.25 MPa, the comparative case is characterized by a cohesion of 25 MPa. Presented in the Fig. 8 is the safety factor, which depends on the relation of shear stress and shear stiffness. The results indicate a moderate influence of the cohesion.

The stresses which result from a variation in the friction angle of the rock are given in Fig. 9. A friction angle of 15° does not lead to a significant change in the results compared with the initial value of 30°.

Regarding the presented model set-up without any injection, production or changes of boundary conditions, there is no difference between a pure mechanical and a coupled hydraulic-mechanical approach. This is also presented in Fig. 9. Here, the simulation results on the left-hand side arise from a pure mechanical calculation while the results on the right side arise from a coupled hydraulic–mechanical simulation. The coupled simulation is characterized by faults, which are filled with water.

The fault is characterized by various parameters given in Table 1. A variation in the friction angle of the fault has a significant influence on the stress field. A shift of the initial value of 20° to a value of 10° leads to the results presented in Fig. 10. Due to the dynamic effects in the system, the resulting stress field varies enormously.

Another significant parameter of the faults is their stiffness. Considering the behaviour of fault systems in context with geothermal applications, this parameter is of special interest. Due to the injection or production of a fluid, the characteristics of the fault zones and the stress and fluid pressure conditions in the fault may change (e.g. Morris et al., 1996; Hillis, 2000; Legarth et al., 2005). By analysing the stiffness of faults and how the system reacts to variations of this parameter, a first insight into this complex interaction can be gained.

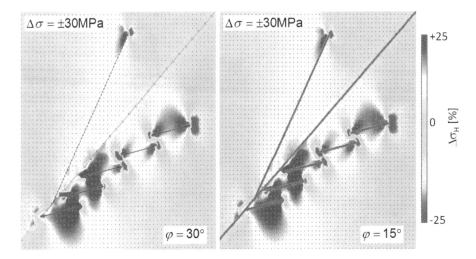

**Figure 9.** Stress field in the vicinity of the fault system. The variation of the friction angle of the rock has essentially no influence on the stress field.

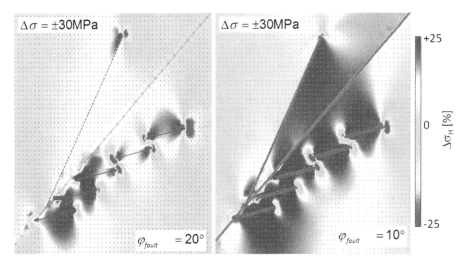

**Figure 10.** Due to the variation in the friction angle of the fault zones, the remaining stress field varies significantly. The area with a significant change in the stress field compared to the initial one is considerably higher due to the lower friction angle of $10°$.

Starting with an assumed shear stiffness of 20 GPa and normal stiffness of 40 GPa, a decrease in both values by 75 % is assumed to occur only in the main fault. This change already has a significant influence on the stress field. This can be seen by comparing Fig. 7 (right) with Fig. 11 (left). Additionally, with the assumption of the same decrease in stiffness for the minor fault, the resulting stresses are shown in Fig. 7 (right). The impact of these changes can be seen immediately – the area which is characterized by a significant decrease in the compressive stresses becomes larger. As a consequence, the equilibrium of forces in the vicinity of the fault system may change.

Nevertheless, these investigations are only a very rough approach to examine the effects within the fault system and their influence on the overall stress field. The key findings of the presented work are shown in Fig. 12. Here, the re-

sulting stress field is presented in combination with the seismic data. Future work might incorporate coupled hydraulic–mechanical effects more precisely. For example, the consideration of the pore pressure in the faults and changes in fault properties such as the aperture of the fault might be of special interest. Furthermore, a detailed investigation of local parameter variations could yield new insights into interactions between faults.

## 3 Conclusions

A significantly simplified 2-D model depicting the fault system in the vicinity of the geothermal well Unterhaching Gt2 is presented. Calibration is carried out in order to achieve a mechanical situation which is very close to the equilibrium state, as this corresponds best to the current state of

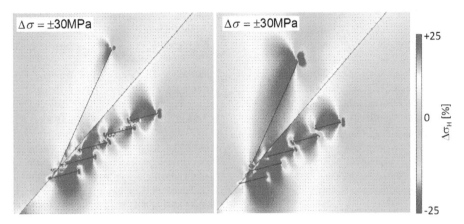

**Figure 11.** Stresses due to the assumption of a decrease in stiffness by 75 % for the main fault only (left) and for the main and the minor fault (right).

**Figure 12.** Seismic data in the area of the Unterhaching well (as given in Fig. 1) combined with the stress field derived by the numerical simulation. The areas shown in green correspond to the initial compressive stress field. However, different colours indicate changes in the stress field compared to the spacial stresses.

ther investigations focus on the impact of various boundary conditions and material parameters. It is necessary to point out that the material properties of the assumed fractures have a significant influence on the movements in the system and consequently on the resulting stress field. In contrast, the material properties of the rock have only a moderate influence. Testing the influence of the initial stress field indicates only a quantitative influence. Higher initially assumed stresses result in higher stresses in the fault system, but the qualitative arrangement remains the same.

Further investigations may focus on the impact of coupled hydraulic–mechanical processes. The influence of the injection on the properties of the fault system will be of special interest. Furthermore, 3-D investigations can incorporate the influence of the non-vertical well and the injection and production of the fluid.

**Acknowledgements.** I would like to express my thanks to the Leibniz Institute for Applied Geophysics (LIAG) for the kind support. The work presented in this paper was financed in the context of a reintegration programme after parental leave.

the system in reality. The presented investigations imply various parameter studies and focus on the impact of minimal movements in the fault system on the stress field. The aim of this model is a better process understanding and better classification of the significant material properties. With this, it is possible to better develop future investigations of coupled thermal–hydraulic–mechanical processes.

The numerical model indicates a relative movement along the main fault. This movement leads to a pressure decrease around the en echelon structures. For the chosen model set-up, the range of all changes in the stress field can be given by approximately 25 % of the initially assumed stresses. Fur-

## References

BStWIVT – Bayerisches Staatsministerium für Wirtschaft, Infrastruktur, Verkehr und Technologie: Bayerischer Geothermieatlas – Hydrothermale Energiegewinnung, 93 pp., München, 2010.

Byerlee, J. D.: Friction of rocks, Pure Appl. Geophys., 116, 615–626, 1978.

Ferrill, D. A. and Morris, A. P.: Dilational normal faults, J. Struct. Geol., 25, 183–196, 2003.

Fossen, H.: Structural Geology. Cambridge University Press, Cambridge, 2010.

Ganz, B., Schellschmidt, R., Schulz, R., and Sanner, B.: Geothermal Energy Use in Germany – Proceedings European Geothermal Congress 2013, 16 pp., Pisa, Italy, 3–7 June 2013.

Gehle, C.: Bruch- und Scherverhalten von Gesteinstrenn- flächen mit dazwischenliegenden Materialbrücken. Schriftenreihe des Institutes für Grundbau und Bodenmechanik der Ruhr-Universität Bochum, 33, 1439–9342, Bochum 2002.

Gudmundsson, A.: Rock Fractures in Geological Processes, Cambridge University Press, Cambridge, 2011.

Hillis, R.: Pore pressure/stress coupling and its implications for seismicity, Explor. Geophys., 31, 448–454, 2000.

Legarth, B., Huenges, E., and Zimmermann, G.: Hydraulic fracturing in sedimentary geothermal reservoirs, Int. J. Rock Mech. Min., 42, 1028–1041, 2005.

Lüschen, E., Dussel, M., Thomas, R., and Schulz, R.: 3D seismic survey for geothermal exploration at Unterhaching, Munich, Germany, First Break, 29, 45–54, 2011.

Lüschen, E., Wolfgramm, M., Fritzer, T., Dussel, M., Thomas, R., and Schulz, R.: 3D seismic survey explores geothermal targets for reservoir characterization at Unterhaching, Munich, Germany, Geothermics, 50, 167–179, 2014.

Mattle, B., John, M., and Spiegl, A.: Numerische Untersuchungen für den Tunnelbau im verkarsteten Gebirge, Felsbau, Rock and Soil Engineering, 21, 29–34, Verlag Glückauf, 2003.

Megies, T. and Wassermann, J.: Microseismicity observed at a non-pressure-stimulated geothermal power plant, Geothermics, in press, doi:10.1016/j.geothermics.2014.01.002, 2014.

Moeck, I. and Backers, T.: Fault reactivation potential as a critical factor during reservoir stimulation, First Break, 29, 73–80, 2011.

Morris, A., Ferrill, D. A., and Henderson, D. B.: Slip tendency analysis and fault reactivation, Geology, 24, 275–278, 1996.

Pollard, D. D. and Fletcher, R. C.: Fundamentals of Structural Geology, Cambridge University Press, Cambridge, 2005.

Reinecker, J., Tingay, M., Müller, B., and Heidbach, O.: Present-day stress orientation in the Molasse Basin, Tectonophysics, 482, 129–138, 2010.

Rinne, M.: Fracture Mechanics and Subcritical Crack Growth Approach to Model Time-Dependent Failure in Brittle Rock. Dissertation, ISBN 978-951-22-9435-0, Helsinki University of Technology, 2008.

Rinne, M., Shen, B., and Lee, H.-S.: Äspö Hard Rock Laboratory, Äspö Pillar Stability Experiment, Modelling of fracture stability by Fracod – Preliminary results. International Progress Report IPR-03-05, Fracom, 2003.

Schumacher, S.: Induzierte Seismizität: Ursachenforschung durch numerische Modellierung am Beispiel Uha Gt2. Der Geothermie Kongress DGK 2013 in Essen, Kongressband ISBN-13:978-3-932570-68-1, Poster, GtV-Bundesverband Geothermie e.V., 2013.

Seithel, R.: Charakterisierung tektonischer Spannungen für ein Geothermieprojekt in der süddeutschen Molasse. M-3, in: Interne Schriftenreihe Institut für Angewandte Geowissenschaften Fachbereich Geothermie, edited by: Kohl, T., 2013.

Shen, B.: FRACOD Version 1.1, User's manual, Fracom Ltd, 2002.

Shen, B. and Stephansson, O.: Modification of the G-criterion of crack propagation in compression. Int. J. of Engineering Fracture Mechanics, 47, 177–189, 1993a.

Shen, B. and Stephansson, O.: Numerical analysis of Mode I and Mode II propagation of rock fractures, Int. J. Rock Mech. Min., 30, 861–867, 1993b.

Shen, B., Stephansson, O., and Rinne, M.: Modelling Rock Fracturing Processes – A Fracture Mechanics Approach Using FRACOD. ISBN: 978-94-007-6903-8 (Print) 978-94-007-6904-5 (Online), Springer, Heidelberg, New York, 2014.

Worthoff, R. and Siemes, W.: Grundbegriffe der Verfahrenstechnik, 3. vollständig überarbeitete Auflage, 2012.

Ziefle, G.: Simulation hydromechanischer Effekte in Störungszonen für geothermierelevante Fragestellungen. Der Geothermie Kongress DGK 2013 in Essen, Kongressband ISBN-13:978-3-932570-68-1, Forum 4.2, 12 pp., GtV-Bundesverband Geothermie e.V., 2013.

Zoback, M. D.: Reservoir Geomechanics. Cambridge University Press, Cambridge, 2007.

# 15

# Geothermometric evaluation of geothermal resources in southeastern Idaho

G. Neupane[1,2], E. D. Mattson[1], T. L. McLing[1,2], C. D. Palmer[3], R. W. Smith[3], T. R. Wood[4], and R. K. Podgorney[1]

[1]Idaho National Laboratory, Idaho Falls, Idaho, USA
[2]Center for Advanced Energy Studies, Idaho Falls, Idaho, USA
[3]Office of Research & Economic Development, University of Idaho, Moscow, Idaho, USA
[4]Department of Geological Sciences, University of Idaho – Idaho Falls, Idaho Falls, Idaho, USA

*Correspondence to:* G. Neupane (ghanashyam.neupane@inl.gov)

**Abstract.** Southeastern Idaho exhibits numerous warm springs, warm water from shallow wells, and hot water from oil and gas test wells that indicate a potential for geothermal development in the area. We have estimated reservoir temperatures from chemical composition of thermal waters in southeastern Idaho using an inverse geochemical modeling technique (Reservoir Temperature Estimator, RTEst) that calculates the temperature at which multiple minerals are simultaneously at equilibrium while explicitly accounting for the possible loss of volatile constituents (e.g., $CO_2$), boiling and/or water mixing. The temperature estimates in the region varied from moderately warm (59 °C) to over 175 °C. Specifically, hot springs near Preston, Idaho, resulted in the highest reservoir temperature estimates in the region.

## 1 Introduction

The state of Idaho in the US has high potential of geothermal energy. The US Geological Survey has estimated that there is up to 4900 MWe of undiscovered geothermal resources and 92 000 MWe of enhanced geothermal potential within the state (Williams et al., 2008). Southern Idaho has been regarded to have high geothermal potential for conventional as well as for enhanced geothermal system (EGS) development (Tester et al., 2006). Geologic evidence such as the passage of the Yellowstone hotspot, Pleistocene basaltic flows, young volcanic features, and warm to hot springs (Mitchell, 1976a, b; Ralston et al., 1981; Souder, 1985) in southern Idaho indicate that the area may have economically viable geothermal resources. More direct evidence of a high-temperature regime at depth in the area is provided by a limited number of deep wells with high bottom-hole temperatures (BHTs) such as the King 2-1 well (202 °C, Table 1). Despite this geologic evidence and high BHTs, estimates of reservoir temperature based on traditional geothermometers applied to the chemistry of waters from springs in the region generally suggest a moderate temperature (Mitchell, 1976a, b).

As a part of an effort to assess the geothermal potential of southern Idaho, we assembled chemical composition of waters measured from numerous springs and wells in the region and used them to estimate reservoir temperatures using an inverse geochemical modeling tool (Reservoir Temperature Estimator, RTEst; Mattson et al., 2015). In this paper, we present results of RTEst applied to southern Idaho thermal water measured at a number of wells and springs.

## 2 Geology and geothermal setting of the area

### 2.1 Geology

The study area is located in both the Basin and Range and Rocky Mountains provinces. Specifically, the western part of the area has geographic characteristics of the Basin and Range such as wide and sediment-filled basins separating fault-bound ranges, whereas the eastern part consists of several thrust-bound narrow sub-parallel ridges with thinly filed basins (Mabey and Oriel, 1970). Geologically, the fold–thrust belt (Fig. 1a) in the area is a part of Sevier fold–thrust

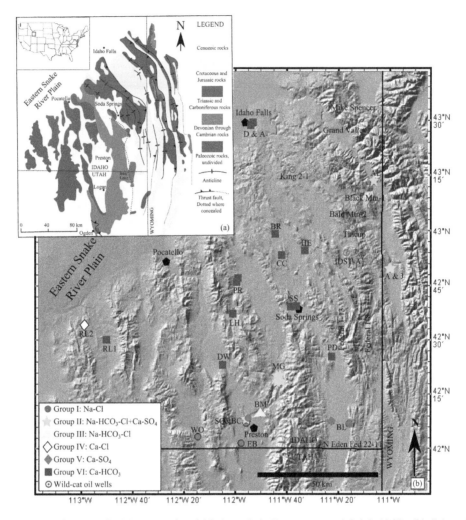

**Figure 1. (a)** Simplified geologic map of Idaho–Wyoming fold–thrust belt (Armstrong and Oriel, 1965). **(b)** Selected springs/wells and general water composition types in southeastern Idaho. The springs/wells codes correspond to the map code given in Table 2. The shaded relief map prepared from NASA 10 m digital elevation model data in GeoMapApp.

zone, locally known as the Idaho–Wyoming fold–thrust belt (Armstrong and Oriel, 1965).

Geology of the area includes thick sequences Paleozoic and Mesozoic carbonate-rich sedimentary rocks deposited in a passive margin setting (Armstrong and Oriel, 1965). During the Jurassic–Cretaceous periods these sedimentary sequences were deformed by compressive stresses associated with the Sevier orogeny resulting in numerous west-dipping low-angled thrust faults (Armstrong and Oriel, 1965). Starting in the Eocene and continuing to the Recent, extensional activities resulted in Basin and Range type topography with normal faults bounding ranges and wide valleys (Armstrong and Oriel, 1965; Dixon, 1982). Quaternary volcanic activity in some areas in the region (McCurry et al., 2011) resulted in volcanic resources (McCurry et al., 2008; Pickett, 2004).

### 2.2   Geothermal setting

The presence of several hot springs and warm springs indicates potential geothermal resources in southeastern Idaho. The western part of study area represents the amagmatic Basin and Range type geothermal system where convective upwelling dominates the thermal discharge along the extensional faults. The discharges of thermal water from springs and seeps in eastern and northern parts of the study area are also controlled by deep normal faults (Dansart et al., 1994). However, some recent works (e.g., McCurry et al., 2011; Welhan et al., 2014) also suggest a deep magmatic geothermal resource in this area. The conceptual model of magmatic-sourced geothermal setting in the fold–thrust belt in southeastern Idaho considers a magmatic geothermal resource at a depth of 12–14 km in an area beneath 58 ka rhyolite domes at China Hat located within the Blackfoot Volcanic Field (BVF) (Welhan et al., 2014). According to this hypothesis, the deep-sourced magmatic hydrothermal fluid

**Table 1.** Depth and corrected bottom-hole temperatures (BHTs) of several wildcat oil exploration wells in southeastern Idaho (Ralston et al., 1981; Souder, 1985; Blackwell et al., 1992; Welhan and Gwynn, 2014).

| Wells | Depth (m) | BHT (°C) |
|---|---|---|
| King 2-1 | 3927 | 202 |
| Grand Valley | 4931 | 140 |
| Mike Spencer Canyon | 4259 | 112 |
| Bald Mountain-2 | 3830 | 140 |
| Black Mountain-1 | 4158 | 120 |
| Federal 1-8 | 5105 | 167 |
| Big Canyon Federal 1-13 | 3551 | 172 |
| IDST-A1 | 4952 | 183 |
| Tincup | 5059 | 180 |
| N Eden Federal 21-11 | 2618 | 84 |

from this zone migrates eastwards along the thrust faults and permeable Paleozoic and Mesozoic layers into a shallower (3–5 km) reservoir. The high-temperature and high-salinity (sodium chloride) thermal fluids encountered at depth in some deep wildcat petroleum wells (such as, the King 2-1 well in Table 1) are possibly associated with these migrated magmatic fluids (Welhan et al., 2014).

## 3 Southeastern Idaho water chemistry data

Chemical compositions of numerous water samples from southeastern Idaho were assembled to assess the potential geothermal reservoir temperatures in the region. Over the last several decades, water samples from springs and wells in southeastern Idaho have been analyzed by several US government agencies and researchers for water quality and management, environmental remediation, and geothermal energy exploration (e.g., Young and Mitchell, 1973; Mitchell, 1976a, b; Ralston et al., 1981; Souder, 1985; Avery, 1987; McLing et al., 2002). A database is compiled of publicly available data for southeastern Idaho springs/wells. From a larger database, 50 water compositions (Table 2, Fig. 1b) were selected for the assessment of deep geothermal temperatures in southeastern Idaho.

## 4 Geothermometry

### 4.1 Approach

Geothermometry is a low-cost but useful geothermal exploration tool that uses the chemical compositions of water from springs and wells to estimate reservoir temperature. The application of geothermometry requires several assumptions: (1) the reservoir minerals and fluid attain chemical equilibrium, and (2) the water that moves from the reservoir to the sampling location retains its chemical signatures (Fournier et al., 1974). The first assumption is generally valid (provided

there is a sufficiently long residence time); however, the second assumption is more likely to be violated. As reservoir fluids move toward the surface, the pressure on the fluid decreases resulting in boiling and subsequent loss of volatiles (e.g., $CO_2$). The fluid temperature will decrease as a result of the associated heat of vaporization as well as thermal conduction. Boiling will increase the concentrations of non-volatile components in the liquid phase. The loss of acid volatiles such as $CO_2$ and $H_2S$ will alter the pH while the loss of redox active species such as $H_2S$ can shift the ratios of redox pairs (e.g., $HS^-/SO_4^{2-}$). These changes in temperature and solute concentrations may result in re-equilibration of the liquid phase with minerals in the zone above the main reservoir. In addition, this altered thermal water may mix with non-thermal waters which will further alter solute concentrations. These processes mask the initial geochemical signature of the reservoir fluid resulting in temperature estimates from traditional geothermometers being diverse and often being inaccurate or inconclusive.

Geothermal temperature predictions using multicomponent equilibrium geothermometry (MEG) provide apparent improvement in reliability and predictability of temperature over traditional geothermometers. The basic concept of this method was developed in the 1980s (e.g., Michard and Roekens, 1983; Reed and Spycher, 1984), and some investigators (e.g., D'Amore et al., 1987; Hull et al., 1987; Tole et al., 1993) have used this technique for predicting geothermal temperature in various geothermal settings. Other researchers have used the basic principles of this method for reconstructing the composition of geothermal fluids and formation brines (Pang and Reed, 1998; Palandri and Reed, 2001). More recent efforts (e.g., Bethke, 2008; Spycher et al., 2011, 2014; Smith et al., 2012; Cooper et al., 2013; Neupane et al., 2013, 2014, 2015a, b; Peiffer et al., 2014; Palmer et al., 2014; Mattson et al., 2015) have been focused on improving temperature predictability of the MEG.

An additional advantage of MEG over traditional geothermometers is that it considers a suite of chemical data obtained from water analyses for temperature estimation. Although MEG has advantages over the traditional geothermometers, it is also subject to the same physical and chemical processes that can violate the basic assumptions of geothermometry. However, MEG also provides a quantitative approach to account for subsurface composition-altering physical and chemical processes through inverse geochemical modeling. Therefore, it is important to reconstruct the composition of geothermal water for estimation of reservoir temperature with a greater certainty.

A newly developed geothermometry tool known as Reservoir Temperature Estimator (RTEst) (Palmer et al., 2014; Mattson et al., 2015) is used to estimate geothermal reservoir temperatures in southeastern Idaho. The RTEst is an inverse geochemical tool that implements MEG with an optimization capability to account for processes such as boiling, mixing, and gas loss. A more detailed description about RTEst can

**Table 2.** Water compositions of selected hot/warm springs and wells in southeastern Idaho used for temperature estimation. Elemental/species concentrations are given in $mg\,L^{-1}$. The pH was measured in the field.

| Springs/wells[a] | T (°C) | pH | Na | K | Ca | Mg | $SiO_{2(aq)}$ | $HCO_3$ | $SO_4$ | Cl | F | Map code[b] | Water type[c] | Data source[d] |
|---|---|---|---|---|---|---|---|---|---|---|---|---|---|---|
| Woodruff WS | 27 | 7.3 | 910 | 87 | 130 | 45 | 29 | 454 | 58 | 1600 | 0.6 | WO | I | 1 |
| E. Bingham W | 63 | 6.2 | 4600 | 770 | 320 | 36 | 68 | 930 | 48 | 7800 | 3.9 | EB | | 1 |
| Squaw HS-1 | 69 | 6.5 | 4184 | 708 | 135 | 23 | 126 | 816 | 27 | 6877 | 4.3 | SQ & BC | | 2 |
| Squaw HS-2 | 73 | 6.6 | 3844 | 533 | 241 | 26 | 126 | 866 | 23 | 6396 | 4.8 | | | 2 |
| Squaw HS W-1 | 82 | 7.8 | 4300 | 880 | 250 | 23 | 130 | 733 | 54 | 7700 | 7 | | | 3 |
| Squaw HS W-2 | 84 | 6.5 | 4368 | 782 | 279 | 24 | 124 | 791 | 35 | 7398 | 4.3 | | | 2 |
| Squaw HS W-3 | 82 | 6.9 | 3996 | 694 | 261 | 21 | 139 | 725 | 35 | 7291 | 4.9 | | | 1 |
| Battle Creek HS-1 | 43 | 6.7 | 3161 | 552 | 174 | 19 | 109 | 696 | 35 | 5241 | 6 | | | 2 |
| Battle Creek HS-2 | 77 | 6.5 | 3071 | 535 | 166 | 15 | 107 | 697 | 29 | 5048 | 6 | | | 2 |
| Battle Creek HS-3 | 81 | 6.5 | 3053 | 533 | 162 | 19 | 109 | 757 | 37 | 5034 | 6 | | | 2 |
| Battle Creek HS-4 | 82 | 6.8 | 4184 | 686 | 215 | 24 | 97 | 610 | 33 | 6967 | 6.4 | | | 2 |
| Wayland HS-1 | 84 | 7 | 3100 | 660 | 160 | 16 | 80 | 699 | 50 | 5400 | 12 | | | 3 |
| Alpine WS | 37 | 6.5 | 1500 | 180 | 560 | 100 | 40 | 880 | 1000 | 2800 | 2.7 | AL | II | 1 |
| Wayland HS-2 | 77 | 6.9 | 499 | 77 | 82 | 22 | 64 | 454 | 323 | 585 | 1 | SQ & BC | | 1 |
| Treasurton WS-1 | 35 | 6.6 | 563 | 127 | 265 | 68 | 54 | 704 | 788 | 632 | 2.2 | MG | | 2 |
| Treasurton WS-2 | 40 | 6.4 | 542 | 110 | 336 | 48 | 54 | 726 | 735 | 629 | 2 | | | 1 |
| Cleavland WS | 55 | 6.2 | 444 | 90 | 259 | 41 | 62 | 565 | 517 | 574 | 1.7 | | | 1 |
| Maple Grove HS-1 | 72 | 7.3 | 490 | 110 | 89 | 24 | 55 | 491 | 260 | 630 | 1.1 | | | 3 |
| Maple Grove HS-2 | 60 | 6.8 | 501 | 82 | 93 | 29 | 85 | 495 | 261 | 601 | 1.1 | | | 2 |
| Maple Grove HS-3 | 76 | 6.8 | 492 | 80 | 93 | 25 | 86 | 494 | 251 | 584 | 1 | | | 2 |
| Maple Grove HS-4 | 71 | 7.8 | 494 | 76 | 69 | 31 | 52 | 424 | 255 | 595 | 0.9 | | | 4 |
| Maple Grove HS-5 | 78 | 6.6 | 492 | 82 | 85 | 30 | 84 | 494 | 256 | 596 | 1.1 | | | 2 |
| Maple Grove HS-6 | 75 | 6.3 | 550 | 71 | 132 | 24 | 66 | 466 | 282 | 586 | 0.3 | | | 1 |
| Auburn HS | 57 | 6.4 | 1327 | 162 | 509 | 76 | 68 | 822 | 996 | 1737 | 0.6 | A & J | | 1 |
| Johnson S | 54 | 6.4 | 1494 | 176 | 454 | 45 | 88 | 973 | 1129 | 1947 | | | | 1 |
| Ben Meek W-1 | 40 | 7.4 | 348 | 20 | 23 | 5 | 90 | 526 | 5 | 321 | 11 | BM | III | 1 |
| Ben Meek W-2 | 45 | 7.3 | 360 | 24 | 25 | 7 | 80 | 524 | 15 | 320 | 10 | | | 1 |
| Ben Meek W-3 | 40 | 6.9 | 368 | 22 | 24 | 7 | 89 | 513 | 13 | 322 | 9.6 | | | 1 |
| Rockland W-2 | 20 | 7.3 | 60 | 24 | 120 | 22 | 70 | 220 | 26 | 280 | 0.2 | RL2 | IV | 5 |
| Bear Lake HS-1 | 40 | 7 | 155 | 48 | 230 | 41 | 43 | 263 | 769 | 72 | 4.2 | BL | V | 1 |
| Bear Lake HS-2 | 39 | 7.2 | 151 | 44 | 227 | 41 | 46 | 255 | 791 | 75 | 4.2 | | | 1 |
| Bear Lake HS-3 | 33 | 7.1 | 163 | 43 | 227 | 41 | 40 | 271 | 758 | 74 | 4 | | | 1 |
| Bear Lake HS-4 | 48 | 6.6 | 180 | 61 | 210 | 55 | 35 | 256 | 800 | 79 | 7.1 | | | 1 |
| Downata HS | 43 | 6.7 | 20 | 9 | 43 | 15 | 29 | 214 | 18 | 20 | 0.4 | DW | VI | 1 |
| Black River WS | 26 | 6.2 | 147 | 217 | 674 | 245 | 33 | 2357 | 1132 | 110 | 3.7 | BR | | 6 |
| Pescadero WS | 26 | 6.4 | 63 | 14 | 188 | 65 | 31 | 658 | 225 | 83 | 1.8 | PD | | 1 |
| Henry WS | 20 | 6.4 | 25 | 8 | 284 | 44 | 40 | 870 | 145 | 32 | 1 | HE | | 1 |
| Steamboat HS | 51 | 7 | 28 | 27 | 645 | 248 | 84 | 2380 | 472 | 8 | 0.3 | SS | | 7 |
| Soda Springs G | 28 | 6.5 | 12 | 23 | 851 | 193 | 35 | 2613 | 801 | 6 | 1.6 | | | 1 |
| Lava HS-1 | 45 | 6.6 | 170 | 39 | 120 | 32 | 32 | 542 | 110 | 190 | 0.7 | LH | | 3 |
| Lava HS-2 | 43 | 6.7 | 176 | 37 | 103 | 29 | 35 | 528 | 91 | 179 | 0.7 | | | 1 |
| Portneuf R WS-1 | 34 | 6.2 | 81 | 62 | 280 | 64 | 38 | 1060 | 270 | 62 | 0.8 | PR | | 8 |
| Portneuf R WS-2 | 41 | 6.3 | 85 | 60 | 275 | 48 | 47 | 1060 | 259 | 53 | 0.7 | | | 1 |
| Corral Creek W-1 | 42 | 6.5 | 101 | 237 | 701 | 263 | 28 | 2845 | 898 | 41 | 2.3 | CC | | 6 |
| Corral Creek W-12 | 41 | 6.8 | 97 | 242 | 620 | 246 | 30 | 2763 | 908 | 43 | 3.5 | | | 6 |
| Corral Creek W-13 | 41 | 6.6 | 101 | 233 | 697 | 263 | 30 | 2723 | 896 | 40 | 2.4 | | | 6 |
| Corral Creek W-14 | 36 | 6.6 | 99 | 233 | 649 | 253 | 30 | 2803 | 884 | 40 | 2.5 | | | 6 |
| Dyer W | 21 | 7.7 | 50 | 3 | 50 | 13 | 68 | 188 | 1 | 61 | | D & A | | 1 |
| Anderson W | 20 | 7.7 | 45 | 7 | 50 | 10 | 111 | 199 | 0 | 45 | | | | 1 |
| Rockland W-1 | 20 | 7.6 | 27 | 13 | 37 | 8 | 160 | 180 | 15 | 28 | 0.6 | RL1 | | 5 |

[a] Well/spring types – W: well, HS: hot spring, WS: warm spring, S: spring, G: geyser; [b] these map codes are used to define the springs/wells in Fig. 2, [c] water types are – I: Na-Cl (12 samples), II: $Na-HCO_3-Cl + Ca-SO_4$ (13 samples), III: $Na-HCO_3$-Cl (3 samples), IV: Ca-Cl (1 sample), V: $Ca-SO_4$ (4 samples), and VI: $Ca-HCO_3$ (17 samples); [d] data sources – 1: Ralston et al. (1981), 2: Mitchell (1976a), 3: Young and Mitchell (1973), 4: Dion (1969), 5: Parliman and Young (1992), 6: Mitchell (1976b), 7: Souder (1985), 8: Mitchell et al. (1980).

be found elsewhere (e.g., Palmer et al., 2014; Neupane et al., 2014; Mattson et al., 2015).

## 4.2 Missing components

The MEG approach requires that measured water composition include all components present in the reservoir mineral assemblage (RMA). For aluminosilicate minerals, this requires measured values of Al that are often not available in historical composition database. For water compositions without measured Al, an Al-bearing mineral (e.g., K feldspar) was used as a proxy for Al during geochemical modeling as suggested by Pang and Reed (1998).

## 4.3 Reservoir mineral assemblage

Based on the geology of southeastern Idaho and literature assessment of secondary minerals generally associated with the dominant rock and water types, we assumed reservoir mineral assemblages (RMAs) consisting of idealized clays, zeolites, carbonates, feldspars, and silica polymorphs (chalcedony) to estimate equilibrium temperatures using RTEst. Recently, Mattson et al. (2015) reported that the RTEst results in similar temperature estimates for the same water compositions when applied with slightly different RMAs. Selection of one or two unrepresentative minerals in the RMAs produced larger uncertainties in estimated temperatures than the estimated temperatures themselves because of poor convergence (Mattson et al., 2015). Therefore, while selecting the RMAs for RTEst, it is recommended to consider local geology, water chemistry (e.g., pH), and expected range of the reservoir temperatures. For more detailed information on selecting RMAs, see Palmer et al. (2014).

Using an appropriate RMA and measured water composition, RTEst estimates an equilibrium reservoir temperature (as well as a fugacity of $CO_2$ and mixing with non-thermal water or boiling) by minimizing an objective function ($\Phi$) that is the weighted sum of squares of the saturation indexes for the selected equilibrium minerals:

$$\Phi = \sum (SI_i w_i)^2, \tag{1}$$

where $SI_i = \log(Q_i/K_{i,T})$ for the $i$th equilibrium mineral ($Q_i$ and $K_{i,T}$ are the ion activity product and temperature-dependent equilibrium constant, respectively for the $i$th mineral) and $w_i$ is the weighting factor for the $i$th mineral.

The weighting factors ensure that each mineral that contributes to the equilibrium state is considered equally and the results are not skewed by reaction stoichiometry or differences in analytical uncertainty. There are three options for weighting factors in RTEst: inverse of variance, normalization, or unit weights. They are discussed in more detail by Palmer et al. (2014). In this paper, we used a normalization method for weighting, which assumes that the analytical errors for all thermodynamic components expressed as basis species are equal and that the thermodynamic activity of

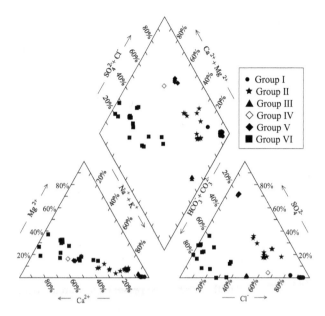

**Figure 2.** Reported chemistry of waters measured from several hot/warm springs and wells located in southeastern Idaho.

water is unity and is invariant. Examples of weighting factors (normalization factors) for several minerals are given in Palmer et al. (2014).

## 5 Results and discussion

### 5.1 Southeastern Idaho spring/well waters

Compositions of waters from hot/warm springs and wells in southeastern Idaho are presented in Table 2. The pH values of southeastern Idaho thermal waters are circum-neutral, ranging from 6.2 to 8.1, with arithmetic mean, median, and standard deviation of 6.87, 6.70, and 0.51, respectively; the field temperatures range between 20 and 84 °C. The aqueous chemistry of these thermal waters shows a large range in total dissolved solids (TDS) from about 250 mg L$^{-1}$ (Downata Hot Spring) to more than 14 000 mg L$^{-1}$ (E. Bingham Well).

The dominant cations in the southeastern Idaho thermal waters are Na and Ca with minor amounts of Mg (Fig. 2). The thermal waters include samples dominated by $Cl$, $HCO_3$, or $SO_4$ while others appear to be dominated by more than one anion. Hierarchical cluster analysis using Ward's (1963) method as implemented in SYSTAT 13 (SYSTAT Software, Inc.) was performed using the 6 Piper diagram (Fig. 2) end-members ($Ca^{2+}$, $Mg^{2+}$, $Na^+ + K^+$, $Cl^-$, $HCO_3^- + CO_3^{2-}$, $SO_4^{2-}$) for classifying water compositions. Six compositional groups were identified within the 50 thermal water samples – Group I: Na-Cl (12 samples), Group II: Na-HCO$_3$-Cl + Ca-SO$_4$ (13 samples), Group III: Na-HCO$_3$-Cl (3 samples), Group IV: Ca-Cl (1 sample), Group V: Ca-SO$_4$ (4 samples), and Group VI: Ca-HCO$_3$ (17 samples) (Table 2). These groups likely reflect differences in sources of water, water–

rock interactions, and structural control of the local geothermal systems.

The Na-Cl and Ca-SO₄ (Group I and V, respectively) type waters may have originated via the water–rock interactions involving pockets of evaporites in the area. Oriel and Platt (1980) have reported the presence of evaporites (e.g., halite, gypsum, and alum) in Middle Jurassic sequences (Preuss Redbeds) in southeastern Idaho. Recently, Welhan et al. (2014) indicated that the high-salinity waters in some deep wildcat petroleum wells may be related to either dissolution of salts from the Preuss evaporites or magmatic waters from a zone as deep as 12–14 km under the BVF in the fold–thrust belt. However, all Na-Cl (Group I) type waters considered in this study are from the surface expressions (hot/warm springs) or from shallow wells and may have originated via water–rock interactions involving evaporites. This type of water is also reported from the Raft River geothermal area (RRGA) located to the west of the present study area (Ayling and Moore, 2013). All Ca-SO₄ (Group V) type waters are from hot springs near Bear Lake, located near the Idaho–Wyoming–Utah border. Deep sourced water from a nearby deep wildcat petroleum well (N Eden Federal well with depth > 2500 m) has a very high SO₄ concentration; however, this water has low Ca concentration and high Na concentration (Souder, 1985). The Ca-SO₄ (Type V) type waters that the Bear Lake hot springs issue may have separate sources of Ca and SO₄, or there may be cation exchange reactions involving Ca and Na along the flow path.

The Ca-HCO₃ (Group VI) type waters are scattered throughout the area. These waters typically exhibit low Cl concentrations (Table 2). With some exceptions (e.g., Black River warm spring, Corral Creek wells, Soda Geyser, Pescadero Warm Spring), these waters also have a low SO₄ concentration. This type of water is generally regarded as a product of the interaction of groundwater with Ca-rich rocks at shallower depth. In the adjoining eastern Snake River Plain (ESRP), the Ca-HCO₃ (Group VI) type water represents the water in the active part of the ESRP aquifer whereas the deeper waters in ESRP area are Na-HCO₃ type (Mann, 1986; McLing et al., 2002).

Only one sample represents Ca-Cl (Group IV) type water, the Rockland W-2 well located in the westernmost part of the study area (Fig. 1). Although this water has some similarity with the Ca-HCO₃ (Group VI) and Ca-SO₄ (Group V) types of water in terms of high Ca content compared to its Na and K concentrations, its high Cl and low Na concentrations make it difficult to assign it as a direct product of a particular type of water–rock interaction.

The remaining two types of waters are mixed waters – Na-HCO₃-Cl (Group III) and Na-HCO₃-Cl + Ca-SO₄ (Group II). The cluster analysis did not identify a separate Na-HCO₃ type water, as it is a representative of deep waters in the adjoining ESRP. It is more likely that these waters are Na-Cl type waters that have interacted with carbonate with or without gypsum/anhydride.

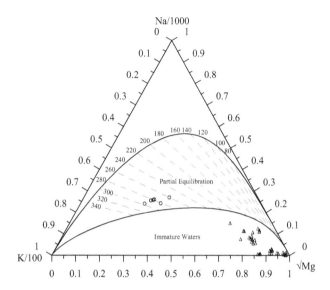

**Figure 3.** Southeastern Idaho waters from hot springs and wells plotted on Giggenbach diagram (Giggenbach, 1988). All partially equilibrated waters are of Na-Cl (Group I) type waters.

## 5.2 Southeastern Idaho geothermal reservoir temperatures

### 5.2.1 Giggenbach diagram

When plotted on a Giggenbach diagram (Giggenbach, 1988), the majority of the southeastern Idaho waters in this study are in the immature zone with some waters in the zone of partial equilibration (Fig. 3). The partially equilibrated waters in Fig. 3 are from hot springs and wells near Preston, Idaho (Battle Creek and Squaw hot springs), and distribution of these waters on Giggenbach diagram indicates that these waters could have interacted with rock at a temperature range of 260–300 °C. The immaturity reflects their low Na content as well as their higher Mg content (Giggenbach, 1988). Waters with high Mg are deemed to be unsuitable for some traditional solute geothermometry, although there have been efforts to develop a Mg correction (e.g., Fournier and Potter, 1979).

### 5.2.2 Temperature estimated by multicomponent equilibrium geothermometry

Estimates of reservoir temperatures for southeastern Idaho thermal waters shown in Table 2 were made using RTEst and several conventional geothermometers (Table 3). The RMAs that were used in RTEst consisted of representative minerals (Mg-bearing minerals – clinochlore, illite, saponite, disordered dolomite; Na-bearing minerals – paragonite, saponite; K-bearing minerals – K feldspar, mordenite K, illite; Ca-bearing minerals – calcite, disordered dolomite; and chalcedony). For thermal waters that do not have measured Al

**Table 3.** Temperature (°C) estimates for southeastern Idaho thermal waters RTEst and other geothermometers.

| Springs/wells[a] | $T^b \pm \sigma^c$ | Quartz[d] | Chalcedony[e] | Silica[f] | Na-K-Ca[g] | Types[h] |
|---|---|---|---|---|---|---|
| Woodruff HS | 97 ± 3 | 78 | 47 | 49 | 56 | I |
| E. Bingham W | 161 ± 4 | 117 | 88 | 88 | 193 | |
| Squaw HS-1 | 179 ± 9 | 151 | 125 | 123 | 204 | |
| Squaw HS-2 | 157 ± 6 | 151 | 125 | 123 | 183 | |
| Squaw HS W-1 | 175 ± 5 | 152 | 127 | 125 | 229 | |
| Squaw HS W-2 | 174 ± 6 | 150 | 124 | 122 | 217 | |
| Squaw HS W-3 | 171 ± 7 | 156 | 132 | 129 | 216 | |
| Battle Creek HS-1 | 169 ± 5 | 142 | 116 | 114 | 205 | |
| Battle Creek HS-2 | 175 ± 6 | 141 | 115 | 113 | 215 | |
| Battle Creek HS-3 | 170 ± 5 | 142 | 116 | 114 | 202 | |
| Battle Creek HS-4 | 171 ± 4 | 136 | 109 | 107 | 204 | |
| Wayland HS-1 | 175 ± 5 | 125 | 97 | 97 | 230 | |
| Alpine WS | 98 ± 9 | 92 | 61 | 63 | 92 | II |
| Wayland HS-2 | 144 ± 7 | 114 | 85 | 85 | 84 | |
| Treasurton WS-1 | 111 ± 3 | 105 | 76 | 77 | 78 | |
| Treasurton WS-2 | 111 ± 9 | 105 | 76 | 77 | 113 | |
| Cleavland WS | 119 ± 7 | 112 | 83 | 84 | 106 | |
| Maple Grove HS-1 | 126 ± 4 | 106 | 77 | 78 | 97 | |
| Maple Grove HS-2 | 123 ± 4 | 128 | 101 | 100 | 73 | |
| Maple Grove HS-3 | 124 ± 3 | 129 | 101 | 101 | 82 | |
| Maple Grove HS-4 | 115 ± 7 | 104 | 74 | 75 | 54 | |
| Maple Grove HS-5 | 126 ± 6 | 128 | 100 | 99 | 67 | |
| Maple Grove HS-6 | 122 ± 5 | 115 | 86 | 87 | 97 | |
| Auburn HS | 107 ± 9 | 117 | 88 | 88 | 104 | |
| Johnson S | 116 ± 13 | 130 | 103 | 102 | 134 | |
| Ben Meek W-1 | 106 ± 7 | 131 | 104 | 103 | 86 | III |
| Ben Meek W-2 | 106 ± 4 | 125 | 97 | 97 | 72 | |
| Ben Meek W-3 | 109 ± 4 | 131 | 103 | 102 | 73 | |
| Rockland-W2 | 110 ± 7 | 118 | 90 | 90 | 93 | IV |
| Bear Lake HS-1 | 113 ± 7 | 64 | 66 | 73 | | V |
| Bear Lake HS-2 | 111 ± 7 | 98 | 68 | 69 | 94 | |
| Bear Lake HS-3 | 107 ± 8 | 92 | 61 | 63 | 92 | |
| Bear Lake HS-4 | 121 ± 4 | 86 | 55 | 57 | 90 | |
| Downata HS | 97 ± 3 | 78 | 47 | 49 | 49 | VI |
| Black River WS | 103 ± 3 | 83 | 52 | 55 | 85 | |
| Pescadero WS | 68 ± 8 | 81 | 49 | 52 | 41 | |
| Henry WS | 60 ± 16 | 92 | 61 | 63 | 89 | |
| Steamboat HS | 96 ± 11 | 100 | 99 | 46 | | |
| Soda Springs G | 59 ± 15 | 86 | 55 | 57 | 88 | |
| Lava HS-1 | 94 ± 6 | 82 | 51 | 53 | 67 | |
| Lava HS-2 | 94 ± 5 | 86 | 55 | 57 | 64 | |
| Portneuf R WS-1 | 100 ± 6 | 89 | 59 | 61 | 92 | |
| Portneuf R WS-2 | 101 ± 9 | 99 | 69 | 70 | 111 | |
| Corral Creek W-1 | 98 ± 3 | 77 | 45 | 48 | 98 | |
| Corral Creek W-2 | 100 ± 4 | 79 | 48 | 51 | 97 | |
| Corral Creek W-3 | 98 ± 3 | 79 | 48 | 51 | 99 | |
| Corral Creek W-4 | 98 ± 3 | 79 | 48 | 51 | 100 | |
| Dyer W | 121 ± 3 | 117 | 88 | 88 | 57 | |
| Anderson W | 144 ± 4 | 143 | 117 | 115 | 74 | |
| Rockland-W1 | 31 ± 4 | 165 | 142 | 138 | 88 | |

[a] HS: hot spring, WS: warm spring, W: well; [b] RTEst-estimated temperature; [c] $\sigma$ is standard error in each RTEst optimized temperature; [d] quartz no steam loss, Fournier (1977); [e] Fournier (1977); [f] Arnórsson et al. (1983); [g] Fournier and Truesdell (1973), Mg correction applied according to Fournier and Potter II (1979); [h] water types are – I: Na-Cl (12 samples), II: Na-HCO$_3$Cl + Ca-SO$_4$ (13 samples), III: Na-HCO$_3$-Cl (3 samples), IV: Ca-Cl (1 sample), V: Ca-SO$_4$ (4 samples), and VI: Ca-HCO$_3$ (17 samples).

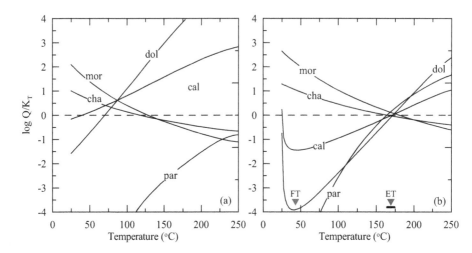

**Figure 4.** Temperature estimation for Battle Creek hot spring near Preston, Idaho. (a) The $\log Q/K_T$ curves for minerals calculated using original water chemistry with K feldspar used as a proxy for Al, (b) optimized $\log Q/K_T$ curves (FT: field temperature (43 °C); ET: estimated temperature (169 °C), the dark horizontal bar below ET represents the $\pm$ standard error for the estimated temperature ($\pm 5$ °C); cal: calcite, cha: chalcedony, dol: disordered dolomite, mor: mordenite K, and par: paragonite).

concentration, a value determined from assumed equilibrium with K feldspar was used in the calculations.

In MEG, the reservoir temperature is estimated by assuming a representative RMA with which the fluid in the reservoir is believed to have equilibrated. Next, the activities of the chemical species in solution are determined and the saturation indices [SI = $\log(Q/K_T)$] calculated using the laboratory measured temperature of the sample. This calculation is repeated for a range of temperatures and the resulting SIs recalculated. The apparent reservoir temperature is the one at which all minerals in the RMA are in equilibrium with the reservoir fluid as indicated by near-zero $\log Q/K_T$ values on a $\log(Q/K_T)$ versus temperature plot [$\log(Q/K_T)$ plot] (Reed and Spycher, 1984; Bethke, 2008). In other words, reservoir minerals in equilibrium with the fluid at depth should yield a common equilibrium temperature with a near-zero $\log(Q/K_T)$ value for each mineral at the same temperature; this common equilibrium temperature is assumed to be the reservoir temperature. If all the $\log(Q/K_T)$ curves do not show a common temperature convergence, then it suggests that there exists errors in analytical data, the selected mineral assemblage does not represent the actual mineral assemblage in the reservoir, or the sampled water must have been subjected to composition-altering physical and chemical processes during its ascent from the reservoir.

Figure 4a shows typical $\log(Q/K_T)$ curves of the RMA (calcite, chalcedony, disordered dolomite, mordenite K, and paragonite) used for the reported Battle Creek hot spring-1 water compositions. The $\log(Q/K_T)$ curves of these minerals intersect the $\log(Q/K_T) = 0$ at a wide range of temperatures, ranging from 40 °C (calcite) to over 250 °C (paragonite), making the $\log(Q/K_T)$ curves derived from the reported water chemistry minimally useful for estimating tem-

perature. The range of equilibration temperature for the assumed RMA is a reflection of physical and chemical processes that may have modified the Battle Creek hot spring-1 water composition during its ascent to the sampling point.

Three common composition-altering processes are degassing, mixing, and boiling. In particular, the loss of $CO_2$ from geothermal water due to degassing has a direct effect on pH, and it is often indicated by oversaturation of calcite (Palandri and Reed, 2001). Similarly, dilution of thermal water by cooler water or enrichment of solutes by boiling is indicated by lack of convergence of $\log(Q/K_T)$ curves over a small temperature range. In principle, these composition-altering processes can be taken into account by simply adding them into the measured water composition and looking for convergence of the saturation indices of the chosen mineral assemblage, but such a graphical approach becomes cumbersome even for two variables (e.g., temperature and $CO_2$).

To account for possible composition-altering processes, RTEst was used to simultaneously estimate a reservoir temperature and optimize the amount of $H_2O$ (to account for mixing or boiling) and the fugacity of $CO_2$ (to account for degassing) (Palmer et al., 2014). The optimized results for Battle Creek hot spring-1 are shown in Figure 4b. Compared to the $\log(Q/K_T)$ curves calculated using the reported water compositions (Fig. 4a), the optimized curves (Fig. 4b) converge to $\log(Q/K_T) = 0$ within a narrow temperature range (i.e., $169 \pm 5$ °C).

The optimized temperatures and composition parameters for the other southeastern Idaho waters reported in Table 2 were estimated using RTEst in the same manner. The estimated reservoir temperatures and associated standard errors are summarized in Table 3.

**Table 4.** Mean and standard deviation[a] of estimated temperature for each group of water.

| Geothermometer | Group 1[b] | Group 2[c] | Group 3[d] | Group 4[e] | Group 5[f] | Group 6[g] |
|---|---|---|---|---|---|---|
| RTEst | $165 \pm 22$ | $119 \pm 11$ | $107 \pm 1$ | 110 | $113 \pm 6$ | $98 \pm 22$ |
| Chalcedony[h] | $110 \pm 24$ | $85 \pm 13$ | $102 \pm 4$ | 90 | $62 \pm 5$ | $67 \pm 28$ |
| Na-K-Ca[i] | $196 \pm 46$ | $91 \pm 21$ | $77 \pm 8$ | 93 | $87 \pm 10$ | $79 \pm 21$ |

[a] Standard deviation for RTEst temperatures are calculated using RTEst temperatures of each group without incorporating standard error associated with estimated temperature of individual sample; [b] Na-Cl type water ($n = 12$); [c] Na-HCO$_3$-Cl + Ca-SO$_4$ type water ($n = 13$); [d] Na-HCO$_3$-Cl type water ($n = 3$); [e] Ca-Cl type water ($n = 1$), since this water type is represented by one sample, no standard deviations were calculated; [f] Ca-SO$_4$ type water ($n = 4$); [g] Ca-HCO$_3$ type water ($n = 17$); [h] Fournier (1977); [i] Fournier and Truesdell (1973), Mg correction applied according to Fournier and Potter II (1979).

### 5.2.3 Temperature estimates with traditional geothermometers

In addition to RTEst, other traditional geothermometers were also compared (Table 3, Fig. 5). Because most of the waters from thermal springs and wells in southeastern Idaho are geochemically immature (Fig. 3), the use of traditional geothermometers to estimate their temperatures is unreliable. Temperatures obtained with silica polymorphs and Na-K-Ca geothermometers can be quite variable, compared with the RTEst temperatures. As shown in Table 4, group-wise mean chalcedony-based reservoir temperatures are consistently cooler than the mean RTEst-calculated reservoir temperature.

Chalcedony-based reservoir temperatures were calculated using the observed silica concentrations. On the other hand, RTEst reservoir temperatures were calculated with reconstructed solute concentrations (i.e., optimized for water gain/loss). For the majority of samples, the chalcedony-based temperatures are lower than the RTEst-estimated temperatures (Fig. 5a). In general, wherever RTEst indicates that the sample contains an appreciable fraction of additional water, the RTEst temperature is higher than the chalcedony-based temperature for that sample.

Mg-corrected Na-K-Ca temperatures are relatively similar to the RTEst temperatures; however, the trend between mean RTEst and Na-K-Ca temperature varies with groups. In general, Na-K-Ca geothermometry predicts in cooler temperatures in the lower temperature range and hotter temperatures in the upper temperature range compared to the RTEst temperatures (Fig. 5b). The main weakness of Na-K-Ca geothermometer is poor reliability in waters with a significant amount of Mg. Even the Mg-corrected Na-K-Ca temperature estimates have poor reliability if the Mg concentration in thermal waters is high and is controlled by non-chlorite minerals. In southeastern Idaho, the Mg concentration in thermal waters is likely to be controlled by carbonate minerals (limestone/dolomite) as these waters may have interacting with carbonate sequences in the reservoir or along the flow path. Compared to the RTEst temperatures, Na-K-Ca temperatures are lower for all but Group 1 waters. The overprediction of temperature for Group 1 waters is likely

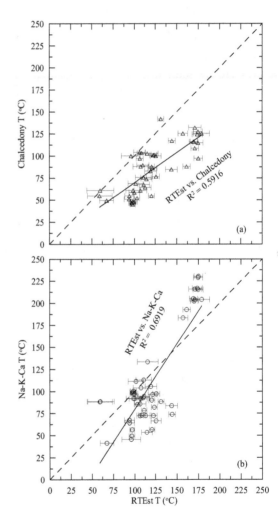

**Figure 5.** RTEst temperatures versus chalcedony (**a**) and Na-K-Ca (**b**) temperatures for the southeastern Idaho thermal waters. Each horizontal bar represents 1 standard error on each side in RTEst temperature estimate.

caused by the disproportionate (relative to Na and K) loss of Ca to calcite precipitation as suggested by Young and Lewis (1981).

### 5.2.4 Estimated temperature versus bottom-hole temperature of wildcat petroleum wells

As reported in Table 1, some of the wildcat petroleum wells in the fold–thrust belt in southeastern Idaho have measured BHTs. In general, RTEst-calculated reservoir temperatures appear positively correlated with nearby BHTs, supporting the argument that MEG can be used to predict deep geothermal reservoir temperatures. The North Eden Federal 21–11 well (2618 m) is located east of the Bear Lake, near the border of Idaho, Utah, and Wyoming. This well has slightly lower BHT (84 °C) than the RTEst temperature estimates (107–121 with standard error ±4 to ±8) for nearby Bear Lake hot springs (represented by letter code BL in Fig. 1). Similarly, the RTEst temperature estimate for Alpine Spring (letter code AL in Fig. 1) (98 ± 9 °C) is similar to the BHT of the nearest deep well, Black Mountain-1 (4158 m, 120 °C). However, there are some thermal features that have different RTEst-calculated reservoir temperatures than the BHTs of some nearby wells. For example, the BHTs measured for Federal 1–8 (167 °C) and Federal 1–13 (172 °C) are significantly higher than the estimated temperature (68 ± 8 °C) for the closest spring (Pescadero Warm Spring with PD letter code in Fig. 1). In this case, it is important to note that BHTs of some nearby wells also varied slightly. For example, Bald Mountain-2 (3830 m) has a BHT of 140 °C whereas the nearby Black Mountain-1 well (4158 m) has a BHT of 120 °C. Such variation in temperature at depth in nearby wells may suggest that some deep temperatures measured in wells are not equilibrated (i.e., disturbed by drilling) or could reflect variable proximity to the thermal water flow paths along fault or other heterogeneities.

The highest BHT was recorded for the King 2-1 well (3927 m, 202 °C), but there are no RTEst temperature estimates in the vicinity of this well. Similarly, there is no deep measured temperature in the vicinity of Battle Creek and Squaw hot springs near Preston. However, these hot springs discharge hot waters (up to 84 °C), and some of the recent shallow wells in the area are reportedly producing water with a temperature over 100 °C. New initiatives (e.g., Wood et al., 2015) would help further assess the geothermal potential of the Preston area system.

### 6  Conclusions

The geological setting coupled with the direct evidence of thermal expressions such as hot/warm springs in the area suggests that southeastern Idaho has good potential for geothermal resources. Our temperature estimates using RTEst with water compositions from southeastern Idaho thermal springs and wells indicate the presence of geothermal reservoirs at depth. Specifically, thermal waters of the Battle Creek hot springs and the Squaw Hot Springs suggest a promising geothermal prospect near Preston, Idaho. The US Department of Energy-sponsored new initiative in the Preston area

with geological, geochemical, and geophysical approaches is expected to further assess the geothermal potential. In several other areas, oil and gas wildcat wells indicate presence of high temperature at depth; however, the moderate RTEst temperature estimates from nearby thermal springs and shallow wells might reflect mixing of local groundwater with deep thermal water and/or re-equilibration of high temperature thermal waters in a shallow low temperature zone.

**Acknowledgements.** Funding for this research was provided by the US Department of Energy, Office of Energy Efficiency & Renewable Energy, Geothermal Technologies Program. We appreciate the help from Will Smith and Cody Cannon for this study. Reviews by John Welhan and an anonymous reviewer greatly improved the quality of this paper.

### References

Armstrong, F. C. and Oriel, S. S.: Tectonic development of Idaho-Wyoming thrust bel, B. Am. Ass. Petrol. Geol, 49, 1847–1866, 1965.

Arnórsson, S., Gunnlaugsson, E., and Svavarsson, H.: The chemistry of geothermal waters in Iceland – III Chemical geothermometry in geothermal investigations, Geochimica et Cosmochimica Acta, 47, 567–577, 1983.

Avery, C.: Chemistry of thermal water and estimated reservoir temperatures in southeastern Idaho, north-central Utah, and southwestern Wyoming, 38th Field Conference, Wyoming Geological Association Guidebook, 347–353, 1987.

Ayling, B. and Moore, J.: Fluid geochemistry at the Raft River geothermal field, Idaho, USA: New data and hydrogeological implications, Geothermics, 47, 116–126, 2013.

Bethke, C. M.: Geochemical and Biogeochemical Reaction Modeling, Cambridge University Press, Cambridge, UK, 564 pp., 2008.

Blackwell, D. D., Kelley, S., and Steele, J. L.: Heat flow modeling of the Snake River Plain, Idaho, US Department of Energy Report for contract DE-AC07-761DO1570, 1992.

Cooper, D. C., Palmer, C. D., Smith, R. W., and McLing, T. L.: Multicomponent equilibrium models for testing geothermometry approaches, Proceedings of the 38th Workshop on Geothermal Reservoir Engineering Stanford University, Stanford, California, 11–13 February, 2013, SGP-TR-198, 10 pp., 2013.

D'Amore, F., Fancelli, R., and Caboi, R.: Observations on the application of chemical geothermometers to some hydrothermal systems in Sardinia, Geothermics, 16, 271–282, 1987.

Dansart, W. J., Kauffman, J. D., and Mink, L. L.: Overview of Geothermal Investigations in Idaho, 1980–1993, Idaho Water Resources Research Institute, University of Idaho, Moscow, Idaho, 1994.

Dion, N. P.: Hydrologic reconnaissance of the Bear River Basin in southeastern Idaho: Idaho Dept. of Water Resources, Water Inf. Bull. 13, 1969.

Dixon, J. S.: Regional structural synthesis, Wyoming salient of western overthrust belt, AAPG Bulletin, 66, 1560–1580, 1982.

Fournier, R. O.: Chemical geothermometers and mixing models for geothermal systems, Geothermics, 5, 41–50, 1977.

Fournier, R. O. and Truesdell, A. H.: An empirical Na-K-Ca geothermometer for natural waters, Geochimica et Cosmochimica Acta, 37, 1255–1275, 1973.

Fournier, R. O. and Potter II, R. W.: Magnesium correction to the Na-K-Ca chemical geothermometer, Geochimica et Cosmochimica Acta, 43, 1543–1550, 1979.

Fournier, R. O., White, D. E., and Truesdell, A. H.: Geochemical indicators of subsurface temperature – 1: Basic assumptions, US Geol. Surv. J. Res. 2, 259–262, 1974.

Giggenbach, W. F.: Geothermal solute equilibria. Derivation of Na-K-Mg-Ca geoindicators, Geochimica et Cosmochimica Acta, 52, 2749–2765, 1988.

Hull, C. D., Reed, M. H., and Fisher, K.: Chemical geothermometry and numerical unmixing of the diluted geothermal waters of the San Bernardino Valley Region of Southern California, GRC Transactions, 11, 165–184, 1987.

Mabey, D. R. and Oriel, S. S.: Gravity and magnetic anomalies in the Soda Springs region southeastern Idaho, US Geological Survey, Washington, DC, Geological Survey Professional Paper 646-E, 1970.

Mann, L. J.: Hydraulic properties of rock units and chemical quality of water for INEL-1: A 10,365-foot-deep test hole drilled at the Idaho National Engineering Laboratory, Idaho, US Geological Survey, Idaho Falls, Idaho, Geological Survey Water Resources Investigations Report 86-4020, p. 23, 1986.

Mattson, E. D., Smith, R. W., Neupane, G., Palmer, C. D., Fujita, Y., McLing, T. L., Reed, D. W., Cooper, D. C., and Thompson, V. S.: Improved Geothermometry Through Multivariate Reaction-path Modeling and Evaluation of Geomicrobiological Influences on Geochemical Temperature Indicators: Final Report No. INL/EXT-14-33959, Idaho National Laboratory (INL), Idaho Falls, Idaho, 2015.

McCurry, M., Hayden, K. P., Morse, L. H., and Mertzman, S.: Genesis of post-hotspot, A-type rhyolite of the Eastern Snake River Plain volcanic field by extreme fractional crystallization of olivine tholeiite, Bull. Volcanol. 70, 361–383, 2008.

McCurry, M., Welhan, J., Polun, S., Autenrieth, K., and Rodgers, D. W.: Geothermal potential of the Blackfoot Reservoir-Soda Springs Volcanic Field: A hidden geothermal resource and natural laboratory in SE Idaho, GRC Transactions, 35, 917–924, 2011.

McLing, T. L., Smith, R. W., and Johnson, T. M.: Chemical characteristics of thermal water beneath the eastern Snake River Plain, GSA Special Paper 353, 205–211, 2002.

Michard, G. and Roekens, E.: Modelling of the chemical components of alkaline hot waters, Geothermics, 12, 161–169, 1983.

Mitchell, J. C.: Geothermal Investigations in Idaho – Part 5, Geochemistry and geologic setting of the thermal waters of the northern Cache valley area, Franklin County, Idaho, Idaho Dep. Water Resources, Water Inf. Bull., No. 30, 1976a.

Mitchell, J. C.: Geothermal Investigations in Idaho – Part 6, Geochemistry and geologic setting of the thermal and mineral waters of the Blackfoot reservoir area, Caribou County, Idaho, Idaho Dep. Water Resources, Water Inf. Bull., No. 30, 1976b.

Mitchell, J. C., Johnson, L. L, and Anderson, J. E.: Geothermal Investigations in Idaho – Part: 9, Potential for direct heat application of geothermal resources, Idaho Dep. Water Resources, Water Inf. Bull., No. 30, 1980.

Neupane, G., Smith, R. W., Palmer, C. D., and McLing, T. L.: Multicomponent equilibrium geothermometry applied to the Raft River geothermal area, Idaho: preliminary results, Geological Society of America, 45, p. 859, 2013.

Neupane, G., Mattson, E. D., McLing, T. L., Palmer, C. D., Smith, R. W., and Wood, T. R.: Deep geothermal reservoir temperatures in the Eastern Snake River Plain, Idaho using multicomponent geothermometry, Proceedings of the 39th Workshop on Geothermal Reservoir Engineering Stanford University, Stanford, California, 24–26 February 2014, SGP-TR-202, 12 pp., 2014.

Neupane G., Baum, J. S., Mattson, E. D., Mines, G. L., Palmer, C. D., and Smith, R. W.: Validation of multicomponent equilibrium geothermometry at four geothermal power plants, Proceedings of the 40th Workshop on Geothermal Reservoir Engineering Stanford University, Stanford, California, 26–28 January 2015, SGP-TR-204, 17 pp., 2015a.

Neupane G., Mattson, E. D., McLing, T. L., Dobson, P. F., Conrad, M. E., Wood, T. R., Cannon, C., and Worthing, W.: Geothermometric temperature comparison of hot springs and wells in southern Idaho, GRC Transactions, 39, 495–502, 2015b.

Oriel, S. S. and Platt, L. B.: Geologic map of the Preston 1° × 2° quadrangle, southeastern Idaho and western Wyoming, Miscellaneous Investigation Series, United States Geological Survey, Washington, DC, Department of Interior, Map I-1127, 1980.

Palandri, J. L. and Reed, M. H.: Reconstruction of in situ composition of sedimentary formation waters, Geochimica et Cosmochimica Acta, 65, 1741–1767, 2001.

Palmer, C. D., Ohly, S. R., Smith, R. W., Neupane, G., McLing, T., and Mattson, E.: Mineral selection for multicomponent equilibrium geothermometry, GRC Transactions, 38, 453–459, 2014.

Pang, Z. H. and Reed, M.: Theoretical chemical thermometry on geothermal waters: Problems and methods, Geochimica et Cosmochimica Acta, 62, 1083–1091, 1998.

Parliman, D. J. and Young, H. W.: Compilation of selected data for thermal-water wells and springs, 1921 through 1991, US Geological Survey, Boise, Idaho, Open-File Report 92-175, 1992.

Peiffer, L., Wanner, C., Spycher, N., Sonnenthal, E., Kennedy, B. M., and Iovenitti, J.: Optimized multicomponent vs. classical geothermometry: insights from mod-eling studies at the Dixie Valley geothermal area, Geothermics, 51, 154–169, 2014.

Pickett, K. E.: Physical volcanology, petrography, and geochemistry of basalts in the bimodal Blackfoot volcanic field, southeastern Idaho, MS Thesis, Idaho State University, Pocatello, Idaho, 2004.

Ralston, D. R., Arrigo, J. L., Baglio, J. V. Jr., Coleman, L. M., Souder, K., and Mayo, A. L.: Geothermal evaluation of the thrust area zone in southeastern Idaho, Idaho Water and Energy Research Institute, University of Idaho, 1981.

Reed, M. and Spycher, N.: Calculation of pH and mineral equilibria in hydrothermal waters with application to geothermometry and studies of boiling and dilution, Geochimica et Cosmochimica Acta, 48, 1479–1492, 1984.

Smith, R. W., Palmer, C. D., and Cooper, D.: Approaches for multicomponent equilibrium geothermometry as a tool for geothermal resource exploration, AGU Fall Meeting, San Francisco, 3–7 December, 2012.

Souder, K. C.: The hydrochemistry of thermal waters of southeastern Idaho, western Wyoming, and northeastern Utah, MS Thesis, University of Idaho, Moscow, Idaho, 1985.

Spycher, N. F., Sonnenthal, E., and Kennedy, B. M.: Integrating multicomponent chemical geothermometry with parameter estimation computations for geothermal exploration, GRC Transactions, 35, 663–666, 2011.

Spycher, N., Peiffer, L., Sonnenthal, E. L., Saldi, G., Reed, M. H., and Kennedy, B. M.: Integrated multicomponent solute geothermometry, Geothermics, 51, 113–123, 2014.

Tester, J. W., Anderson, B. J., Batchelor, A. S., Blackwell, D. D., DiPippo, R., Drake, E. M., Garnish, J., Livesay, B., Moore, M. C., Nichols, K., Petty, S., Toksöz, M. N., and Veatch, R. W.: The future of geothermal energy – impact of enhanced geothermal systems (EGS) on the United States in the 21st century, Massachusetts Institute of Technology, Cambridge, Massachusetts, 372 pp., 2006.

Tole, M. P., Ármannsson, H., Pang, Z. H., and Arnórsson, S.: Fluid/mineral equilibrium calculations for geothermal fluids and chemical geothermometry, Geothermics 22, 17–37, 1993.

Ward, J. H., Jr.: Hierarchical grouping to optimize an objective function, J. Am. Stat. Assoc., 58, 236–244, 1963.

Welhan, J. A. and Gwynn, M.: High heat flow in the Idaho thrust belt: A hot sedimentary geothermal prospect, GRC Transactions 38, 1055–1066, 2014.

Welhan, J. A., Gwyunn, M., Payne, S., McCurry, M., Plummer, M., and Wood, T.: The Blackfoot volcanic field, southeast Idaho: a hidden high-temperature geothermal resource in the Idaho thrust belt, Proceedings of the 39th Workshop on Geothermal Reservoir Engineering Stanford University, Stanford, California, 24–26 February 2014, SGP-TR-202, 13 pp., 2014.

Williams, C. F., Reed, M. J., Mariner, R. H., DeAngelo, J., and Galanis Jr., S. P.: Assessment of moderate- and high-temperature geothermal resources of the United States, US Department of the Interior, US Geological Survey, Menlo Park, California, Fact Sheet 2008-3082, 2 pp., 2008.

Wood, T. R., Worthing, W., Cannon, C., Palmer, C., Neupane, G., McLing, T. L., Mattson, E., Dobson, P. F., and Conrad, M.: The Preston Geothermal Resources; Renewed Interest in a Known Geothermal Resource Area, Proceedings of the 40th Workshop on Geothermal Reservoir Engineering Stanford University, Stanford, California, 26–28 January 2015, SGP-TR-204, 14 pp., 2015.

Young, H. W. and Lewis, R. E.: Application of a Magnesium Correction to the Sodium-Potassium-Calcium Geothermometer for Selected Thermal Waters in Southeastern Idaho, GRC Transactions, 5, 145–148, 1981.

Young, H. W. and Mitchell, J. C.: Geothermal investigations in Idaho – Part 1: Geochemistry and geologic setting of selected thermal waters (No. NP-22003/1), US Geological Survey and Idaho Dept. of Water Administration, Boise, Idaho, 43 pp., 1973.

# PERMISSIONS

# LIST OF CONTRIBUTORS

**S. Schumacher and R. Schulz**
Leibniz Institute for Applied Geophysics, Stilleweg 2, 30655 Hanover, Germany

**A. Santilano, A. Manzella, G. Gianelli, A. Donato, G. Gola, I. Nardini, E. Trumpy and S. Botteghi**
National Research Council, Institute for Geosciences and Earth Resources, Pisa, Italy

**M. A. Grant**
MAGAK, 14A Rewi Rd, Auckland 1023, New Zealand

**G. Falcone**
Department of Geothermal Engineering and Integrated Energy Systems, Institute of Petroleum Engineering, Clausthal University of Technology, Clausthal-Zellerfeld, Germany

**S. Homuth**
Züblin Spezialtiefbau GmbH, Ground Engineering, Europa Allee 50, 60327 Frankfurt a. M., Germany

**A. E. Götz**
University of Pretoria, Department of Geology, Private Bag X20, Hatfield, 0028 Pretoria, South Africa

**I. Sass**
Technische Universität Darmstadt, Geothermal Science and Technology, Schnittspahnstraße 9, 64287 Darmstadt, Germany

**D. Reyer and S. L. Philipp**
Georg August University of Göttingen, Geoscience Centre, Department of Structural Geology and Geodynamics, Germany

**M. M. Lahan, R. T. Verave and P. Y. Irarue**
Geological Survey Division, Mineral Resources Authority, P.O. Box 1906, Port Moresby, NCD, Papua New Guinea

**K. Breede, K. Dzebisashvili and G. Falcone**
Dept. of Geothermal Engineering and Integrated Energy Systems, Institute of Petroleum Engineering, Clausthal University of Technology, Clausthal, Germany

**V. Chauhan**
Reykjavik University, Reykjavik, Iceland
UNU Geothermal Training Programme, Reykjavík, Iceland

**Á. Ragnarsson**
IcelandGeoSurvey (ISOR), Reykjavik, Iceland

**J. Limberger**
Department of Earth Sciences, Utrecht University, Utrecht, the Netherlands

**P. Calcagno**
BRGM, Orléans, France

**A. Manzella and E. Trumpy**
Institute of Geosciences and Earth Resources, CNR, Pisa, Italy

**T. Boxem and M. P. D. Pluymaekers**
TNO – Geological Survey of the Netherlands, Utrecht, the Netherlands

**J.-D. van Wees**
Department of Earth Sciences, Utrecht University, Utrecht, the Netherlands
TNO – Geological Survey of the Netherlands, Utrecht, the Netherlands

**Y. Cherubini**
University of Potsdam, Institute of Earth and Environmental Science, Potsdam, Germany
Helmholtz Centre Potsdam – GFZ German Research Centre for Geosciences, Potsdam, Germany

**M. Cacace, M. Scheck-Wenderoth and V. Noack**
Helmholtz Centre Potsdam – GFZ German Research Centre for Geosciences, Potsdam, Germany

**M. Nakagawa and Y. Koizumi**
Colorado School of Mines, Golden, Colorado, USA
anow at: Kajima Corporation, Tokyo, Japan

**A. Fauzi**
PT. Geo Power Indonesia, Menara Palma #15-02A-B Kuningan, 12950 Jakarta, Indonesia

**G. Ziefle**
Leibniz Institute for Applied Geophysics (LIAG), Stilleweg 2, 30655 Hannover, Germany
Federal Institute for Geosciences and Natural Resources (BGR), Stilleweg 2, 30655 Hannover, Germany

**E. D. Mattson and R. K. Podgorney**
Idaho National Laboratory, Idaho Falls, Idaho, USA

**G. Neupane and T. L. McLing**
Idaho National Laboratory, Idaho Falls, Idaho, USA
Center for Advanced Energy Studies, Idaho Falls, Idaho, USA

**C. D. Palmer and R. W. Smith**
Office of Research & Economic Development, University of Idaho, Moscow, Idaho, USA

**T. R. Wood**
Department of Geological Sciences, University of Idaho – Idaho Falls, Idaho Falls, Idaho, USA

# Index

Printed in the USA
CPSIA information can be obtained
at www.ICGtesting.com
JSHW051446221024
72173JS00006B/1592

9 781682 864623